PUBLIC-PRIVATE PARTNERSHIPS

PUBLIC-PRIVATE PARTNERSHIPS

Managing risks and opportunities

EDITED BY

Akintola Akintoye, Matthias Beck and
Cliff Hardcastle

School of the Built and Natural Environment
Glasgow Caledonian University

Blackwell
Science

© 2003 by Blackwell Science Ltd,
a Blackwell Publishing Company
Editorial Offices:
9600 Garsington Road, Oxford OX4 2DQ, UK
  Tel: +44 (0)1865 776868
Blackwell Science, Inc., 350 Main Street, Malden,
MA 02148-5018, USA
  Tel: +1 781 388 8250
Iowa State Press, a Blackwell Publishing Company,
2121 State Avenue, Ames, Iowa 50014-8300, USA
  Tel: +1 515 292 0140
Blackwell Publishing Asia Pty Ltd, 550 Swanston
Street, Carlton South, Victoria 3053, Australia
  Tel: +61 (0)3 9347 0300
Blackwell Wissenschafts Verlag, Kurfürstendamm 57,
10707 Berlin, Germany
  Tel: +49 (0)30 32 79 060

First published 2003 by Blackwell Science Ltd

Library of Congress
Cataloging-in-Publication Data is available

ISBN 0-632-06465-X

A catalogue record for this title is available from the
British Library

Set in 10/12 Palatino
by DP Photosetting, Aylesbury, Bucks
Printed and bound in Great Britain by
The Bath Press, Bath

For further information on
Blackwell Publishing, visit our website:
www.blackwellpublishing.com

Contents

Contributors

Akintola Akintoye is a Chartered Quantity Surveyor and a Chartered Builder. He holds the Chair of Construction Economics and Management and is the Director of Research, School of the Built and Natural Environment at Glasgow Caledonian University. Before his academic career, he worked as a quantity surveyor on major building and civil engineering projects. He is the current Chairman of the UK-based Association of Researchers in Construction Management (ARCOM) and the Co-editor of the *Journal of Financial Management of Property and Construction*. In addition, he is a Visiting Professor to the Department of Civil Engineering, Asian Institute of Technology and the Department of Building and Real Estate, Hong Kong Polytechnic University. He has gained international recognition for his scholarly work in the area of construction risk management and procurement, construction estimating and modelling, construction economics, and construction inventory management. He is an active member of two Working Commissions of the International Council for Research and Innovation in Building and Construction (CIB).

Joseph A. Anderson is Managing Partner of the Singapore office of Morrison & Foerster LLP, a major international law firm and a member of the firm's Project Development and Finance Group. The main focus of his practice is development and financing of major infrastructure projects in China, India, Indonesia, the Philippines, Thailand, and Vietnam, representing project sponsors and lenders. He has been actively involved in international legal, trade and energy matters since his graduation from Georgetown University's School of Foreign Service where he received a B.S. in Foreign Service in 1986. He received his Juris Doctorate from Harvard University Law School in 1989.

Darinka Asenova works as a Research Fellow in the Division of Risk at Glasgow Caledonian University. Having a scientific and economics background, her current research interests include a range of risk-related issues such as risk management in PFI projects, investigating the behaviour of the financial service providers in the PFI environment, communication of scientific risks as well as policy implications of the risk communication.

Matthias Beck is Professor of Risk Management at the Division of Risk at Glasgow Caledonian University. He has been a Co-principal Investigator in the DETR/EPSRC Link Project – Standardised Framework for Risk Analysis and Management in PFI Projects – and has acted as consultant to a number of organisations including the ILO, Nestle Cadbury, Marsh, the Association of

British Insurers as well as Lord Cullen's inquiry into the Ladbroke Grove and Southall rail disasters.

Kate Boothroyd is a Chartered Surveyor. She joined Gardiner & Theobald in 1988 and in 1992 set up their Risk Management Group, which provides risk management advice to clients, contractors and consultants on a variety of projects. She was promoted to Associated Director of G&T's Risk Management Group in 1995, and became the first quantity surveyor to be elected a Fellow of the Institute of Risk Management in 1996 – she is now Deputy Chairman. Kate joined AMEC in 1997 as Director of Risk Management, working on a variety of commercial and PFI projects. In January 2000 she become the Total Risk Management (TRM) Implementation Manager responsible for developing the best practice on risk management AMEC–wide.

Paul A Bowen is Professor and Head of the Department of Construction Economics and Management at the University of Cape Town in South Africa. He is a nationally-rated researcher and a member of the Academy of Science of South Africa. His professional and academic experience relating to the construction industry has resulted in more than 125 publications. He serves on the editorial/technical advisory boards of a number of internationally refereed journals. His research interests include project procurement, communication, and the sustainability assessment of buildings.

Lucy Chege is an Infrastructure Investments Specialist at the CSIR, Building and Construction Technology in South Africa. She has worked closely with the South African Department of Public Works in its public-private partnership (PPP) projects. Lucy has authored and co-authored several publications on PPPs in South Africa from the perspective of financing, investment, and the processes involved in these projects. Prior to joining the CSIR, Lucy worked as a quantity surveyor in a construction firm. She has a bachelor's degree in Building Economics from the University of Nairobi, a Master of Science in Project Management from the University of the Witwatersrand and is currently studying for an MBA (specialising in Finance) at Manchester Business School.

Ezekiel Chinyio is currently a Senior Lecturer with the University of Central England in Birmingham, UK. Prior to that, he worked as a Postdoctoral research fellow with Glasgow Caledonian University, where he played an active role in the study of the risk management of 'private finance initiative' projects. His research interests concern the areas of construction contracts and procurement, construction economics, risk management and built environment education.

Peter J. Edwards is an Associate Professor at RMIT University in Melbourne, Australia, with responsibility for offshore delivery of undergraduate and postgraduate degree programs. He has an extensive background in the construction industry, professionally and academically, and has authored over 80 academic publications. He serves as *ad-hoc* referee for a number of international journals and has been the recipient of research funding for a number of projects. His research interests include project risk management and value management and he acts as a consultant in these areas.

Alasdair W. Fergusson has worked in the construction industry over the last 40 years. His experience ranges from the tight discipline of QS practice, through partnership in Bell Ingram to his present position in a large plc – Carillion. A four year spell as a Civil Servant and overseas service in Saudi Arabia, Iraq, Australia and Europe added to his unique blend of knowledge. His most rewarding stint in recent times – Technical Advisor to the FCO for the first overseas PFI project, The British Embassy in Berlin. In addition to PFI and mainstream FM he has become deeply involved in MOD Prime Contracting.

Cliff Hardcastle holds an MSc by Research and a PhD. He is a Member of the Chartered Institute of Building and the Association of Cost Engineers. He is currently Dean of the School of the Built and Natural Environment at Glasgow Caledonian University. His previous research work in construction procurement includes the potential of integrated databases for information exchange in the construction industry, and the analysis of approaches to procurement and cost control of petrochemical works. He is currently involved in a number of applied research projects including cultural change in the construction industry, the building procurement decision-making process and decision making and risk in PFI projects. He is a member of the EPSRC Peer Review College, the RICS Research Advisory Board, an Advisor to the Malaysian Board of Quantity Surveyors, and referee for a number of international research councils, including the United Kingdom, South Africa and Hong Kong.

John Hood is a Lecturer in the Division of Risk at Glasgow Caledonian University. Prior to joining the University, he spent a number of years working in the insurance industry. His main current research interests lie in the areas of public sector risk management, the risks associated with outsourcing of functions and in political risk management in multinational corporations. He has published in a wide range of journals on a number of risk managed-related topics, presented papers at UK and overseas conferences, contributed to risk management guidance documents for local authorities and contributed to several insurance, risk management and public policy textbooks. He is a co-opted member of the Executive Committee of the Association of Local Authority Risk Managers (Scotland).

Caroline Hunter holds an MA from the University of Glasgow, an LLB and a postgraduate diploma in legal practice, both from the University of Dundee. She is currently completing her PhD at the Division of Law at Glasgow

Caledonian University. Her research interests include administrative law and gender issues in the legal profession.

John Kelly is a Chartered Surveyor with industrial and academic experience. His quantity surveying career began with a national contractor moving to a small architects practice specialising in commercial development and later to an international surveying practice. His academic career began at the University of Reading as a Research Fellow, moving to Heriot-Watt University as a Lecturer and later Senior Lecturer. His research into value management began in 1983 and has attracted significant grant income. His numerous publications include the first UK textbook on value management (written with Professor Steven Male of the University of Leeds). Professor Kelly is a firm believer in the principle of putting research into practice and has undertaken value management studies as research consultancy on a variety of construction projects.

Mohan Kumaraswamy is an Associate Professor at the Department of Civil Engineering of The University of Hong Kong, where he teaches and researches in construction engineering and management. He has previously worked with a contractor, a client/consultant and a project management company in Sri Lanka, as well as with a contractor in Nigeria. He has also led study teams on internationally funded assignments. His present research interests include construction procurement systems (including decision support for their selection and sub-systems such as in contractor selection and BOT-type public-private partnerships), project evaluation, knowledge management, technology transfers/exchange, claims and dispute minimisation, quality and productivity issues, and construction industry development.

Bing Li is currently at the School of the Built and Natural Environment of Glasgow Caledonian University undertaking a PhD programme in Construction Management and Economics. He graduated from Tongji University, Shanghai, China, where he received his BSc degree in Building Engineering. He worked for many years as a construction manager with Fujian, the fourth largest Building Construction Company in China. His knowledge in international project management includes working as a member of the China Construction Industry Development Programme on a World Bank-sponsored project. He thereafter did his postgraduate study at the School of Civil and Structural Engineering, Nanyang Technological University, Singapore, where he was awarded a Master of Engineering for research in Construction Management. He has co-authored many journal and conference papers. His research interests are in the areas of risk management, construction joint ventures (JVs) and public-private partnerships.

Tjiamogale Eric Manchidi is an Executive Director of Procurement Dynamics (Proprietary) Limited, a Procurement Advisory Consultancy specifically formed to advise both government and the private sector. He was previously a Director for Asset Procurement and Operating Partnership Systems, a public-private partnership programme in the National Department of Public Works. He served as a member of the Inter-ministerial Task Team on public-private partnerships that developed the current South African Government PPP Framework and also a member of the Municipal Services Partnerships Task Team. He acted as a Chief Negotiator on behalf of Government on the procurement of the first two PPP prison projects in South Africa. He formulated a framework for PPPs in Government Office Accommodation and implemented it on several projects that are at various procurement stages. He has published and presented papers on infrastructure development and procurement issues in both local and international conferences.

Evelyn McDowall is Principal Consultant in the PPP Unit of Turner and Townsend Group. She is a regular contributor to the Facilities Management technical journals on all aspects of PPP procurement and Facilities Management strategy. She is a former Chair of the BIFM in Scotland. She started her career in 1990 at the Centre for Facilities Management at University of Strathclyde where she worked as a Research Officer on a wide range of FM research projects. In 1992 she gained an MSc in Facilities Management. In 1996 she joined the Symonds Group as a Facilities Management Consultant and developed her experience of offering strategic advice on the structure and management of FM services in a range of public and private sector organisations. During this period she gained her first experience of PPP projects, acting as project manager for the company's successful bid for the DSS PRIME portfolio (700 sites). In 1999 she joined Turner & Townsend Group where she specialised in education PPP projects, advising clients at all stages of procurement on all technical aspects of managing and delivering FM services.

Andrew Mills joined a risk management consultancy in 1977, following careers with a UK insurer and broker. Then followed ten years in industry with multinational conglomerates before joining Merrett Health Risk Management, where he was Director of Operations and was responsible for the technical content of the NHS risk management manual. Andrew moved to Hartlepool in 1995 as Head of Risk Management for the local acute NHS Trust and has helped them to survive two mergers, continuing his strategic role in the new organisation. Qualified as a Chartered Insurance Practitioner and a Fellow of the Institute of Risk Management, Andrew has been Senior Vice Chairman of that Institute and Chair of their education and training function. He was the founder of a virtual federation of risk-related groups in healthcare and has assisted the NHS Executive in several steering and advisory groups related to risk management and controls assurance.

Stephen Ogunlana is an Associate Professor of Construction Engineering and Management at the School of Civil Engineering, Asian Institute of Technology, in Pathumthani, Thailand. He obtained bachelor and master degrees from Obafemi Awolowo University in Nigeria. His doctoral degree is from Loughborough University, UK. Dr. Ogunlana's research interests are in project management, construction process simulation, cost estimating and risk analysis, organisational learning, training, productivity improvement and project improvement through stakeholder participation. He consults for several governments and Habitat in project management and training.

Jirapong Pipattanapiwong is a doctoral candidate in Infrastructure Systems Engineering at Kochi University of Technology, Japan. He received Bachelor of Engineering in Construction Engineering from King Mongkut Institute of Technology, Ladkrabang (KMITL), Thailand in 1998 and Master of Engineering in Construction Engineering and Management from the Asian Institute of Technology (AIT) in 2000. Before pursuing a doctoral degree, he served as a Research Associate at AIT under the Japanese Government on Regional Education and Development Project. His work involved conducting researches, publishing papers, organising symposia, and assisting in graduate teaching. His past research works were related to the development and application of risk management process and quality management in construction. Currently, he is undertaking doctoral research in the area of risk management, particularly risk perception and risk allocation.

Pantaleo D. Rwelamila is a Professor of Project Management at the Graduate School of Business Leadership, University of South Africa. He graduated in Building Economics at the University of Dar Es Salaam in Tanzania, received his MSc Degree in Project Management from Brunel University and a PhD in Project Management and Procurement Systems from the University of Cape Town. He was a Senior Partner, Quantum Consultants (Pty) Ltd in Botswana, an Associate Professor in the Department of Construction Economics and Management at the University of Cape Town. He has worked in a number of countries, including Tanzania, Kenya, Uganda, Botswana and Sweden. He was a full time secretariat member of the South African Government Task Team responsible for developing the construction industry policy for two years, 1998 and 1999. He received a joint Centre for Science and Industrial Research outstanding achiever award in 1999.

Kalidindi N. Satyanarayana is currently an Associate Professor in the Department of Civil Engineering at the Indian Institute of Technology, Madras, India. He obtained his MS and PhD degrees in Civil Engineering (Project Management and Construction Engineering) from Clemson University, USA and B.Tech degree in Civil Engineering from the Indian Institute of Technology, Madras. He is actively involved in teaching, research and consulting, related to project management, construction contracts, construction quality management, risk assessment of privately financed infrastructure projects, and information technology in construction.

Michaela M. Schaffhauser-Linzatti works as an Assistant Professor at the Department of Finance and Banking at the University of Vienna, Austria, and is also affiliated as Guest Professor at the Europa Universität Viadrina in Frankfurt/Oder, Germany. She holds an MBA in Business Administration from the Vienna University of Economics and a PhD in Social and Economic Sciences from the University of Vienna. Michaela Schaffhauser-Linzatti concentrates on the research areas of accounting, transportation, healthcare systems, privatisation, and public-private partnership. Her professional membership includes INFORMS, Austrian Working Group on Banking and Finance, Austrian Transportation Society, Schmalenbachgesellschaft and Society of Social Politics.

Arthur L. Smith is President of Management Analysis, Incorporated (MAI), a US-headquartered firm (Vienna, VA) which specialises in consulting services related to public-private partnerships and privatisation. He has 25 years experience in performing and managing analyses of public-private partnerships on five continents, advising governmental agencies on partnership strategies and planning. This experience includes managing the US Government's largest managed competition and development of new cost evaluation methodologies for unique partnership environments. Mr. Smith is also currently serving as Vice President for the US National Council for Public-Private Partnerships. He holds a Master of Science degree in Technology Management from the University of Maryland.

Bill Stein is a Senior Lecturer in the Division of Risk at Glasgow Caledonian University. Following a career as legal liability and property underwriter for a major insurer, he has developed an interest in the management of risk where insurance solutions play only a small part, such as in the NHS. He is interested generally in the development of risk management in the NHS and has a particular interest in the mental health service. He has contributed to several textbooks including those published by the Chartered Insurance Institute and the Institute of Risk Management. He is an Honorary Consultant in Risk Management at Renfrewshire & Inverclyde Primary Care NHS Trust, a graduate of the University of Strathclyde and of Glasgow Caledonian University, Fellow of the Chartered Insurance Institute, Chartered Insurance Practitioner, Member of the Institute of Risk Management, and Member of the Institute of Learning and Teaching.

A.V. Thomas is Assistant Professor in Civil Engineering at D.D. Institute of Technology (Deemed University), Nadiad, India and is currently doing his PhD at the Indian Institute of Technology, Madras. He obtained a M.Tech degree in Building Technology and Construction Management from the Indian Institute of Technology, Madras. His current research interest is in the area of risk management in private infrastructure projects.

Robert Tiong is an Associate Professor in the School of Civil & Environmental Engineerng, Nanyang Technological University, Singapore. He is the Coordinator of the MSc programme in International Construction Management and Lecturer for the Project Financing subject. His research and consulting interests focus on construction project management, project financing, risk management and structuring of BOT concession contracts for privately financed infrastructure projects. He has published extensively, including a monograph on 'The Structuring of BOT Projects' and a research report on 'Evaluation of Risks in BOT Projects'. He was also a reviewer for the UNIDO Guidelines for the Development, Negotiaion, and Contracting of BOT Projects.

Andrew Walsh is a Partner in the Projects & Commercial Group of Pinsent Curtis Biddle the sale or privatisation of public sector businesses or assets, one of the UK's largest national law firms. From a corporate background for the last ten years or so, Andrew has specialised in UK project finance work including PFI and PPP projects, outsourcing and contracting-out typically involving some public sector element, and with a particular focus on the transport, local government and education sectors. He has acted for a wide range of public authorities, including several government departments, private sector consortia and debt funders in relation to PFI and PPP projects.

Tsunemi Watanabe is an Associate Professor in the Department of Infrastructure Systems Engineering at the Kochi University of Technology. He received bachelor and master's degrees from the Hokkaido University and a doctoral degree from the Department of Geography and Environmental Engineering at Johns Hopkins University. He has taught environment, construction, and infrastructure management at the University of Tokyo, the Kochi University of Technology, and the Asian Institute of Technology. His past research works include stochastic programming models for environmental decision making under uncertainty, design of procurement systems of public works, bidding analysis, safety and quality management in construction, risk analysis and risk management process in a construction project and sustainable construction.

Xue-Qing Zhang worked as a Research Assistant at the Department of Civil Engineering of The University of Hong Kong, examining BOT-type procurement modalities, while studying the 'Procurement of Privately Financed Infrastructure Projects' in general. He has worked with province and ministry level governmental departments in China, assisted in the preparation and management of a group corporation, and served as an editor of the *Journal of Soil and Water Conservation in China*.

Introduction: public-private partnership in infrastructure development

Akintola Akintoye, Matthias Beck and Cliff Hardcastle

Background

Public-private partnerships (PPP) in facilities development involve private companies in the design, financing, construction, ownership and/or operation of a public sector utility or service. Such partnerships between the public and private sector are now an accepted alternative to the traditional state provision of public facilities and services. Arguably, the joint approach allows the public sector client and the private sector supplier to blend their special skills and to achieve an outcome, which neither party could achieve alone.

The use of PPP is not straightforward. There are complex issues that should be addressed by governments in order to embrace this procurement method for infrastructure development. Malhotra (1997) has argued that governments involved in private finance initiatives (PFI) need to concern themselves with issues such as transparency of the process, competitiveness of the bids, appropriate allocation of risk, developer returns commensurate with risks, government guarantees and credit enhancements. According to Malhotra (1997), risk sharing among the government, utility, lenders and developers is often at the heart of most reservations or debate about private sector BOT/BOO (build-operate-transfer/build-own-operate) projects.

Examples of facilities developed through PPP project funding abound worldwide. It has been used in industrialised countries, such as the United States, in Eastern Europe, the Pacific Rim and in countries with tremendous new infrastructure demands, such as in Latin America. Where the energy generation has been privatised, this has encouraged the private development of new electrical production. The capital-intensive nature of road projects, in a time of intense competition for limited governmental resources, have made PPP road projects based on either shadow or toll revenues attractive. Other projects in which PPP infrastructure developments are being used include waste disposal and telecommunications.

In the UK, the most popular type of PFI contract is design, build, finance and operate (DBFO) or design, construct, manage and finance (DCMF). Currently, most PFI schemes operate under a DBFO contract arrangement in which the public sector makes monthly, quarterly or annual payments for the use of privately owned facilities over the lifetime of the concession. For example, the major PFI developments for the National Health Service (NHS) are DBFO schemes with the primary concession period ranging between 25 and 40 years. The PFI prison building programme is being advanced under the DCMF label (as opposed to DBFO) for which Kent (1998) has extensively documented the extent of the government's commitment. Elsewhere, PFI is to be found embodied in contracts labelled 'design, build, operate and maintain' (DBOM) and 'build, own, operate and transfer' (BOOT).

All over the world where PPP procurement has been used in one form or another, the way in which associated risks are handled or treated has become an important issue. This book addresses issues associated with the risk management of PPP projects based on contributions from eminent scholars and practitioners from different parts of the world. The book is unique in the sense that it focuses on practical experiences. In addition, the book also draws on materials from the research project 'Standardised Framework for Risk Assessment and Management of PFI Projects' which was conducted at Glasgow Caledonian University and funded by the UK Engineering and Physical Sciences Research Council and the Department of the Environment, Transport and Regions (now called the Department of Trade and Industry).

Structure and summary of the chapters

This book is structured into four parts, namely issues of communication, stakeholder perspectives, international perspectives and frameworks for the management of PFI risks.

Risk, value and communication in PPP procurement

Part one covers issues dealing with risk, value and communication in relation to PPP procurements and discusses the general issues of PPP uptake around the world. This part has four chapters.

Chapter 1 by Bing Li and Akintola Akintoye presents an overview of the involvement of the private sector in the delivery of public services across the world. The chapter notes that the level of development of PPPs across the world differs widely. Typically, in the developed countries, PPPs are seen in all areas of public service provision including education, health services, waste management and public buildings. In addition, there is a strong regional and sector concentration in the adoption of PPP procurement. Areas where PPP is fast growing are French-speaking Africa, Eastern Europe, Central Asia (particularly Turkey) and South Asia. PPP is least developed in Sub-Saharan African countries, although there is potential for it to grow if there is an appropriate political will.

Chapter 2 by Cliff Hardcastle and Kate Boothroyd explores the nature of risks associated with PFI projects. The chapter suggests that when viewed from the perspective of the participants, many of the risk issues are common to the main parties but their importance is variable, while some risk issues are particular to a specific party. Some of the key risks encountered by PFI participants include risks associated with availability, commissioning, construction, credit, cost, demand, demographic changes, design, environment, finance, land, legislative changes, legal, market, operation, performance, planning permission, political/social issues, specification, sponsor, technical, technological obsolescence, time and volume.

Chapter 3 by John Kelly addresses the management of value for money in PFI projects. In this context, value for money gains are defined as 'improvements in the combination of whole-life costs and quality that meets the user's requirements' (OGC, 2000). The value management service described is one that maximises the functional value of a project by managing its development from concept to use through the audit of all decisions against a value system determined by the client.

Chapter 4 by Peter Edwards and Paul Bowen is concerned with risk perception and communication in PPP projects. The chapter commences with a description of the communication environment of project stakeholders. Having provided a communication framework against which to explore risk perceptions and communication in PPPs, risk and risk management are defined and a social perspective of risk explored. This perspective is predicated upon a view of risk as part of a social construct (and thus residing in people rather than in projects). It is concluded that in PPP projects risk arises from the decision-making of project stakeholders engaged in pursuing project objectives at procurement, functional or strategic levels. Perceptions of these risks vary among the different stakeholders, thereby creating the need for effective risk communication.

Stakeholder perspectives on PPP risks and opportunities management

Part two details stakeholder perspectives on PPP including the views of those involved in construction, legal services, finance, facilities management and local administration.

Chapter 5 by Ezekiel Chinyio and Alasdair Fergusson covers PPP risk from a construction perspective. Since PFI aims to transfer many risks from the public to the private sector, the authors call for prudent risk management to ensure that risks are identified, assessed, mitigated and/or priced. Citing an investigation involving PFI practitioners, the chapter discusses the ways in which risks are currently managed in PFI projects. The chapter emphasises an iterative approach to risk management, involving the tasks of risk identification, mitigation, evaluation and control.

Chapter 6 by Darinka Asenova and Matthias Beck provides an investigation of the risk management practices of UK financial services providers. The chapter discusses findings from an empirical analysis of the financial risk management measures utilised by 14 financial companies in the UK. This empirical analysis indicates that the main risk categories routinely examined by financiers include performance-related risk, financial risk, time and cost overruns and other construction risks. In terms of their risk assessment methodology, financial companies place heavy reliance on previous experience, advice from external consultants, checklists, site visits and case studies. The chapter concludes that it may be necessary for companies to place greater emphasis on the reputational risks associated with PPPs.

Chapter 7 by Andrew Walsh discusses how the risk allocation objectives of PFI projects can be attained through a combination of competitive procurement processes, contractual structures and contract terms. The chapter describes how various techniques can be used to identify, allocate and control project risks. In addition, it discusses how wider economic and political developments are likely to impact on the risk management of PFI projects.

Chapter 8 by Evelyn McDowall offers a strategic overview of PPP projects that incorporate facilities management (FM). It assesses how risk is managed throughout the life of the project using tools such as risk allocation matrices, risk registers, risk calculation methodologies, and management practice. The chapter suggests that improvements can be made to the PPP process which reduce risks and decrease the cost of projects in the future.

Chapter 9 by John Hood, Andrew Mills and William Stein details how in recent years practice of risk management has developed in the UK public sector. The introduction of PPP/PFI, with its emphasis on risk transfer, has raised the profile of risk management in the public sector. Nonetheless, expertise within the public sector is frequently fragmented across a number of departments. The authors suggest that unless the public sector develops systems to co-ordinate its risk management expertise, an imbalance between public sector and private sector practices will continue to exist.

International perspectives on PPP risks and opportunities management

Part three provides an international perspective on risk management within the context of PPPs in countries such as Singapore, Austria, Hong Kong, Thailand, the USA, India and South Africa.

Chapter 10 by Robert Tiong and Joseph A. Anderson explores privately-developed infrastructure projects in Asian countries. Asia's track record with respect to effective PPPs in the infrastructure sector is mixed. In this chapter, the application of risk management strategies in several privately-financed projects is discussed. The experiences of investors and lenders are likely to provide a strong foundation for the success for future projects.

Chapter 11 by Michaela M. Schaffhauser-Linzatti explores PPP in Austria. In Austria, large-scale privatisation of nationalised firms has been reducing the dominant share of public enterprises since 1985. Nevertheless, public involvement is still an important element of Austrian economic policy, among others, for providing infrastructure. So far, high volume infrastructure projects have been exclusively financed by federal and local public institutions. Budgetary shortages as well as EC guidelines force them to co-operate with private partners. This chapter describes the existing legal framework and presents two successfully implemented PPP projects, Climate-Wind-Channel Vienna and Cargo-Terminal Werndorf.

Chapter 12 by Mohan M. Kumaraswamy and Xue-Qing Zhang discusses risk management approaches utilised in BOT-type PPPs in China and Hong Kong. Relevant developments and trends of BOT-type PPP projects in China in general, and in Hong Kong in particular, are presented, with specific reference to the evolving and improving practices in risk management and their related strengths and weaknesses.

Chapter 13 by Arthur L. Smith covers PPPs in the USA. This analysis suggests that PPPs offer significant opportunities for improved service delivery, but that they also entail an array of risks. The chapter explores the role of risk analysis in structuring a successful partnership for both low-risk and high-risk projects. It is argued that infrastructure projects, with their requirements for capitalisation and construction, typically have an inherently higher level of risk than service-oriented partnerships.

Chapter 14 by Pantaleo D. Rwelamila, Lucy Chege and Tjiamogale E. Manchidi deals with PPP in South Africa, where municipal authorities are facing tremendous service delivery challenges. In 1998, after two years of preparation, the South African National Government paved the way for PPPs by creating the Municipal Infrastructure Investment Unit (MIIU), a non-profit company tasked with providing technical assistance and grant funding to municipalities investigating innovative service delivery partnerships. The authors argue that, while most forms of traditional privatisation include 'transfer of risks' to one party, risk sharing between the public and the private sectors may improve the efficiency and effectiveness of privatisation.

Chapter 15 by Satyanarayana N. Kalidindi and A.V. Thomas discusses the risks associated with Indian PPP-based road projects. In this context, risk allocation provisions in six road project concessions published by the Ministry of Surface Transport (MOST) are reviewed. The perception of BOT participants on risk management capability and preference of risk allocation with respect to critical risks are also investigated.

Framework for risks and opportunities management of PPP infrastructure development

Chapter 16 by Jirapong Pipattanapiwong, Stephen Ogunlana and Tsunemi Watanabe investigates multi-party approaches to risk management in PFIs. The involvement of a large number of parties increases risks, since each party has different objectives. Furthermore, a risk response taken by one party may create risks to other parties. A multi-party risk management process, however, can assist decision-makers in systematically and efficiently managing risks. The procedure has been applied to a public bridge and elevated road construction project financed by the Asian Development Bank as a case study.

Chapter 17 by Matthias Beck and Caroline Hunter discusses the obstacles to PFI uptake in UK local authorities. Based on a quantitative study of 116 British local authorities, the authors conclude that PFI uptake has been concentrated among large authorities, while smaller authorities were less likely to enter the PFI arena. The authors suggest that this is attributable to the technical complexity of PFI schemes which requires a level of expertise that may be beyond many smaller local authorities.

Chapter 18 by Akintola Akintoye, Matthias Beck, Cliff Hardcastle, Ezekiel Chinyio and Darinka Asenova develops a broad-based framework for identifying and addressing PFI risks. The chapter reviews the gateway produced by the Office of Government Commerce (OGC) as a prelude to the presentation of a new PFI risk assessment framework. The goal of the framework is to provide a basis for improved communication between the different PFI participants; greater consistency in PFI risk assessment and management practices; a harmonisation of the risk management terminology used by different participants; and, improved guidance on tools available for risk management, and when they could be used. Leading PFI practitioners and academics assisted in shaping and finalising the model.

References

Kent D. (1998) Secure partnerships. *Private Finance Initiative Journal*, **3**(5), 15–18.

Malhotra, A.K. (1997) Private participation in infrastructure: lessons from Asia's power sector. *Finance and Development*, **December**, 33–35.

Office of Government Commerce (2000) *Value for Money Measurement*. OGC, London.

PART ONE

PPP: Risk, value and communication in public-private partnership procurement

1 An overview of public-private partnership

Bing Li and Akintola Akintoye

1.1 Introduction

Governments worldwide have sought to increase the involvement of the private sector in the delivery of public services. These initiatives have taken many forms, such as the outright privatisation of previously state-owned industries (Ng, 2000), contracting out of services, such as refuse collection (Sindane, 2000) or cleaning to private firms and the use of private finance in the provision of social infrastructure (Tanninen-Ahonen, 2000). Privatisation has occurred in over 100 countries, most notably in the former Communist countries of Central and Eastern Europe. Contracting out of labour-intensive services has also been widespread. Concessions to build and operate large-scale infrastructure networks, such as roads, have been of particular interest to rapidly developing countries in South America and South East Asia (Hall, 1998).

It is generally recognised that a public-private partnerships (PPP) programme offers a long-term, sustainable approach to improving social infrastructure, enhancing the value of public assets and making better use of taxpayer's money. The concept of PPP in the United States and Europe has existed for centuries, but has become more prominent in recent decades in local economic development (Keating, 1998). Partnerships come in all sizes and types which makes it difficult to group them in a consistent fashion. The most important PPPs since the 1990s have been in the sectors of education, health and transportation. There is a considerable range in partnerships, from those dominated by the private sector to those dominated by the public sector (Savitch, 1998). Some types of partnerships are more prevalent in some nations than others.

This chapter discusses the benefits of PPP and documents the implementation of PPPs across the world. The PPP models examined all belong to the project level PPPs.

1.2 Concept and characteristics of PPP

The numbers and types of PPPs are overwhelming, making the definition of a PPP difficult. In some cases, city officials might describe a tax concession for which business promises to create jobs in the future as a partnership. In other instances, hiring a private contractor to manage a parking garage or to collect garbage might be labelled a PPP. A partnership might be as extensive as privatising facilities or services, or it might simply involve applying financing or management techniques from the private sector (McDonough 1998). This idea of bringing in private finance to finance public sector infrastructure originated with the early occurrences of PPP (The World Bank and the International Finance Corporation, 1992). At the time, the terms 'privatisation', PPP, Alternative Service Delivery (Ford & Zussman, 1997) and Municipal Service Partnerships were used to mean the same thing. Carroll and Steane (2000) defined PPPs in broad terms to encompass a very wide diversity of partnerships and the circumstances in which they arise as 'agreed, co-operative ventures that involve at least one public and one private-sector institution as partners'.

According to the government approach (HM Treasury, 2000), PPPs bring public and private sectors together in long-term partnerships for mutual benefit. It covers a wide range of different types of partnership, including:

(1) The introduction of private sector ownership into state-owned businesses, using the full range of possible structures (whether by flotation, or the introduction of a strategic partner), with sales of either a majority or a minority stake.
(2) The private finance initiative (PFI) and other arrangements, where the public sector contracts to purchase quality services on a long-term basis, so as to take advantage of private sector management skills given the incentive of having private finance at risk. This includes concessions and franchises, where a private sector partner takes on the responsibility for providing a public service, including maintaining, enhancing or constructing the necessary infrastructure.
(3) The franchising of government service provision into wider markets, and other partnership arrangements where private sector expertise and finance are used to exploit the commercial potential of government assets.

The National Council for Public Private Partnership of the USA (Norment, 2000) defines PPP along similar terms to its UK counterpart in terms of a 'contractual arrangement between a public sector agency and a for-profit private sector concern, whereby resources and risks are shared for the purpose of delivery of a public service or development of public infrastructure'. The objective of PPP, accordingly, is to utilise the economies of the private sector to deliver more effectively the service or infrastructure. This can include everything from outsourcing of an Operation and Management contract to full privatization (i.e. transfer of assets from the public to the private sector).

The Canadian Council for Public Private Partnerships (1998) defines a PPP as a co-operative venture between the public and private sectors, built on the expertise of each partner, that best meets clearly defined public needs through the appropriate allocation of resources, risks and rewards. The Council does not consider a contracting out arrangement as a true PPP.

The United Nations organisation PPPUE defines PPP broadly to include informal dialogues between government officials and local community-based organisations, to long-term concession arrangements with private businesses, but not privatisation (http://www3.undp.org/pppue/prog.html, August, 2000). This idea is in accordance with United Nations' statements (1995), in which co-management is seen as part of the definition of a true PPP in addition to shared ownership and equal responsibility.

Academic and industrial participants in PPP projects still regard the concept of PPP as being very ambiguous. Some have argued that PPP includes a wide range of co-operation between the public sector and the private sector. Bennett and Krebs (1991) note that partnerships are part of local economic development, while Collin (1998) argues they are part of municipal development. The inclusion of PPP in Local Economic Development (LED) programmes was initiated in the 1980s by such countries as Britain and Germany, with the focus on specific sectors or social groups at a sub-national level. Bennett and Krebs (1991) state that a straightforward example of a PPP in LED might be a situation in which a city provides land and buildings to a business venture, and the business partner provides the labour, raw material, capital, and management expertise. Both parties share in the risks, gains, and losses of the business venture. In other words, both sides take an equity position in the business. More common are partnerships in which the local government waives taxes or fees or relaxes regulations to entice business investment, or the local government acquires the land and makes public improvements in the hope of enticing private business investment.

There are also several academic researchers and industrial practitioners who suggest that a PPP is different from an outright privatisation initiative. Middleton (2000) described PPP as a successor to privatisation. Moore and Pierre (1988), Faulkner (1997) and Collin (1998) claim that PPPs should be regarded as a viable alternative to privatisation and socialisation, because they provide the opportunity to alter the institutional milieu without the loss of municipal influence.

Sindane (2000) argues in favour of PPP as compared to contractual arrangements where a private party takes responsibility for all, or part, of a government's (departments) functions. Such contractual arrangements, accordingly, are different from the total selling off of state assets or the complete transfer of responsibility for relevant services. Spillers (2000) similarly suggests that PPP offers advantages over privatisation and concession schemes.

1.3 Features of PPP

Although there is no unified definition of PPP all definitions have common features or characteristics. This has led Peters (1998) to identify five general defining features of partnerships.

Firstly, a partnership involves two or more actors, at least one of which is public and another from the private business sector, such as in the context of a Build-Operate-Transfer (BOT) project (Tiong, 1992) and joint venture company (Eckel & Vining, 1985). Councils of Government in the US typically involve a number of local governments co-operating to provide some common services (Peters, 1998). Several practitioners (Tarantello & Seymour, 1998) suggest that partnerships between non-profit organisations and local governments should also be counted as PPPs. Here, very often, more actors are involved and more complex relationships exist (Peters, 1998).

Secondly, in a PPP, each participant is a principal, i.e. each of the participants is capable of bargaining on its own behalf, rather than having to refer back to other sources of authority. In some instances, the public sector has to set up a special agency capable of entering into partnership before collaboration becomes possible (Grimsey & Graham, 1997; NHS, 1999).

A third defining feature of partnerships is that they establish an enduring and stable relationship among actors. There are numerous simple, one-off transactions between the public and private sectors. However, even if a government agency should return to the same supplier to purchase goods and services year after year, this does not constitute a partnership (Peters, 1998). In a PPP there is a continuing relationship, the parameters of which are negotiated among the members from the outset (Middleton, 2000), or a process in which such a partnership is created (Moore & Pierre, 1988).

Fourthly, in a PPP, each of the participants brings something to the partnership (Collin, 1998; Peters, 1998). Therefore, for the partnership to be a genuine relationship, each will have to transfer some resources – material or immaterial – to the partnership. The transfer of material resources (money or land, etc.) is rather obvious (Tiong, 1992). The transfer of other resources, such as authority and any other symbolic values (Bennett & Krebs, 1991), can constitute a less obvious form of partnership.

Finally, a partnership implies that there is some shared responsibility for outcomes or activities (Collin, 1998; HM Treasury, 2000). This differs from other relationships between the public and the private sectors in which the public sector retains control over policy decisions after receiving the advice of organisations in the private sector. In contrast, actual partnerships produce mutual shared responsibility which can make accountability for these decisions difficult to ascertain. Thus, often partnerships are separate organisational structures, rather than bargaining relationships which have been established among otherwise autonomous organisations. This view is closely related to the analysis by Grant (1996) who suggested that ideas of shared authority and responsibility, joint investment, sharing liability/risk-taking and mutual benefit stood at the core of a partnership. The South African Government has accepted some of these ideas in that it excludes an agreement between an institution and a private party, where the latter performs an institutional function without accepting the significant risks, from PPP status and defines it as a borrowing transaction (Government Gazette, 2000; viewed on 8 September 2000).

1.4 Benefits of PPP

Generally, PPP procurement can provide a wide variety of net benefits for a government. Chief amongst these is the possibility of more and better projects being built and services being provided (HM Treasury, 1997). A detailed analysis of potential benefits from PPP has been provided by the Nova Scotia Government (NS, 2000), which is discussed below.

1.4.1 Enhance government's capacity to develop integrated solutions

In a conventional procurement process, projects with a broad scope are generally broken down into their component parts and managed as separate units that have to be implemented sequentially due to budget limitations. As a result, the opportunity to develop an integrated solution that effectively addresses a public sector need is often missed. With PPP procurement, the scope for procurement is expanded to reflect a broader context and the focus can shift to developing an integrated solution. For example, without private sector participation, the Channel Tunnel would not have been finished (Finnerty, 1996).

1.4.2 Facilitate creative and innovative approaches

By moving away from the detailed definition of inputs to a description of desired outcomes, the PPP procurement process allows bidders to compete on the basis of their ability to develop unique and creative approaches to the delivery of the required project (Birnie, 1999; NS, 2000). For example, the developer could build facilities that service multiple purposes while being simultaneously used for a variety of specific community purposes (Utt, 1999). To attract private sector investment in sanitation, Sohail (2000) suggested combining sanitation and water supply together as a package for contractors, and at the same time obliging the users to purchase both water and sewerage connections. However, not all innovative solutions are satisfactory to users. Williams (1998) found that in the USA, contracting out solid waste management did not improve levels of government and resident satisfaction with these services.

1.4.3 Reduce the cost to implement the project

A PPP procurement approach offers the potential benefit of reducing costs, often significantly, or delivering higher quality for the same cost, both for the design-build phase of the project and for the operational phase of the project. Arthur Andersen and Enterprise LSE (2000) claimed that the average saving in a PFI project was 17%, and not 20% as argued earlier by National Audit Office (NAO). Hall (1998) argued that recent contracting out arrangements won by the public sector for refuse collection in the UK achieved less than half the cost reductions than were made in similar contracts by the private sector.

Although the issue of cost savings arising from PPP is still being debated, many PPP participants agreed that cost reductions have been obtained in addition to gains associated with faster delivery of the project and the transfer of risk to the private sector. Cost reductions in PPP procurement have been attributed to synergies, economies of scale and reductions in life-cycle costs (NS, 2000).

1.4.4 Reduce the time to implement the project

In a conventional procurement process, the government construction of major infrastructure projects is typically broken down into relatively small pieces and carried out over an extended period, with the initiation of each phase being tied to a multi-year capital plan. Additionally, acquiring the funds for major public construction projects often entails a complicated and lengthy process with an uncertain outcome (Utt, 1999).

According to NS (2000), the total construction period in a PPP can be reduced because a PPP:

- Enables design and construction to be undertaken concurrently rather than sequentially
- Incorporates incentives in the project that reward the private partner for on-time completion of the project
- Reduces the number of times a government project or proposal goes out to tender
- Discourages the temptation to make ongoing changes to the project design, which can cause both delays and create cost overruns.

The experiences associated with the building of the Tate's Cairn Tunnel in Hong Kong (Downer & Porter, 1992) and a PFI prison (Hall, 1998) have indicated that some time reduction can be associated with PPPs.

1.4.5 Transfer certain risks to the private project partner

One of the objectives of PPP in public procurement is to transfer risks from the public sector to the private partner. Thus, an appropriate risk transfer strategy needs to be developed as part of the planning process of the PPP project, in which risks best managed by the private sector partner are transferred to it, and risks best managed by the public sector partners are retained by it (NS, 2000). As a general rule, PFI/PPP schemes should always transfer to the private partner design, construction and operation risks (both cost and performance). In a PFI risk allocation survey within the UK, Akintoye *et al.* (1999) confirmed that the design and construction risks, operation and financing risk are usually assigned to the private sector. Outside the UK, Thobani (1999) and Hambros (1999) have observed that private partners should often bear exchange and interest rate risks. Demand and other risks will vary widely from contract to contract and between different types of services (HM Treasury, 1995). For example, Hall (1998) stated that residual risk has been largely retained by the public sector.

1.4.6 Attract larger, potentially more sophisticated, bidders to the project

By virtue of the size and scope of a PPP project, governments are often able to solicit interest among larger potential bidders and therefore increase the intensity of the competition. In this context, traditional and new bidders are motivated to propose new and more attractive terms to the government. IJ (2001) (viewed in January 2001; http://212.125.86.158/news.asp?story id=5489) reported that top oil companies such as Exxon, Shell and BP joined in bidding for a 30% joint venture with the China National Offshore Oil Corporation (CNOOC) consortium.

Highly competent consortia, however, can also delay the government's selection process. China, for instance, delayed awarding the construction contract of a US$600 million liquefied natural gas terminal in Shenzhen in Southern China. The Chinese Government needed time to choose between four bidding companies or consortia because each was better than its competitor in its own way (IJ, 2001).

1.4.7 Access skills, experience and technology

Governments can gain new skills, technology and knowledge as a result of undertaking a PPP project. The PPP procurement process requires a rigorous analysis of the project including an analysis of opportunities for innovation. This can expand government expertise beyond that associated with a conventional procurement process. Nielsen (1997) claimed that a significant trend in PPP, like design-build, BOT and BOOT projects, has been the requirement for technology transfer to local project participants.

1.5 PPP models

Today the UK Government is endorsing eight types of PPP models for public services and facilities procurement as shown in Table 1.1. There is clearly some overlap, with a number of existing PPP projects fitting into more than one category. PFI is probably the best known approach for attracting private investment to new developments, such as hospitals in the National Health Service (NHS), or schools for the DfEE.

Since PPP approaches come in so many types and forms, an attempt should be made to summarise them in a model. PPPs usually originate from the initiatives of a service provider, purchaser or regulator (Batley, 1996; Corry *et al.*, 1997). Depending on the degree of government control and private economic scale (Gentry & Fernandez, 1997; Savitch, 1998), private sector involvement can vary from the provision of a service, to outright ownership of facilities. Past analyses have recognised five types of private involvement (Figure 1.1), namely service contracts, leasing, joint ventures, concessions and privatisation.

As concerns service contracts, researchers such as Batley (1996),

Table 1.1 *PPP models in the UK.*

Model	Definition
Asset sales	The sale of surplus public sector assets
Wider markets	Introducing the skills and finance of the private sector to help make better use of assets (both physical and intellectual) in the public sector
Sales of business (by flotation or trade sale)	The sale of shares in state-owned business, by flotation or trade sale, with the sale of a minority (e.g. BNFL) or majority (e.g. CDC) stake
Partnership companies (e.g. NATS)	Introducing private sector ownership into state-owned business, while preserving the public interest and public policy objectives through legislation, regulation, partnership agreements, or retention by government of a special share
Private finance initiative	The public sector contracts to purchase quality services, with defined outputs, on a long-term basis from the private sector, and includes maintaining or constructing the necessary infrastructure; the term also covers financially free-standing projects where the private sector supplier designs, builds, finances and then operates an asset
Joint ventures	Partnerships in which the public and private sector partners pool their assets, finance and expertise under joint management, so as to deliver long-term growth in value for both partners
Partnership investments	Partnerships in which the public sector contributes to the funding of investment projects by private sector parties, to ensure that the public sector shares in the return generated by these investments
Policy partnerships	Arrangements in which private sector individuals or parties are involved in the development or implementation of policy

Source: HM Treasury, 2000.

Stonehouse *et al.* (1996), Gidman *et al.* (1998) and Sindane (2000) have suggested that these contracts represent the simplest form of partnership. A lease arrangement involves a situation where the private sector uses public facilities, and pays a rental fee to provide service. According to Batley (1996), Gidman *et al.* (1998) and Sindane (2000), in a leasing arrangement, the service provider is responsible for the operating, repair, and maintenance costs of assets. The service provider can also be responsible for the collection of tariffs from the service consumers, and thus assumes the collection risk. Usually, the service provider is not responsible for making any new capital investments or for the replacement of the leased assets. Generally, the estimated duration of lease contracts is between eight and fifteen years.

In joint ventures, the government and private companies assume

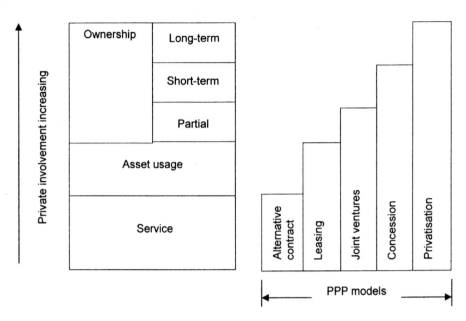

Figure 1.1 *PPP models and private sector involvement level.*

co-responsibility and co-ownership for the delivery of services. Joint venture PPPs provide a vehicle for 'true' public-private partnerships, in which governments, business, non-government organisations and others can pool their resources and generate a shared 'return' (Brodie, 1995; Batley, 1996; Owen and Merna, 1997; Bennett, 1998; Gidman *et al.*, 1998; Keene, 1998; HM Treasury, 2000; Sindane, 2000). The public and private sector partners can either form a new company or assume joint ownership of an existing company, which provides a service (for example, the public sector sells part shares of an existing company to the private sector). Joint ventures are generally used in combination with other types of PPPs (Bennett, 1998; Zhang *et al.*, 1998). For example, the government awards the newly established mixed capital firm with a concession contract for the provision of services.

Under joint ventures, the government acts as the regulator and active shareholder in the operating company. From this position, it may share in the operating company's profits and help ensure the wider political acceptability of its effort. The private sector partner often has the primary responsibility for performing the daily management of the operation. Under a joint venture, the public and private sector partners work together from the earliest possible stages, often forming an institutional vehicle or Project Development Entity (Bennett, 1998) during the pre-investment and development phase of the project. This model provides a forum for direct collaborative dialogue between the public and private sector partners as they work to develop the final project. In a more basic scenario, this can take the form of a working group.

Concession is the most important PPP arrangement for the private sector and is said to contribute to best value service in public services (Tiong, 1992; Finnerty, 1996; Kopp, 1997; Middleton, 2000; Sindane, 2000). The service provider finances, designs and builds a new service facility or substantially improves an existing one. The service provider retains ownership of the completed facility and operates, maintains and repairs it for the duration of the contract, which is typically 20–30 years. The government grants concessions to recover the cost by collecting user charges and tariffs (Tiong, 1992; Tillman, 1997; Bennett, 1998).

Privatisation involves the sale of a state-owned asset either by auction, public stock offering, private negotiation, or outright grant to a private organisation that assumes operating responsibilities. Also known as asset sales (HM Treasury, 2000), this approach involves the complete transfer of equity to the private sector without time limitations. This is a very common practice overseas, as countries race to spin off state-owned enterprises to private investors (Jomo, 1995). Kopp (1997) referred to privatisation as 'divestiture'. New Zealand's sale of its national railroad to Wisconsin Central in return for $60 million in private capital services is an example of divestiture in which governments cash-in money-losing state-owned enterprises for short-term budget relief or investment capital (Wisconsin Central Transportation Corporation, 1995).

1.6 PPP in Europe

It has been suggested that PPP usage in Europe is still very uneven and recent. According to Ball (1999), Spain, Portugal, Finland and Holland have the most advanced PFI-type arrangements. The European Commission July 1997 policy document, AGENDA 2000, makes a reference to the need to develop public-private financing. This referred to the need to improve the regulatory framework 'of public procurement, particularly in respect of the needs of public/private financing, and to promote the implementation of best practice procurement procedures in both the public and private sectors' (Janssen, 2000). In this context, the commission wishes to see an effective co-ordination of its existing financial instruments including the Cohesion Fund, the European Regional Development Fund (ERDF), the European Investment Bank (EIB) and European Investment Fund (EIF).

Joint public/private ownership enterprises have a long history in Europe (Eckel & Vining, 1985). Government ownership can be direct or indirect whereby the government directly either owns shares in the corporation, or a state holding company acquires shares in the company. The most famous example of joint ownership in Germany is Volkswagen. The federal government and the state of Lower Saxony each own 20% of Volkswagen shares, while 60% are owned by private shareholders. Mixed enterprise corporations are also frequent in France (Johnson, 1978). State holding companies are common in many European countries including Italy, Britain and France. In Italy, several large holding companies own shares in a variety of companies, often holding less than a controlling interest (Mazzolini, 1979). In Austria, the major nationalised banks serve as state holding companies for a number of dependent companies (Van der Bellen, 1981).

According to Merna and Smith (1994), the use of BOT contracts is quite popular in Europe as a technique for consortia of constructors and financiers to finance, construct, operate and maintain economic infrastructure. Duffield (1998) stated that BOT could also be applied to small contractors and small projects. The concept of a small contractor partially financing a heating project had been shown to have the potential to benefit tenant, owner and contractor alike.

Stienlet *et al.* (2000) reported how an outsourcing of government service performed in Flanders, Belgium. Outsourcing is defined as the formal contracting out of a general and technical service to a third party. It implies that government enters into a strategic alliance with a third party, while at the same time the steering capacity of the public principal is ensured. Within the Ministry of Flanders, four services were considered to be suitable for outsourcing as a whole: the legal department, the printing office, translation and the IT function.

Reijniers (1994) reported that the parliament coalition in the Netherlands first mentioned PPPs in 1986, in the context of a planned rationalisation of government services. Since then, PPPs have provided the government with a means of reducing the financing deficit, while providing private investment companies with new markets (Reijniers, 1994). Some investment projects undertaken in the Netherlands which could be designated PPP projects are the tunnel under the river Noord, Parkeerschap Den Bosch, Parking Amsterdam, the Westerschelde bank connections, the Betuwe railway, high-speed tracks, Fokkerweg Schiphol-Oost, the Maastricht Sphinx plan, Zeedijk Amsterdam and the Amsterdam IJ bank project.

Spiering (2000) found that PPP projects often play a role in city revitalisation projects. In 1996, the local government of Nijmegen in the Netherlands and ING Real Estate Development made a joint venture agreement. The goal of agreement was the completion of the Mariënburg-project and the disposal of the common real estate.

In France, there is long history of private sector involvement in the provision of public goods. The tradition, going back to the sixteenth century, has been to limit, as much as possible, the involvement of the public sector and to seek the participation of private companies in the building and operation of infrastructure (Frilet, 1996). In the area of water services, several private companies were set up during the last century; for example, the Compagnie Générale des Eaux (in 1853) and the Lyonnaise des Eaux (in 1880). These two companies are also the international leaders in the privatisation of water infrastructure. Since the 1970s, a new wave of PPP is rolling across France. The water distribution sector illustrated this trend with the share of the population receiving private service increasing from 31% in 1854, to close to 60% in 1980 and roughly 75% in 1991 (Frilet, 1996). The involvement of the private sector in France takes various forms, from a BOT-type venture to a simple management service contract. In fact, there is generally a clear interest in the public entity transferring the burden of building and operating the facility to the private company. However, since public entities often want to control the quality and costs of services clauses are usually introduced in order to satisfy these concerns. This contractual scheme is known as a Concession contract.

Ribeiro (2000) claims that BOT was first used in Portugal for infrastructure provision in the mid 1970s. Since then, the BOT model has become accepted as an alternative way of securing private-sector involvement in infrastructure projects. Cardoso (2000), meanwhile, has argued that the Portuguese PPP age began fairly recently in the early 1990s, with a project in the energy sector. In recent years, economic growth in Portugal has resulted in a great demand for basic infrastructures like motorways, bridges, railways, and pipelines. To meet this need for infrastructure, the BOT model is being increasingly used by the Portuguese government in its drive to implement privately financed infrastructure projects (Middleton, 2000).

According to Tanninen-Ahonen (2000), the first Design, Build, Fund and Operate (DBFO) project was started in Finland in 1997. The Lahti motorway project was implemented according to Finnish laws and construction practices, and turned out to be successful in terms of time and money saved. Implementation of the project involved amendments to tax laws.

1.7 PPP in the UK

PPP is a key element in the Government's strategy for delivering modern, high quality services and promoting the UK's competitiveness (HM Treasury, 2000). PPPs cover a range of business structures and partnership arrangements, from the PFI to joint ventures and concessions, outsourcing, and the sales of equity stakes in state-owned business.

HM Treasury (2000) reported that the Labour Government took action to enhance PPPs within days of taking office in May 1997. These initiatives focused on PFI provision and included:

- Contracts with an estimated capital value of over £8 billion which were signed since the election in areas as diverse as hospitals, schools, military helicopter training and water treatment services
- The renegotiation, in June 1998, of the Channel Tunnel Rail Link project, with a capital value of over £4 billion
- Government PPP plans for a further £20 billion of investment. This includes PFI projects currently out to tender and the Government's plans for PPPs for London Underground and National Air Traffic Services (http://www.partnershipsuk.org.uk/).

PFI is the principal model of PPPs in the UK NHS. In the health sector, the NHS is responsible for providing high quality clinical care to patients. PPP in the NHS is not simply about the financing of capital investments, but about exploring the full range of private sector management, commercial and creative skills (NHS, 1999).

Major PFI schemes are typically DBFO, whereby the NHS makes annual payments for the use of privately owned facilities over a primary concession period of 25–40 years. Unlike PFI in other government infrastructure and service schemes, each PFI hospital is tendered by a separate Trust with their own limited budget (Grimsey & Graham, 1997). This means that the private sector partner is responsible for (NHS, 1999):

- Designing the facilities (based on the requirements specified by the NHS)
- Building the facilities (to time and at a fixed cost)
- Financing the capital cost (with the return to be recovered through continuing to make the facilities available and meeting the NHS's requirements)
- Operating the facilities (providing facilities management and other support services).

NHS rules (1999) require that all PFI schemes must demonstrate value for money (VFM) for expenditure by the public sector. This can be achieved if the private sector assumes risks which would otherwise have been borne by the public sector and through cost-effective management by the private sector. The first wave of NHS schemes was studied by Gaffney and Pollock (1999). In this study all outline business cases (OBCs) showed savings of a range between 9% and 229%.

In 1997, the UK Department for Education and Employment began to select proposals for PPPs in schools submitted by Local Education Authorities (LEA) in England and Wales for its financial support. This programme was implemented as a comprehensive nationwide pilot project to demonstrate the feasibility of partnerships for the improvement of school facilities. School PFI projects usually involve the design, building, financing and operation of a package of services for a single school and/ or the maintenance of a number or all school buildings. A key difference from traditional procurement is that PFI requires LEAs to specify outputs as opposed to inputs. This enables the private sector to develop innovative and cost effective solutions within the public sector (DfEE, 2000).

The New Deal for Schools (NDS) of April 1998 gave further support for PFI pilot projects. In total, this involves facilities at over 220 schools with a capital value of close to £200 million (DfEE, 1999). By the end of 1998, more than 70 schools were already scheduled to be replaced or renovated under a PFI approach.

School PPP projects involve buying asset-related services for schools from the private sector. PPP projects will normally involve using assets – such as school buildings and other facilities – including equipment relating to specific services such as heating systems and school meals. The contractor provides and operates these assets, and is often able to generate income from them, for example, from commercial use by the third parties when facilities are not required by the school. Any commercial use is subject to agreement with the LEA and schools.

1.8 PPP in the USA

In the USA, recent years have witnessed an increased interest in the privatisation of government-produced services, particularly at the county and municipal level. Among the services that are historically most commonly contracted out in the USA are solid waste disposal, street construction, management and operating of facilities, building repair, ambulance services, vehicle repair and maintenance, architectural and engineering services, and legal counsel. Small cities are often at the forefront of privatisation (Florestano & Gordon, 1980; Hirsch & Osborne, 2000).

At the same time, leading companies in the USA are beginning to become involved in the social sector – in public schools, welfare-to-work programmes, and inner city redevelopment. Some examples are Bell Atlantic providing computer networks in Union City schools; Marriott International refining its training programme and the Bank of Boston launching the Community Bank programme (Kanter, 1999). Public buildings are also successfully developed under the arrangement of PPP. The Modesto revitalisation project includes a public plaza, a multi-screen cinema, a parking garage and a city/county administrative building with retail shops scheduled for completion in 1999. According to McDonough (1998), these arrangements are probably better characterised as a public–public-private partnership, since city and county officials have to agree on the details before they can negotiate with a private sector developer.

According to Utt (1999), the Public Schools Partnership Act, which was introduced by Senator Bob Graham (D-FL) as S.2397, amends the federal tax code to allow the use of tax-exempt private activity bonds for the construction of privately owned school facilities, leased back to the public school system. Utt (1999) notes that if this bill were broadened to enhance its versatility, it could serve as the foundation for a legislative plan that encourages the use of PPPs to build public schools more rapidly and at a lower cost across the country.

Bloomfield *et al.* (1998) reported how a correctional facility had been financed and built through a complex series of non-competitive contracting arrangements using lease-purchase financing and design-build construction in the USA. In 1991, the Commonwealth of Massachusetts enacted special legislation authorising Plymouth County to enter into a long-term financing lease for a new correctional facility to house federal, state, and county inmates. The special legislation also exempted all project-related expenditures from state bidding and oversight laws that normally govern publicly funded construction projects in Massachusetts.

Martin (1996) noted that hundreds of thousands of housing units have been developed as a result of PPPs in the USA. From a centralised accounting and maintenance system, Dade County, Florida, has structured a portfolio of public-housing properties that are individually managed by private companies as market rate housing, while retaining the social responsibilities of the public sector.

According to Donnellon (1997), the majority of the USA's wastewater treatment operations are still currently run by public entities. However, there is a major push underway toward privatisation of these facilities. Donnellon reported many examples of the initiatives in wastewater treatment operations including Wheelabrator Environmental System, Inc. and Treated Water Outsourcing (TWO), a Nalco/US Filter joint venture, who formed a strategic alliance that pursued industrial outsourcing projects for wastewater treatment. TWO's sole purpose is to design, build, own and operate the treatment facilities and treat water and wastewater on customer sites.

Since the mid 1990s, PPPs (P3s in Canada) have been used selectively to implement large infrastructure projects in Nova Scotia, Canada (NS, 2000). PPP procurement was driven by the twin provincial imperatives of eliminating deficits and a pressing desire to provide new infrastructure. Most notably, the Government was interested in the off-balance sheet treatment of new projects by having them counted as operating leases as opposed to capital leases. The new infrastructure projects included:

- The Construction of 39 schools
- The Highway 104 Western Alignment, linking Truto to Amherstt
- An adult correctional/forensic institution.

According to Grant (1996) PPPs in Canada usually take the form of a partnership between government and the private business sector to deliver public services. Such PPPs produce achievements in two ways. At the most basic level, the partnership creates a dynamic new environment for change within entrenched government bureaucracies. At the strategic level, it allows governments to implement change without losing sight of the true business of government.

The Government of New Brunswick is among the leaders in responding to the need to re-engineer social systems and has chosen a public-private sector partnership as a key element in its strategy. One example in Grant's (1996) study is that of Andersen Consulting who worked with the New Brunswick Human Resources Development Department to redesign business processes in a way that will improve services to citizens and save money. It involves an up-front investment by the firm and a value-based compensation arrangement under which payment of the firm's fee is contingent on the delivery of benefits to the government.

Since experiencing pervasive financial shortfalls, Nova Scotia's Government implemented ambitious programmes to use PPPs to facilitate the construction of new schools. The first phase of Nova Scotia's PPP programme encompassed as many as 41 new schools within three years. Eight have already been completed and are now in service. Under the PPP programme, Nova Scotia's Ministry of Finance requests bids from qualified developers to provide one or several school facilities built to the ministry's specifications in a designated district. Completed projects are provided on a 'turnkey' basis – the developer furnishes the desks, telephones, blackboards and computers, while the school system provides teachers, principles and students. Prospective qualified bidders compete on the price, and the cost of the project is converted into a 20-year lease with annual rent payments equal to 85% of the capitalised cost of the project. In order for the developer to make up the difference in cost and earn a profit on his investment, the contract is structured so that the school system leases the building for specific hours (such as 8:30 AM to 3:30 PM, Monday to Friday, September to June, as well as select off-hour periods). During the hours and days in which the public school system is not using the facility, the developer can rent its space to other approved and compatible organisations and businesses (Utt, 1999).

Stonehouse *et al.* (1996) reported how in the autumn of 1993, the Toronto Hospital (TTH) embarked on an ambitious programme of establishing relationships with a number of private sector organisations with a view to reducing the hospital's operating costs and sustaining (if not improving) quality and productivity. TTH, independently owned and operated as a public hospital, is the major University of Toronto teaching hospital. The complexity of the hospital's operations, in conjunction with severe financial pressures, suggested that the organisation should rely on external skills and resources for functions that are not the hospital's central businesses. A strategic decision was made to alter the traditional pattern of inhouse services, and an assortment of PPPs has emerged. Four types of outsourcing are at the disposal of hospitals: products, services, contract management, and the contracting out of management and the means of production (functional outsourcing). Each involves different degrees of process redesign, product outsourcing, management and labour outsourcing, and capital equipment procurement. Each is predicated on different expectations with respect to savings and new revenue, function by function. Each relationship is tailored to the hospital's requirements and takes into account myriad human resource issues, statutory obligations and internal performance expectations. In addition, by using private sector expertise, the hospital elected to pay more attention to a couple of revenue-generating assets in its property management portfolio such as its parking lots and its 400-apartment residence building (Stonehouse *et al.*, 1996).

The private sector participated very actively in infrastructure development in Latin America and the Caribbean (Roger, 1999). Acevedo (2000) decribed a PPP programme in Brazil which involved a professional training programme for low-income young people, which was created through a partnership of the public sector, private sector and non-governmental organisations (NGOs). Another kind of PPP spreading quickly in Brazil in the last ten years involves the transformation of privately owned natural areas into privately owned protected natural areas (De Araujo, 2000). The Private Reserves of Natural Heritage (PRNHs) are private nature reserves in which the landowner decides if he wants to turn his/her property into a PRNH, without losing his property rights.

Kopp (1997) reported that Mexico has tendered development franchises for over 5400 km of a new limited access motorway between 1989 and 1994. The stated objective of the project was to bring the majority of the service into use before the end of President Salinas' term in 1994. Attracting private capital was critical to the success of the programme. The Ministry of Communications and Transportation selected and divided the feasible areas into discrete projects, prepared preliminary alignments, cross-sections, and structural designs, assembled the right-of-way, developed traffic forecasts and cost estimates, and tendered each project to interested developers.

Bennett (1998) described a joint venture in water service provision in Colombia. Facing huge inefficiencies and poor services, the Government of Cartagena liquidated the public water and sewerage utility and created a mixed-capital company – Acuacar – to serve the city's 750 000 inhabitants. Acuacar is jointly owned by the Government of Cartagena and Aguas de Barcelona, a Spanish provider of water services. It has been awarded a 26-year operation and maintenance contract. The city of Cartagena continues to act as the sole owner of the system, with sole responsibility for funding expansion. Aguas provides operating services and receives a fixed percentage of total revenues and divided distributions from Acuacar's profits.

The Manizales project originated from wide-ranging attempts to address critical issues related to water supply and quality, including the problems caused by coffee producers using a traditional, highly polluting, coffee washing process. With the assistance of Sustainable Project Management (SPM) and UNDP, a new company called Agua Pura SA was set up in late 1994. The shareholders include the Government of Caldas State, the departmental water supply and distribution companies, the electricity utility, the local co-operative of coffee growers, and a provincial financial institution. This group developed a business plan for a full domestic solid waste collection, disposal, and recycling operation, covering 21 municipalities and 5000 commercial businesses, with an initial investment of $3.5 million. Agua Pura SA expected sales in year one of $1.4 million. The Manizales project has been considered worthy of replication.

1.9 PPP in Asia-Pacific

PPPs have been used in many Asian Pacific countries including China, Thailand, Australia, Vietnam, Malaysia, the Philippines, Sri Lanka and Japan.

PFI is arousing intense interest in Japan where it is supported by an Act of Parliament. The most important measure of that Act is the exemption of PFIs from the five-year limitation on central government. Nakamura (2000) noted that local governments are enthusiastic about PFI because of their severe financial difficulties. Some Japanese construction companies already have experience of build-operate-transfer schemes overseas, though no PFI projects are as yet operational in the country. According to Ball (1999), the initiative is deemed suitable for a variety of sectors within the Japanese economy, including transport, health and waste disposal.

China is actively investigating ways to introduce project financing, especially through the BOT schemes to meet the needs for the country's infrastructure and to be attractive to foreign investors and lenders (Wang *et al.*, 2000). Several state-approved BOT projects have been awarded since late 1996, such as the Shanghai Da Chang water project, Changsha power project and Chengdu water project, and Laibin B power plant projects.

Lu *et al.* (2000) have reported that Taiwan plans to engage in PFI developments over the next years. The Government's Council of Economic Planning and Development (CEPD) statistics reported that the country will invest nearly US$40 billion in infrastructure projects, of which US$10 billion is expected to be financed internationally. These projects will include the largest BOT project ever, the North/South High Speed Rail Project. According to Lu *et al.* (2000) current BOT projects in Taiwan can be classified into three categories: construction-stage projects, bidding-stage projects and planning-stage projects. Lu *et al.* (2000) noted that these projects include high-speed rail, mass transit, highways, cable cars, airport terminals, harbours, bridges, office buildings, gyms, resorts, shopping centres, incinerators, parks, and land developments.

Stein (1994) reported on Vietnam's involvement in BOT. Following Vietnam's amended Foreign Investment Law of December 1992, the amended foreign investment regulations of April 1993 and the BOT regulations in November 1993, investors can enter into an implementation agreement, or BOT Contract, with an agency or entity of the Government of Vietnam. The BOT company can be either 100% foreign owned or a joint venture company with Vietnamese partners.

According to the *Daily News* (1995), the Government of Sri Lanka expressed its desire to conduct future investments for new infrastructure projects in conjunction with private sector participants via BOT, or BOO arrangements (Ranasinghe, 1999).

In 1995, the Clark Development Corporation (CDC) used the joint venture model, a variant of the BOT model, to attract private funds for utility development in the Clark Economic Zone in the Philippines. The principal elements of the joint venture model were capital investment by a strategic partner and eventual sale to the public of the CDCs share of the joint venture corporation (Gavieta, 1999).

The Tate's Cairn Tunnel in Hong Kong is the longest automobile tunnel in the territory. The four-lane facility connects densely-populated Kowloon with the rapidly developing new towns to the north through a 4-km twin-tube tunnel. The project was tendered to a development consortium which saved 17 months from the Government's 54-months planned construction programme. According to Downer and Porter (1992), heavy traffic volumes and substantial travel timesaving have resulted in a remarkably quick repayment of the initial construction capital.

In Malaysia, privatisation has been used as a means for affirmative action in redressing ethnic income differences (Ng, 2000). Privatisation in the Malaysian context includes the sale or divestment of state concerns; public issue or sale of shares in state-owned public enterprises; placement of shares with institutional investors; sale or lease of physical assets; joint public-private ventures; schemes to draw private financing into construction projects; 'contracting out' of public services previously provided within the public sector; and allowing private competition where the public sector previously enjoyed a monopoly (Jomo, 1994).

Korea has tried to address its infrastructure shortage through the introduction of private funding and more efficient project management. Park (1998) reported that the Korean Government passed 'The Private Capital Inducement Act for Infrastructure Development' (The Act), which served as enabling legislation for private infrastructure development in 1994. The Act states basic objectives, such as increased supply of infrastructure facilities, their efficient management, and balanced development of the whole economy. It also sets forth the methods, procedures, and rules relating to government supervision for private infrastructure providers, and provides tax, financial and other incentives for the private sector to invest in infrastructure projects. The Act classified infrastructure facilities into two types: primary facilities and secondary facilities. In general, primary facilities are more sensitive to public interest and require more investment than secondary facilities. They include such infrastructure facilities as road, railway, harbour, airport, sewage, and telecommunications. Examples of secondary facilities are power plants, distribution complexes, bus terminals, and parking lots. Primary facility projects are developed on a build-transfer-operate (BTO) mechanism and secondary facility projects are developed through a BOO mechanism.

PPP is also becoming popular in Australia. New South Wales (NSW), like other states of Australia, has faced increasing demands for infrastructure of all types. Meeting these demands has been difficult due to budgetary restraints. Raneberg (1994) has described the four key reform objectives of the NSW Government as the optimal allocation of scarce public sector resources, efficiency, better service, and accountability for performance. A range of market-orientated initiatives have been employed to secure these objectives. These include measures ranging from wholesale privatisation to contracting out in-house service needs, as well as private sector participation in infrastructure projects.

Laurie (2000) described the Melbourne City Link project which was financed by the private sector as a BOOT scheme with a concession period of 34 years. The Concession Deed is the primary contract between the State and the City Link developer Transurban City Link Limited. City Link is a $2 billion ($US1.2 billion) privately funded electronic tollroad in the heart of Melbourne. The project joins together three of the city's freeways, creating a 22-km expressway linking the major routes between Melbourne Airport, the port and industrial centres in the Southeast.

According to an NSW Government report (1993), contracting out services from road maintenance to hospital catering achieved average savings of 20%. At least 34 major projects of privately provided infrastructure have either been completed or are planned in NSW. The total cost of these amounts to over A$13 billion. Virtually all these projects are BOO, BOT or BOOT models.

1.10 PPP in Eastern Europe

PM (1995) reported how Bulgarian municipal governments have been looking for ways to divest themselves of their landholdings in a cost-effective, fair, and transparent manner. Several international developers are providing training for Bulgarian public officials in structuring PPPs. These international developers include the United States Agency for International Development, the International City/Country Management Association (ICMA), and the Planning and Development Collaborative International (PADCI).

Bulgarian cities and towns are encouraging private housing development through PPPs which concede to private developers the right to build housing units on municipally owned land. In exchange for these land development rights, the municipalities receive from the contractors a pre-determined proportion of units (approximately 25%) that they can use for social housing. Now the ICMA and the Bulgarian Municipal Development Programme (BMDP) are working to expand the development process beyond the three cities (Strara Zagora, Rousse and Bourgas) and beyond housing development to include large, complex and mixed-use (residential/commercial/industrial) projects (PM, 1995).

Plewik (2000) reported on an amendment to the project of the Motorway Construction Program which allowed several different financial systems to be included in the programme, including a license system, BOT and public-private joint financing programmes. The PPP system involves state partici-pation in construction in terms of the direct financing of certain sections or otherwise, by creating a guarantee system and securing the financial stability of the project. In practice this means that the state is ready to cover possible current deficit in the first years of operations, as well as being willing to minimise potential deficit through reduced taxes. Government assistance mechanisms were intended to enhance different types of relief and exemptions. The amendment of income tax regulations provides for tax exemption from subsidies coming from the state and the independent budgets of local autonomies

In Slovenia the Kozjansko-Obsoelje Centre has become the initiator of the co-development and chief promoter of tourist development projects. According to Strukelj and Potocan (2000), the combination of the simulta-neous implementation of public and private interest in the development proved to be successful in the first year of the Centre's operation.

In the Czech Republic, the Spisska Regional Environmental and Energy Company (SREEC), has become the vehicle for PPP. It is a joint venture between the municipality (40%) and a Slovak private company, Pluralité-Mega (60%). Supported by SPM and the Swiss and Canadian Governments, SREEC created subsidiaries with local and international partners and investors to implement projects in district heating and energy efficiency, forest management linked with housing development, a capacity building centre for community development and solid waste management.

1.11 PPP in Africa

African states are now recognising that PFI/PPP is probably the most effective way for them to go forward.

PPP is very much a buzzword in South Africa today, where the prison sector in particular has become an area for PPP (Ball, 1999). PPP agreements need prior written approval by the national treasury or the relevant provincial treasury (Government Gazette, 2000) (viewed on 8 September 2000). A build-operate-train-transfer (BOTT) programme has been implemented in a water system in South Africa. However, according to Gentry and Fernandez (1997), this kind of PPP has not yet been widely applied worldwide.

Botswana is said to be an excellent example of a country that has exploited its natural resources through PPP arrangements (Ball, 1999).

A wastewater service is provided on a small scale in Addis Ababa, Ethiopia. Gentry and Fernandez (1997) noted that non-profit making organisations played the most important role in this programme. Forty-one different donor entities, including international charities, foreign aid agencies and NGOs helped fund the projects.

Gidman *et al.* (1998) claimed that in Zimbabwe, private sector participation is now being increased on the initiative of government to rationalise the local government service. A project in Harare envisages the creation of an energy-environment management enterprise to bring eco-efficient technologies to the Willowvale Industrial Park, in order to improve energy and water management practices.

Lavigne (1995) and Franceys (1997) reported on a water supply lease contract in Guinea. In 1986 the Guinean National Water Company awarded a ten-year leasing contract for a Guinean certified company, with 51% of shares held by SAUR and Compagnie Générale des Eaux of France, following an international call for tenders. The leasee collected the fee to cover operation and investment costs.

Franceys (1997) described a Ugandan pilot water management system project, funded by a French Government grant, which is expected to pave the way for the privatisation of Uganda's water systems management.

From the mid 1980s, Cameroon was besieged by a serious economic crisis, which was further aggravated by political and social uprisings during the early 1990s. Public finances dropped considerably, making it difficult to pursue with vigour its rural development strategy. Many local communities initiated their own development projects and sought foreign assistance directly, either through their own local NGOs or through foreign NGOs based in Cameroon. Tafah and Asondoh (2000) reported on one such local initiative – the Niger Integrated Rural Development Project (NIRDP) which received the approval and support of the European Union and the Netherlands Development Organisation (SNV). These two organisations work in partnership with the local community to ensure a better standard of living for the Niger people.

1.12 Conclusions

In most developed countries, PPPs are utilised to some degree or another in the provision of services or infrastructure. In the Western European hemisphere, the UK has taken a lead position in PFI procurement, although other countries experiment with a wide range of PPPs. In the US, there is a wide diversity of PPPs which involve almost all sectors of government, whereby contracting out waste management to private companies is very popular. Moreover, the Government strongly supports the business sector in joint capital investment in sectors such as energy, water, transportation, etc.

In the developing world, there is a strong regional concentration of PPP contracts, principally in Latin America, and followed by South East Asia. Areas in which there is growing private participation include French-speaking Africa, Eastern Europe, Central Asia (particularly Turkey), and South Asia. The regions with the least PPP are Sub-Saharan African countries, where PPP projects are normally on a small scale.

References

Acevedo C. (2000) A training program in Brazil: an example of public-private partnership. In *Public and Private Sector Partnerships: the Enabling Mix* (eds L. Montanheiro & M. Linehan), pp. 11–20. Sheffield Hallam University Press, Sheffield.

Akintoye A., Fitzgerald E. & Hardcastle C. (1999) Risk management for local authorities. In: *Private Finance Initiative Projects*, pp. 81–91. RICS *Cobra*, London.

Arthur Andersen and Enterprise LSE (2000) *Value for Money Drivers in the Private Financial Initiative*. The Treasury Taskforce, http//www.treasury-projects-taskforce.gov.uk/series_1/andersen/7tech_contents.html.

Ball J. (1999) Marketing a concept abroad. *Project Finance*, **193**, 18–19.

Batley R. (1996) Public-private relationships and performance in service provision. *Urban Studies*, **33**(4–5), 723–751.

Bennett E. (1998) *Public-Private Cooperation in the Delivery of Urban Infrastructure Services (Water and Waste)*. PPPUE Background Paper, UNDP/Yale Collaborative Programme. http://www.undp.org/pppue/.

Bennett R. & Krebs G. (1991) *Local Economic Development Public-private Partnership Initiatives in Britain and Germany*. Belhaven Press, London.

Birnie J. (1999) Private finance initiative (PFI) – UK construction industry response. *Journal of Construction Procurement*, **5**(1), 5–14.

Bloomfield P., Westerling D. & Carey R. (1998) Innovation and risk in a public-private partnership: financing and construction of a capital project in Massachusetts. *Public Productivity and Management Review*, **21**(4), 460–471.

Brodie M. (1995) Public/private joint ventures: the government as partner – bane or benefit? *Real Estate Issues*, **20**(2), 33–39.

Cardoso N. (2000) Portuguese public-private partnerships. *The Newsletter of the International Project Finance Association*, **1**(2), 5–8.

Carroll P., & Steane P. (2000) Public-private partnerships: sectoral perspectives. In: *Public-Private Partnerships: Theory and Practice in International Perspective* (ed. S. Osborne), pp. 36–56. Routledge, London.

Collin S. (1998) In the twilight zone: a survey of public-private partnerships in Sweden. *Public Productivity and Management Review*, **21**(3), 272–283.

Corry D., Le Grand J. & Radcliffe R. (1997) *Public/Private Partnerships: a Marriage of Convenience or a Permanent Commitment?* IPPR, London.

Daily News (1995) President's Policy Statement. *Daily News*, 7 January, **12**, 15.

De Araujo A. (2000) Private reserves of natural heritage: a public-private partnership for nature conservation in Brazil. *Public and Private Sector Partnerships: the Enabling Mix* (eds L. Montanheiro & M. Linehan), pp. 83–93. Sheffield Hallam University, Sheffield.

DfEE (1999) *New Deal for Schools PPP Pilots – Interim Progress Report.* Department for Education and Employment, London.

DfEE (2000) *Public Private Partnerships: a Guide for School Governor.* Department of Education and Employment, London.

Donnellon T. (1997) Privatisation of wastewater treatment facilities: promising opportunities, risk management challenges. *Water Engineering and Management*, **144**(11), 12–15.

Downer J. & Porter J. (1992) Tate's Cairn Tunnel, Hong Kong: South East Asia's longest road tunnel. In: *Proceedings of the 16th Australian Road Research Board Conference* (7), pp. 153–165. Perth.

Duffield C. (1998) Commercial viability of privately financed heating systems in Europe – a case study. *Engineering, Construction and Architectural Management*, **5**(2), 3–8.

Eckel C. & Vining A. (1985) Elements of a theory of mixed enterprise. *Scottish Journal of Political Economy, Edinburgh*, **32**(1), 82–93.

Faulkner H. (1997) Bridges to sustainability: engaging the private sector through public-private partnerships. In: *Bridges to Sustainability: Business and Government Working Together for a Better Environment* (ed. L. Gomez-Echeverri). Yale School of E&ES Bulletin 101. http://www.undp.org/pppue/.

Finnerty J. (1996) *Project Financing: Asset-based Financial Engineering.* Wiley, New York.

Florestano P. & Gordon S. (1980) Public vs. private: small government contracting with the private sector. *Public Administration Review*, **40**(1), 29–34.

Ford R. & Zussman D. (1997) *Alternative Service Delivery: Sharing Governance in Canada.* Institute of Public Administration of Canada, Ottawa.

Franceys R. (1997) *Private Sector Participation in the Water and Sanitation Sector.* http://www.lboro.ac.uk/well/occpaps/nos.htm.

Frilet M. (1996) Underlying contractual and legal conditions for a successful private/public partnership in the water sector. *The International Construction Law Review*, **13**(3), 281–290.

Gaffney D. & Pollock A. (1999) Pump-priming the PFI: why are privately financed hospital schemes being subsidised? *Public Money and Management*, **19**(1), 55–62.

Gavieta R. (1999) 'Private funds, utility development, and government policy: the Clark experience. *Journal of Project Finance*, **4**(4), 51–55.

Gentry B. & Fernandez L. (1997) Evolving public-private partnerships: general themes and urban water examples. In: *Globalisation and the Environment: Perspectives from OECD and Dynamic Non-Member Economies*, pp. 19–25, OECD, Paris. http://www.undp.org/pppue/.

Gidman P., Blore I., Lorentzen J. & Schuttenbelt P. (1998) *Public-Private Partnerships in Urban Infrastructure Services*. UMP Working Paper Series 4, Nairobi: UNDP/Habitat/World Bank. pp. 12–36, http://www.ndp.org/pppue/.

Government Gazette (2000) http://www.gov.za/gazette/regulation/2000/21082.pdf. p. 54.

Grant T. (1996) Keys to successful public-private partnerships. *Canadian Business Review*, **23**(3), 27–28.

Grimsey D. & Graham R. (1997) PFI in NHS. *Engineering, Construction and Architectural Management*, **4**(3), 215–231.

Hall J. (1998) Private opportunity, public benefit? *Fiscal Studies*, **19**(2), 121–140.

Hambros S. (1999) *Public-Private Partnerships for Highways: Experience, Structure, Financing, Applicability and Comparative Assessment*. Council of Deputy Ministers Responsible for Transport and Highway Safety, Canada.

Hirsch W. & Osborne E. (2000) Privatisation of government services: pressure group resistance and service transparency. *Journal of Labour Research*, **21**(2), 315–326.

HM Treasury (1995) *Private Opportunity, Public Benefit: Progress in the Private Finance Initiative*. HMSO, London.

HM Treasury (1997) *Bates Review*. HMSO, London.

HM Treasury (2000) *Public Private Partnerships – the Government's Approach*. HMSO, London. http://www.hm-treasury.gov.uk/docs/2000/ppp.html.

IJ (2001) *Infrastructure Journal*, http://www.infrastructurejournal.com.

Janssen J. (2000) The European construction industry and its competitiveness: a construct of the European Commission. *Construction Management and Economics*, **18**, 711–720.

Johnson C. (1978) *Japan's Public Policy Companies*. American Enterprise Institute for Public Research, Washington.

Jomo K. (1994) Privatisation. In: *Malaysia's Economy in the Nineties* (ed. K.S. Jomo). Pelanduk Publications, Petaling Jaya.

Jomo K. (1995) *Privatisation Malaysia: Rents, Rhetoric, Realities*. Westview Press, Boulder.

Kanter R. (1999) From spare change to real change. *Harvard Business Review*, **77**(2), 122–132.

Keating M. (1998) Commentary: public-private partnerships in the United States from a European perspective. In: *Partnerships in Urban Governance: European and American Experience* (ed. J. Pierre), pp. 163–186. St Martin's Press, New York.

Keene W. (1998) Reengineering public-private partnerships through shared-interest ventures. *The Financier*, **5**(2–3), 55–59.

Kopp J. (1997) *Private Capital for Public Works: Designing the Next-generation Franchise for Public-Private Partnerships in Transportation Infrastructure.* Master Thesis, Department of Civil Engineering, Northwestern University, USA. http://iti.acns.nwu.edu/clear/infr/kopp/index.htm.

Laurie J. (2000) *Melbourne City Link.* ATSE. http://www.atse.org.au/publications/focus/focus-laurie.htm.

Lavigne J. (1995) Two African experiences. In: *Water Management in Cities* (ed. D. Lorrain). Economica, Paris.

Lu Y., Wu S., Chen D. & Lin Y. (2000) BOT projects in Taiwan: financial modelling risk, term structure of net cash flows, and project at risk analysis. *Journal of Project Finance*, **5**(4), 53–63.

McDonough K. (1998) Constructive financing. *The American City and County*, **113**(6), 18–27.

Martin J. (1996) The management of public housing: forging new partnerships. *Journal of Property Management*, **61**(2), 24–29.

Mazzolini R. (1979) *Government Controlled Enterprise.* Wiley, New York.

Merna A. & Smith N. (1994) *Projects Procured by Privately Financed Concession Contracts.* Project Management Group, UMIST Publications, Manchester.

Middleton N. (2000) *Public Private Partnerships – a Natural Successor to Privatisations?* http://www.pwcglobal.com/uk/eng/about/svcs/pfp/ppp.html.

Moore C. & Pierre J. (1988) Partnership or privatisation? The political economy of local economic restructuring. *Policy and Politics*, **16**(3), 169–178.

Nakamura T. (2000) The PFI situation in Japan. *The Newsletter of the International Project Finance Association*, **2**(1), 5–7.

Ng B. (2000) Privatisation of ethnic wealth redistribution in Malaysia: experiences and lessons for developing countries. *Public and Private Sector Partnerships: the Enabling Mix* (ed. L. Montanheiro & M. Linehan), pp. 461–467. Sheffield Hallam University, Sheffield.

NHS (1999) *Public Private Partnerships in National Health Service. The Private Financial Service: Good Practice.* HMSO, London.

Nielsen K. (1997) Trends and evolving risks in design-build, BOT and BOOT projects. *The International Construction Law Review*, **14**(2), 188–197.

Norment R. (2000) Executive director of NCPPP, email communication.

NS (2000) *Review of Public Private Partnership Processes.* http://www.gov.ns.ca/finance/index.htm (visited July 2001).

NSW Government (1993) *Budget Paper No. 2.* Treasury, Sydney.

Owen G. & Merna A. (1997) The private finance initiative. *Engineering, Construction and Architectural Management*, **4**(3), 163–177.

Park T. (1998) Private infrastructure development in Asia. *Journal of Project Finance*, **4**(3), 25–33.

Peters B. (1998) With a little help from our friends: public-private partnerships as institutions and instruments. In: *Partnerships in Urban Governance: European and American Experience* (ed. J. Pierre), pp. 11–33. Macmillan, London.

Plewik M. (2000) Public-private partnership: toll motorways construction in Poland. In: *Public and Private Sector Partnerships: the Enabling Mix* (ed. L. Montanheiro & M. Linehan), pp. 499–506. Sheffield Hallam University, Sheffield.

PM (1995) Public/private partnerships for land development in Bulgaria. *Public Management*, **77**(11), 30–31.

Ranasinghe M. (1999) Private sector participation in infrastructure projects: a methodology to analyse viability of BOT. *Construction Management and Economics*, **17**, 613–623.

Raneberg D. (1994) Innovations in the public-private provision of infrastructure in the Australian State of New South Wales. In: *Public Management – New Ways of Managing Infrastructure Provision*, pp. 27–54. OECD, Paris.

Reijniers J. (1994) Organisation of public-private partnership projects: the timely prevention of pitfalls. *International Journal of Project Management*, **12**(3), 137–142.

Ribeiro F. (2000) Appraisal of BOT system for infrastructure projects in Portugal. In: *CIB W92 Procurement System Symposium* (ed. A. Serpell), pp. 621–630. Department of Construction Engineering and Management, Pontificia Universidad Católica de Chile, Santiago.

Roger N. (1999) Recent trends in private participation in infrastructure. *Public Policy for the Private Sector*. The World Bank Group, Note 196, Washington, DC.

Savitch H. (1998) The ecology of public-private partnerships: Europe. In: *Partnerships in Urban Governance: European and American Experience* (ed. J. Pierre), pp. 175–186. Macmillan, London.

Sindane J. (2000) Public-private partnerships: case study of solid waste management in Khayelitsha-Cape Town, South Africa. In: *Public and Private Sector Partnerships: the Enabling Mix* (ed. L. Montanheiro & M. Linehan), pp. 539–564. Sheffield Hallam University, Sheffield.

Sohail M. (2000) *PPP and the Poor in Water and Sanitation – Interim Finds*. Water, Engineering, and Development Centre, Loughborough University, Leicestershire.

Spiering W.D. (2000) Public private partnership in city revitalisation – a Dutch example. In: *Public and Private Sector Partnerships: the Enabling Mix* (ed. L. Montanheiro & M. Linehan), pp. 565–576. Sheffield Hallam University, Sheffield.

Spillers C.A. (2000) Airport privatisations: smooth flying or a crash landing? *The Journal of Project Finance*, **5**(4), 41–47.

Stein S.W. (1994) Build-operate-transfer (BOT): a re-evaluation. *The International Construction Law Review*, **11**(2), 101–113.

Stienlet G., van Sprundel P. & De Ryck W. (2000) Public-private partnerships: what about outsourcing? In: *Public and Private Sector Partnerships: the Enabling Mix* (ed. L. Montanheiro & M. Linehan), pp. 577–590. Sheffield Hallam University, Sheffield.

Stonehouse J., Hudson A. & O'Keefe M. (1996) Private–public partnerships: the Toronto hospital experience. *Canadian Business Review*, **23**(2), 17–20.

Strukelj T. & Potocan V. (2000) Partnership of public and private sectors in the development of tourism in Slovenia. In: *Public and Private Sector Partnerships: the Enabling Mix* (ed. L. Montanheiro & M. Linehan), pp. 591–598. Sheffield Hallam University, Sheffield.

Tafah E.E.O. & Asondoh R.T. (2000) Partnership mix in assisting local communities in sustainable development in Cameroon. In: *Public and Private Sector Partnerships: the Enabling Mix* (ed. L. Montanheiro & M. Linehan), pp. 599–609. Sheffield Hallam University, Sheffield.

Tanninen-Ahonen T. (2000) PPP in Finland: developments and attitude. In: *CIB W92 Procurement System Symposium* (ed. A. Serpell), pp. 631–639. Santiago, Chile.

Tarantello R. & Seymour J. (1998) Affordable housing through non-profit/private–public partnerships. *Real Estate Issues*, **23**(3), 15–17.

The Canadian Council for Public Private Partnerships (1998) http://home.inforamp.net/~partners/awardfaq.htm (visited November, 2001).

The World Bank and the International Finance Corporation (1992) *IFC Investing in the Environment*. The World Bank, Washington, DC.

Thobani, M. (1999) Private infrastructure, public risk. *The Newsletter of the International Project Finance Association*, **1**(1), 5–7.

Tillman R. (1997) Shadow tolls and public-private partnerships for transportation projects. *Journal of Project Finance*, **3**(2), 30–37.

Tiong R.L.K. (1992) *The Structuring of Build-Operate-Transfer Construction Projects*. CACS, Nanyang Technological University, Singapore.

United Nations (1995) Comparative experiences with privatisation: policy insights and lessons learned. *United Nations Conference on Trade and Development*. United Nations, New York/Geneva.

Utt R.D. (1999) *How Public-Private Partnerships Can Facilitate Public School Construction*, http://www.heritage.org/library/backgrounder/bg1257.htm (visited November 2001).

Van Der Bellen A. (1981) The control of public enterprises. *Annals of Public and Co-operative Economy*, **52**(1–2), 73–100.

Wang, S.Q., Tiong, R.L.K., Ting S.K. & Ashley D. (2000) Evaluation and management of foreign exchange and revenue risks in China's BOT projects. *Construction Management and Economics*, **28**, 197–207.

Williams R.L. (1998) Economic theory and contracting out for residential waste collection: how 'satisfying' is it? *Public Productivity and Management Review*, **21**(3), 259–271.

Wisconsin Central Transportation Corporation (1995) *1995 Annual Report*. WCTC, Rosemont, Illinois.

Zhang W.R., Wang S.Q., Tiong R.L.K., Ting S.K. & Ashley D. (1998) Risk management of Shanghai's privately financed Yan'an Donglu tunnels. *Engineering, Construction and Architectural Management*, **5**(4), 399–409.

2 Risks overview in public-private partnership

Cliff Hardcastle and Kate Boothroyd

2.1 Introduction

The scope of this chapter is a pedagogic view of risks in private finance initiative (PFI) projects, to include types and perceptions of risks. However, the chapter does not go into risk assessment and mitigation, as these issues are covered in other chapters in this book. An overview of risks and the PFI form of procurement are made from a theoretical perspective, before contemporary risks in PFI projects are examined from a practical dimension.

Recent research, funded by the Engineering and Physical Science Research Council (EPSRC) and the Department of the Environment, Transport and the Regions (DETR) (now the Department of Trade and Industry), and conducted by Glasgow Caledonian University, has informed the practical considerations of this chapter.

2.1.1 Public-private partnerships

Public-private partnerships (PPP) can be described as a contractual relationship where a private party takes responsibility for all or part of a government's (departments) functions. In essence, it is a contractual arrangement between a public sector agency and private sector concern, whereby resources and risks are shared for the purpose of delivering a public service, or for developing public infrastructure. Some project schemes may not be very robust or their financial burdens may be unbearable for the private sector alone, forcing public bodies to share in the financing and risk-bearing of such projects, or vice versa (Merna & Smith, 1999). Various measures, like outsourcing, deregulation, privatisation, etc., have been used by government to enhance PPP. The intent, in any case, is to combine the resources of the public and private sectors, in the quest of providing services at optimal levels to the public.

The extent of the roles and responsibilities of both public and private sectors could vary in different PPP projects. However, the public sector (always) retains responsibility for deciding on the nature of services to be provided, the quality and performance standards of these services to be attained, and taking corrective action if performance falls below expectation (Smith, 2000).

Lam (1999) reports that, despite some casualties, BOT (build, operate, transfer) projects, which are a subset of PPP, have been completed and have started reaping expected revenue. However, such procurement strategies are not completely devoid of risks, albeit secondary and residual risks. For instance, Akintoye and Black (1999) identified some risks associated with partnering arrangements, namely; managers being unwilling to relinquish control, partners becoming complacent, partners reverting to adversarial relationships, etc. Comerford and Puryear (1995) also pointed out that if the services to be provided, the service provider(s), the goals of the project, the guarantees and the relationships between project participants have been ill defined, then things can go wayward. Therefore, co-operative arrangements should not be used blindly.

2.2 Private finance initiative

PFI is a type of PPP where project financing rests mainly with the private sector. This initiative represents the strategy in the UK through which government contracts to purchase quality public sector services, on a long-term basis from the private sector, and includes maintaining and possibly constructing the necessary infrastructure. This is fundamentally about the delivery of a service rather than the procurement of construction assets. However, such assets may be required to facilitate the delivery of the required service.

Having passed through the learning curve, the use of PPP and PFI for public service delivery is gaining ground all over the world. This increasing usage is partly because citizens in many countries are demanding ever higher levels of services from their governments, and funds are usually insufficient to meet all these demands (Goldsmith, 1997). Therefore many governments are turning to private finance as an alternative means of funding some of their service delivery needs.

In the UK context, PFI overshadows other forms of PPP because the Government is currently promoting it. In this regard therefore, it may not be untrue, in the context of public service delivery, that the terms PFI and PPP are used interchangeably. PFI is discussed here at length, partly because of the foregoing consideration, and partly given insights gained into its practice through research.

2.2.1 PFI in the UK market

In the UK, the stage for introducing PFI was set in the 1980s when different forms of procurement such as outsourcing, privatisation, BOT etc., became manifest (Mustafa, 1999). The aim of these alternative forms of procurement was to enhance or replace direct government involvement in the provision of public services. Their eventual usage has been reported to have delivered cost savings in many project schemes (PFP, 1996).

The Ryrie rules of the 1980s also served as an impetus for introducing PFI (Birnie, 1999). However, Pickering (1999) pointed out that the leasing industry has *a priori* used PFI before it re-emerged in its present form. Although PFI has been implemented in other countries, like the USA, Australia, New Zealand, etc., it took until November 1992 before the scheme was launched in the UK for the provision of public services.

It is a fundamental requirement in PFI procurement that appropriate risks are transferred to the private sector. It has in this light been commented that:

> PFI has heightened the awareness of project risks in ways that public procurement hitherto has not been able to do, so that the identification, allocation and management of risks has grown to become an essential part of the PFI process.' (ICE and FIA, 1998; p. 46)

Although the aim of PFI is to transfer many risks to the private sector, the ideal objective is to allocate risks optimally such that each party bears those risks it is best able to cope with (PFP, 1995). Negotiations are used as a vehicle for reaching a compromise view on how best to apportion the project risks between its main participants.

2.2.2 Components of PFI contracts

In PFI, there is the design aspect and the consequent construction contract. There is also the operation and maintenance, or facilities management contract (Mustafa, 1999). Environmental concerns too, should not be over-looked (Mellish, 2000). When the socio-political dynamics of PFI are equally considered, the scheme could be said to be complex.

The nature of the formation of the Special Purpose Vehicle (SPV) and the direct relationships between its members can vary. Figure 2.1 shows a typical set-up of a PFI contract. If a typical PFI scheme is examined in its entirety, different parties can be linked with the scheme. Each of these parties will view the project risks with different objectives (Wynant, 1980). It is thus helpful to decide at an early stage who will fill the role of project leader, to pool all the parties together effectively (HM Treasury, 1999).

From Fig. 2.1 it can be seen that the key players in PFI projects would include the sponsors, client, constructors, facilities managers and financiers. In addition, there are the designers who form part of the SPV. These constituent parties and their functions are examined below to see how they influence the risk profile of a PFI project.

The design aspect

The SPV is responsible for designing any physical structure that pertains to the PFI project. Since specifications are not provided by the client, innovative inputs are allowed and encouraged in the design. Typical sources of risk attributable to design include incomplete and delayed drawings. These sources of risk are not the responsibility of the client under a PFI contract.

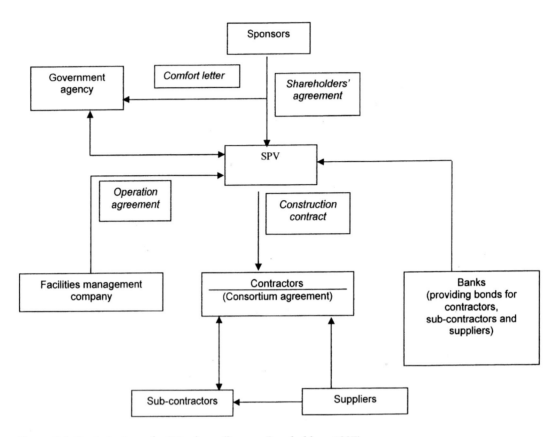

Figure 2.1 *Typical set-up of a PFI scheme* (Source: Beenhakker, 1997).

The construction aspect

In PFI procurement, the client specifies a needed service or services and the bidding consortia are given a free hand to evolve their own ways of providing the specified services. Thus, construction facilities serve as a means to an end towards delivering the client's expected services. This leeway, which contrasts with traditional construction procurement, allows bidders to innovate in designing any needed facilities that will enhance the provision of services. Examples of construction facilities in PFI schemes include: residential buildings for the provision of reformation (prison) services; roads and bridges for the provision of transportation; school buildings for educational activities, etc.

Tiong (1990) suggested that in BOT projects, and thus PFI schemes, this construction aspect is relatively high risk.

The operations/facilities management aspect

Facilities management (FM) is of greater importance and prominence in PFI schemes. Even at inception stages, sponsors are expected to consider the life-cycle cost commitments of a project (HM Treasury, 1999). As service contracts in PFI schemes are of long durations, an element of uncertainty will always surround FM in PFI deals.

Tiong (1990) suggested that BOT projects, and thus PFI schemes, consist of a relatively low-risk utility aspect.

2.3 Project procurement

Procurement is discussed at this juncture because PFI is used as an alternative procurement approach. The way PFI fits into procurement in general and its relationship with other forms of procurement are portrayed.

According to McDermott (1999) the CIB W92 produced a working definition of procurement at its meeting in 1991 as 'the framework within which construction is brought about, acquired or obtained'. Procurement thus encompasses a wide range of activities, including:

- Land acquisition for a project
- Design of buildings or facilities
- Construction of facilities
- Commissioning of projects or facilities
- Management of facilities.

There are thus different activities involved in the procurement of a project, especially if it is very big or complex. Such activities must be carefully co-ordinated and managed, otherwise the project could go wayward and become a burden, or even unsuccessful.

There are equally different forms of procurement that could be used for a particular project. For example, Franks (1998) categorised procurement systems in the following manner (Table 2.1).

Depending on the subject under consideration, procurement systems could be classified in different ways. Therefore the foregoing classification is not stereotypical but illustrative. However, it remains that several procure-

Table 2.1 *Procurement systems.*

Main type of system	Procurement strategies
Designer-led competitive tender	Traditional approach
Designer-led construction managed for a fee	(1) Management contracting (2) Construction management
Package deal	Turnkey Design-and-build British Property Federation Design, build, finance and operate (and/or private finance initiative)
Partnering	A stand alone (partnering) strategy

ment approaches abound, and new concepts occasionally emerge in the market. A detailed description of the different forms of procurement is outside the scope of this chapter. The overview serves merely to bring out the relationship between risks and procurement.

Given the circumstances and peculiarities that may surround certain projects, not all forms of procurement would be suitable for a project. Some procurement approaches may be more effective and beneficial for certain projects. Therefore, an *ad hoc* choice must be made to decide on the preferable form of procurement that befits a project. Factors to be considered in the choice of a procurement approach for a project would include:

- Objectives of the client
- Size and nature of the building programme
- Complexity of the work
- Availability of expertise
- Nature of funding required
- Nature of technology involved.

The decision on the type of procurement to be employed on a particular project should precede the (detail) briefing phase, since the type of procurement assists in determining who should participate in the brief (Franks, 1998).

Each type of procurement has a bearing on the allocation of risks to the different parties involved in the project. While the client bears more responsibilities, and thus risks, in the traditional form of procurement, his/her responsibilities and risks are diminished in the 'package-deal' types of procurement.

Privatised projects are usually large and complex and thus have their associated problems. Particularly, the interrelationships between the project participants are very complex, giving room for conflicts. Thus, procurement strategies that encourage co-operation like partnering, BOT, PFI and DBFO are held to augur well for risk management (Simon *et al.*, 1997), because they seek to bring about a congruence in objectives between the parties, thereby smoothing the flow of activities/events (Chapman, 1997).

Both clients and contractors should analyse and manage risks separately since they have different objectives. Almost every other party involved in a project should analyse and manage the risks relevant to its participation (Touran *et al.*, 1994). As such, input from one party to the other is beneficial for overall project risk analysis and management, buttressing the need for co-operative strategies in procurement (Chapman, 1997).

2.4 Theoretical overview of risks

A risk occurs where either the outcome or consequence of an activity or decision is less than certain (Boothroyd & Emmett, 1996). Both the outcome and consequence of a decision could simultaneously be uncertain. The presence of risks in a project could hinder the achievement of its objectives. Therefore, risks must be managed carefully, in order to achieve any set of project objectives.

2.4.1 A financial view of risks

The finance of large-scale PFI schemes involves thorough scrutiny, especially by the financiers. They do this to ensure that risks associated with a project, and their potential consequences, are not left unchecked. Financiers are mostly risk averse, and often aim to ensure that all possible benefits of a project are balanced with the attendant risks. Therefore, risk management by financial institutions assumes very high prominence in PFI projects.

There are different classifications of risks identified in the financial literature. Sharpe (1964), for instance, considered *systematic* and *specific* risks. Similarly, King (1999) distinguished between *systematic* and *non-systematic* risks. Wilmott *et al.* (1995) have classified risks into *specific* and *non-specific* risks. Apart from market function and specificity, risks can be examined from other perspectives. For instance, Oldfield and Santomero (1997) distinguished four generic risk categories, namely: credit risk, counterparty risk, operational risk, and legal risk. These different risks are explained below.

2.4.2 Systematic (or market) risks

These are also known as non-specific risks (Wilmott *et al.*, 1995). They concern changes in broad economic conditions that affect a whole market. This, for example, may relate to changes in asset values as a result of systematic environmental factors. Other examples of market risks include changes in consumer spending, level of industrial output, interest rates, exchange rates, energy prices, high-impact weather effects, etc. Market risks affect all equities to some extent.

Systematic risks cannot be completely avoided and are considered to be undiversifiable. Banks are known to place capital at risk in order to generate transactions between different market participants and to pursue profit through the efficient supervision of investments. While this function does not alter the structure of the systematic risks in the market, financial institutions allegedly make the capital formation process more efficient and reduce inefficient risk taking. This risk redistribution effort encourages more investments in real assets and contributes to the creation of wealth (Greenspan, 1999).

2.4.3 Non-systematic (or specific) risks

These are associated with a particular asset, company or segment of the market. They are known as 'specific risks' because they do not exert an impact on the whole, but rather on specific components of the market. Examples of activities which can introduce specific risks to a particular company include the introduction of a new product, changes in management, etc. Since specific risks do not affect the entire market, investors that are affected by specific risks can diversify into a range of other activities to mitigate their impact.

2.4.4 Credit risk

Credit risk arises from the possible default of a debtor, with respect to settling a credit facility. When debtors fail to fulfil their contractual obligations, the interest and principal on their loans are not paid within the agreed time. In some cases, credit risks are occasioned by systematic risk.

2.4.5 Counterparty risk

This risk concerns the trading process rather than the investment portfolio. It occurs when one of the trading parties does not perform its obligations as a result of either unexpected systematic factors, or legal or political risks.

2.4.6 Operational risk

Operational risk arises in the course of processing, confirming and reconciling transactions. It can be triggered by a broad range of factors including human errors, inadequate control, systems failure, varying measurement units, inconsistent standards, etc.

2.4.7 Legal risk

A legal risk comes about when new legislation and regulations are introduced with adverse consequences on existing transactions. Legal risks are also associated with actions of fraud or non-compliance with security laws. The consequence of some legal risks could particularly be a big problem for some transactions because the parties affected may not be able to perform their obligations.

Suffice to note that financial risks are multi-faceted. The definitions, however, show that the types of risks emphasised by the financial sector are different from those of the construction sector. According to Chicken (1994), banks are mainly concerned with economic risks.

2.5 Checklist of risks in PFI

The following have been listed as risks in PFI schemes (Private Finance Panel, 1995; Gallimore *et al.*, 1997; Jones, 1998; Birnie, 1999; Salzmann & Mohamed, 1999; Tiffin & Hall, 1999):

- Site acquisition (possibility of obtaining the wrong land, or the right land at the wrong price)

- Feasibility studies (failure to identify key downsides with the intended project)
- Acquiring planning approval (unusual delays could arise, or permission may be denied for ill-defined schemes)
- Design (the technical solution may be unworkable or inefficient)
- Construction (there could be cost and/or time overruns, as well as poorly constructed solutions)
- Commissioning (may be delayed due to several unmet targets)
- Operating risks (including maintenance; malfunctions and delays are key issues here)
- Demand (revenue) risk and its change may render facilities under-utilised
- Occupation and usage risks over time could overstretch the capability limits of resources
- Obsolescence/technology risk could render a scheme unfruitful
- Residual value risk (achieving a high standard of facilities/services at the end of the concession period can be difficult)
- Economic risks (including fall in revenue; financiers pulling out, etc.)
- Legislative/regulation risks (e.g. future planning regulations, health and safety features, etc. may affect the project adversely)
- Taxation risks (change in taxes/laws)
- Bid process/complicated negotiations, being lengthy and costly
- Political (governmental support of *international* projects may not be forthcoming)
- Corruption
- Consortium structure (partners could be mismatched)
- Local partners (could pose interface problems or could use different systems/procedures)
- Project management ability (may be inadequate for the present task)
- Existing infrastructure
- Raw material (supply, availability, etc.)
- Financing (foreign exchange)
- *Force majeure* (circumstances beyond one's control)
- Market competition (could erode the potential gains of a project)
- Revenue tariffs (may be lower than projections)
- Project performance (may be lower than projections)
- Foreign exchange
- Inflation
- Financing risks.

Lam (1999) infers that the greatest risks of BOT and thus PFI projects occur at the later part of construction and the early part of operation of the facility.

2.5.1 Prioritisation of risks by PFI participants

Some clients and contractors who were surveyed by Akintoye *et al.* (1998) opined that government has a mistaken belief that the private sector will-

Table 2.2 *Ranking of PFI risks by contractors, clients and lenders.*

Risks	Ranking or risks			
	Contractors	Clients	Lenders	All
Design risk	1	5	10	1
Construction cost risk	2	6	6	2
Performance risk	4	2	8	3
Risk of delay	7	3	7	4
Risk of cost overrun	3	9	3	5
Commissioning risk	17	1	5	6
Volume risk	8	10	2	7
Risk of operating/maintenance cost	9	4	13	8
Payment risk	10	14	1	9
Tendering cost risk	6	17	9	10
Contractual risk	5	11	15	11
Legal risk	11	19	12	12
Market risk	14	16	11	13
Residual value risk	16	12	14	14
Planning risk	13	18	19	15
Environmental risk	15	8	23	16
Safety risk	21	7	20	17
Financial risk	12	22	18	18
Credit risk	25	24	4	19
Possible change in government	20	20	16	20
Project life risk	19	13	26	21
Changes in European legislation	24	15	22	22
Development risk	18	21	24	23
Bankers' risk	23	26	17	24
Debt risk	22	25	21	25
Land purchase risk	26	23	25	26

NB: 1 = most important (Source: Akintoye *et al.*, 1998).

ingly accepts risks. The interviewees further suggested that government should be considered as a party that could also bear more risks associated with PFI projects. Scores in Table 2.2 show that different PFI participants prioritise project risks differently. Their differing conception, mode of identification, assessment, evaluation, reporting and management of risks may well be attributable to these differences.

2.6 The multi-disciplinary dimension of risk management

Although risk analysis and management is practised in construction, its concepts were adopted rather than devised. It would seem that circumstances did lead the construction industry into embracing risk analysis and management. The field of construction is unique in a number of accounts. Firstly, construction contracts are loose and not rigid. Although standard forms of contract have been formulated, several variants are always developed. For instance, the Joint Contracts Tribunal (JCT, 1980) has six variants, which are (Murdoch & Hughes, 1996):

- Local authority edition with quantities
- Private edition with quantities
- Local authority edition with approximate quantities
- Private edition with approximate quantities
- Local authority edition without quantities
- Private edition without quantities.

A standard form of contract, such as the JCT variant 5, allows a client and contractor to enter into a contract where construction outputs are ill-defined, i.e. the full extent of work being contracted is not exactly known, and thus the project completion price cannot be accurately ascertained. Uncertainty surrounds such contracts. Even in situations where full construction quantities have been defined, accurate pricing and forecasting of project costs over the construction phase is almost impossible. It has thus become a tradition that contingency sums are added to bid prices to cover for price and other uncertainties (risks). The contingency allowance, which is added to construction project estimates or bids, is a risk response strategy to ensure that construction risks will be accommodated should they materialise in the course of a project.

The complexity, dynamism and fragmented nature of construction undertakings often combine to amplify the number and at times magnitude of construction risks. The contingency premium alone may be insufficient to counter these diverse risks, so the construction industry has been adopting risk assessment and management techniques used in other industries. Although elaborate risk analysis was embraced very late in the construction sector, some quantitative techniques, like Monte Carlo simulation, have been utilised albeit to a limited extent (Burchett *et al.*, 1999).

Since the construction industry adopted risk analysis and management, it has followed the lead of the pacesetters. In this regard, it has been suggested that insurance and other financial sectors hold the leading edge in terms of risk analysis and management (Raftery, 1994; Boothroyd & Emmett, 1996). Apart from their pioneering role in developing risk evaluation techniques, insurance and financial sectors have been, and are, involved in construction production through the provision of capital and security. The insurance and financial sectors do assess (the riskiness of) clients, contractors and projects before assisting any scheme. Their risk evaluation techniques thus play a valuable role in construction production. To improve or even optimise its performance, the construction industry will benefit by understanding fully the risk evaluation tools employed by the financial sector.

Also, the advent of PFI has encouraged financial institutions, in conjunction with the construction industry, to play a more involving role in the provision of public services. For PFI schemes to succeed, construction, finance and other sectors involved in PFI undertakings need to understand each other in terms of practices, techniques etc., and to harmonise their approaches. This strengthens the need for an examination of PFI risks from the perspective of all the participants involved.

2.7 Practical risk considerations in PFI

This section discusses the perception of risks by participants in PFI projects. These participants include constructors, designers, financiers, facilities managers, clients and independent advisers. Their understanding of risks is covered, as well as the current key risks prevalent in the PFI market.

2.7.1 Understanding risks

PFI participants associate risks with events which cause something to go wrong. A risk is seen as the uncertain possibility of something happening in the future. Risks concern potential problems, i.e. the possibility of something going wrong that can result in increased cost or cause delay. It could be something unforeseen happening in the way a project deal is structured, or it might be foreseen but unclear *ab initio*.

Under any perspective, what PFI participants refer to as risk, in commercial terms, comes down to what it costs if that risk event should occur. If construction, design, finance and service delivery are considered, their risks are quantifiable in financial terms. For instance, if an organisation is taking the risk of 'designing and building' a particular facility and it designed a building that started to structurally fail, it would be responsible for rectifying that failure and associated loss. The effort involved in rectifying the defect(s) comes down to money. Equally, if that organisation failed to deliver certain services, it would incur financial penalties. Essentially then, risks can be represented in financial terms.

The foregoing perspective assumes a negative dimension, risk as a downside event. Many PFI participants look at the downside, i.e. what could go wrong, and what it would cost. They thus assess the chance of a risk happening, and its magnitude. They then use these two pieces of data to work out the cost to them of a risk occurring. Mathematically, it is the product of probability and consequence, which are the two main features of a risk. So, if a risk has been identified, its estimation would first involve an assessment of how likely it is that the risk would materialise, and what the consequential impact would be. The two are sometimes multiplied together and their product is used as a basis for pricing the risk.

Few PFI participants though recognise that there is an upside to risks. If the cost of a project feature were less than expected, that would result in a higher than planned profit, which is an upside. That is also risk. Risks thus present both exposures to unwanted threats, and opportunities for gain.

It is more realistic to think of a risk as an event that has an impact, which could be positive or negative. However, despite the potential gains inherent in project risks, many PFI participants continue to find it helpful to think of risks as downside events.

2.7.2 Contemporary key risks in PFI

By its nature, PFI exhibits certain features that distinguish it from other forms of procurement. Therefore, while PFI may share risks that are similar with other forms of procurement, it has risks that are peculiarly prominent in it.

Effectively PFI participants look at risks in the different phases of their projects: conception, inception, design, construction, commissioning, operation and termination. The sponsors, by their disposition, look at two principal risk issues: (1) who bears the increase in cost; and (2) who bears the consequences of time delay(s) in the project?

Discussed below are the key risks encountered by contemporary PFI participants.

(1) **Availability**: this relates to facilities being on hand for use by the client (either owner or ultimate end-users). On road projects, there are usually availability risk issues in terms of road closures. Other PFI schemes (for example, involving hospitals, schools, accommodation facilities, etc.) could face different situations that could render them unavailable.

(2) **Commissioning**: pertains to the transition from construction to operation of facilities. Inability to supply power on time could, for example, delay the commissioning of a project. In one PFI hospital project that is operating smoothly now, a hiccup was encountered, in that the new wards could not receive any hot water supply when the patients were due to move in. An emergency measure had to be adopted temporarily to supply these wards with hot water before the fault with the main water system could be identified and rectified. Although penalties are stipulated against the SPV for such defaults, the embarrassment and inconvenience caused to clients (owners and users) could be high. Failure to commission a project on time could trigger-off protests from concerned organisations and members of the public, another form of risk, which is discussed below as 'social risk'.

(3) **Construction**: this is fundamental. Construction works in PFI schemes are usually sublet under fixed-price contracts. It has been a long-term view that construction projects are beset by cost and time overruns. In addition, quality standards, health and safety issues remain of para-mount concern. These concerns are not automatically removed in PFI projects and furthermore the default of a sub-contractor can tarnish the reputation of the SPV, in its bid to win future contracts.

(3) **Credit**: this involves corporate credit from sponsors, operators and construction companies. As PFI schemes are developed using non-recourse financing, the organisations involved must be reputable to raise the funds needed for each development. If such organisations are not credit-worthy, project financing could be curtailed.

(5) **Cost**: the real issue of PFI is that the private sector is contracting with the public sector for a long period, say 30 years. The public sector effectively pays an agreed price over that concession period, which is indexed in some form. The challenge for the private sector is to deliver the output specification to the required performance within the cost assumptions made at the outset of the transaction. So the private sector has to deliver the building to cost and to time, and then it has got to provide the underlying services within that building to within the cost they have specified in their model, including the cost of repair and maintenance for a facility. Although the costs are usually index-linked, predicting the cost of repair and maintenance over 25 or 30 years is subject to uncertainty that can work out negatively for a project.

(6) **Demand**: this deals with the demand for the services provided. In earlier PFI schemes, this was a serious issue, but more recently the public sector tends to bear this risk, so the private sector is less concerned about the level of usage of services. Thus, for example, in a prison project, the SPV will not be concerned if all the cells are filled or not, as that risk will rest with HM Prisons Service. However, some PFI schemes, by their peculiar nature, still tilt the demand risk to the private sector, as it is best placed to bear that risk in such circumstances. For example, in a waste management project, the volume (or demand) risk tends to lie with the private sector that would predict the amount of wastes to be conveyed and treated over the years. Although the client will provide initial input data to predict the volume of wastes, the ultimate responsibility may belong to the private sector, because the project is of a specialist nature. Only specialist organisations (utility firms) tend to be involved in such schemes and are better placed to predict the wastes to be conveyed, especially the varying strengths of such wastes.

(7) **Demographic changes**: these can affect the extent to which certain services are utilised. A waste management scheme will be used again to illustrate this point. Such schemes are usually designed to support a certain density of a population in a city or region. If there is a massive increase in the number of people or factories supported by a particular scheme, the pipeline conveying the wastes may become inadequate. Demographic variations could also trigger-off environmental risks on certain projects.

(8) **Design**: this is fundamental, in that the private sector has to design a functional facility to deliver the services desired by the client. Design is more of a problem for hospitals than for roads. Roads tend to be fairly substantially designed when projects are let whereas the hospitals are usually designed as part of the letting process from scratch. The inability of clients to fully understand design concepts compounds this risk. For example, the quality, size, aesthetics, etc. of facilities are usually not fully comprehended by clients until after construction when it is impossible or expensive to change.

(9) **Environment**: this relates to the impact of the facilities on the

environment, or vice versa. In terms of construction, it is not only the way the project is built that impacts on the environment, but the long-term effect of operating it as well.

(10) **Finance**: this is often as a consequence of inflation. PFI parties have to take a view on interest rates and inflation due to the long-term nature of projects. The SPV will in particular consider the risk of being able to make the project pay and cover the cost of the funding, given the fluctuations that will occur.

(11) **Land**: this is a primary concern in PFI. 'Who is responsible for acquiring the land?' can be a major question. After that, who takes the risk that the land does not have antiquities on it, or is not going to be subject to subsidence? Added to these, the location of the land becomes vital as it also affects the acceptability of the project by those near it. Demographic and environmental considerations can compound the difficulties surrounding the choice of land. In some PFI projects, nearby residents and other concerned citizens have protested the choice of certain sites, which they wanted preserved as green sites.

(12) **Legislative changes**: relates to changes in law, regulations, etc. It is an important factor because PFI deals are comprised of a complex collection of individual contracts. If the government establishes a law that changes the way some things in the health sector are carried out, this may expose some SPVs involved in hospital projects to high risks. Regulations from/concerning the EU come into effect intermittently, which could have an effect on some PFI contracts. Generally the private sector is covered for legislative changes that affect an entire market. However, changes that only affect specific projects are borne by the private sector. Although risks from legislative changes may be infrequent, SPVs should be aware of them and forecast their occurrence and impact as much as possible.

(13) **Legal**: these are different from legislative risks. They deal with whether, for example, a client is actually empowered to enter into a particular contract. Can the person who signed the contract on behalf of a particular council commit that council to the contract? There have been cases where local authorities had signed contracts, when subsequently they were found not to be empowered to do so. That is a legal risk. Legal risks were a particular issue when the PFI began. For example, at that time it was not certain that an NHS Trust was a legal entity entitled to sign a contract. The changes in the law, in this regard, were held up due to the impending elections and this particular issue drove up the legal costs.

(14) **Market**: this concerns many sectors in the PFI arena, like health, education, IT, housing, etc. Some sectors do present their own peculiar risks, for example, IT projects are noted for being very complex and risky. Most other types of projects do, however, have an element of market risk. Private organisations may avoid markets they are not familiar with. For example, some construction outfits that are used to

delivering educational projects will not bid for hospital projects, because they are not experienced in that sector.

(15) **Operation**: that the facility would operate within cost and within the constraints of the concession agreement. This is a big risk that sub-sumes other miniature risks like security, energy consumption, welfare, communications, etc.

(16) **Performance**: this is related to operational risk but concerns the service being delivered. In a housing scheme for instance, the SPV may be expected to provide houses to comply with the Housing Corporation standards for the construction of new homes, and to contain substantially all the features of the 'lifetime homes' standard published by the Joseph Rowntree Foundation. If the SPV does not comply, it could be penalised.

(17) **Planning permission**: this can permit or deny a scheme. Planning permission must be obtained before a project can proceed in full. As a consequence of planning permission delays PFI schemes may be delayed to the point where they become non-viable. Social risk can have an impact on the acquisition of planning permission. If a notice needs to be served, or people have to be displaced and compensated, or wayleaves are involved in a project, then securing planning permission may be delayed.

(18) **Political**: this is a rare occurrence, but it can affect projects. An example is used to illustrate political risk. A project was originally let by the Highways Agency (HA), the HA then novated the project to the Greater London Authority and the Transport London Authority. There was a risk in the transfer, in that the stature of the client effectively changed from central government to local government. There was thus a political aspect. Also in terms of political risk, for example in school projects, consortia would look at the sustainability of schools in terms of whether they thought the schools would still be required in 25 years. Their considerations would partly be political, i.e. what is the situation if the local government structure or the government structure changes?

(19) **Residual value**: the risk the private sector sees in residual value is that the road, hospital, or whatever meets an agreed specification for a period of time after it has been handed back to the client. For example, a road typically needs at least 10 years worth of life left in it when it is handed back. There is thus a residual specification risk. Schools must be handed back in a defined condition. The SPV must therefore determine how much to expend in order to ensure that a school is at a particular quality level when it is handed over to the client. Residual values of facilities are critical to public sector clients who will inherit many of the facilities that are used to deliver services. In some schemes, however, the facilities do not revert back to the client, but remain with the private sector, who will continue to maintain and use them for service delivery.

(20) **Social issues**: this is often equated with 'protester risk', where some section of the community is not in favour of a project or an aspect of it.

A dimension of protester risks is that of vandalism. The construction of the A38 road was attended by many protests. It was a major issue for the consortium responsible which had to manage the protests, and be prepared to act in that situation. Social risk is an issue that must be considered in PFI schemes. It is clearly important that the project should be socially acceptable and that there should not be any adverse consequences as a result of the project. Clearly the private and even the public sector would not wish to be pursuing a project that was going to be deemed to be unpopular with the electorate or have an adverse impact on society.

(21) **Specification**: this is a function of the design. Clients normally state their requirements for a PFI in output terms. The private sector is then responsible for developing the requirements to achieve the detailed needs of the client. The specifications have to be right in order to meet the standards of the client and to fully satisfy the client. In addition, the specifications must meet health and safety and other statutory regulations. All of this burden is transferred to the private sector.

(22) **Sponsor**: this is the level of commitment of the sponsors to the transaction. A lack of commitment could delay or even abort a PFI scheme. Funding in terms of equity will not be forthcoming if the sponsors are not seen to be willing and committed.

(23) **Technical**: concerns the effective and efficient functioning of equipment, materials, processes, etc. In a hospital setting, for instance, technical risks are primary, as theatre and other medical equipment have to operate effectively. The refrigeration has to work, the laundry machines have to work, the communication system has to work, etc.

(24) **Technological obsolescence**: this is easily apparent with IT. The technology underpinning a certain service can become obsolete. Meanwhile, one of the ambitions of clients is to improve the level of their services and to keep pace with advances in the global market. If technological advances are not addressed in a project, the clients may become disenchanted with the scheme.

(25) **Time**: this concerns the delays in a PFI project. Sources of delay in the construction aspect could be adverse weather; change of mind by the authorities; strikes in construction etc. The design could be delayed, so could the acquisition of planning permission. Many features could give rise to project delay, and PFI participants need to be wary of them.

(26) **Volume**: this is related to demand risk. However, volume risk is about capacity, while demand risk is associated more clearly with usage. For example, on a road, there is a traffic volume risk which must be forecast. In this case, the road must be designed to sustain the number and types of vehicles that will use it. The demand risk on the road will pertain to whether the road is actually used by vehicles or not. Volume risk thus feeds back directly into the design.

2.8 A sectoral consideration of the major risks in PFI

Research carried out at Glasgow Caledonian University has identified the major risks considered by the different PFI contributing sectors and these are shown in Figs 2.2 and 2.3. There are risks which are considered by all five sectors such as performance-related risk, financial risk, cost overrun risk, construction risks, volume risk, legislative changes, etc. (see Fig. 2.2). Each sector also considers other risks that may specifically inhibit its performance (see Fig. 2.3).

In Figs 2.2 and 2.3, a suffix is added to some risk types to distinguish the attitudes of the different sectors. The suffix 'F' is used to denote views from the financial sector. 'CS' denote views from consultants; 'C' for answers from the construction sector, while 'FM' and 'CL' refer to operators (or facilities managers) and client organisations, respectively. A checklist of all the risks in Figs 2.2 and 2.3 is produced in Table 2.3 where the meaning of each risk is explained.

Construction companies often consider the adverse effects of weather, availability of resources within the locality of a project, environmental perils, access roads, etc. Financial organisations are mainly concerned with financial performance, but also consider planning permission, i.e. whether it will be obtained or delayed; the credit worthiness of the organisations involved in the project; and availability of land for the project.

Clients pay particular attention to the quality of the asset and/or quality of services to be delivered by the SPV. They also keep a close watch on their affordability, to make sure it is not exceeded by the SPV. The demand/volume risk is also important to clients who often bear this risk. For example, a university that is building a hostel for students would be concerned about unoccupied rooms both during term time and holidays. Where facilities would be handed back to clients at the end of the concession period, the residual value of the assets becomes a prominent issue with them.

The risks in Table 2.3 number 40, which is relatively large. They seem to suggest that current PFI participants look at many risks in their projects. Diverse dimensions of risks are critically examined. The planning and general management of the project is usually considered. The design, construction and operation of facilities are also usually considered. Even the tail of the project (residual value of facilities) is not left out. The interfaces between different project phases, possible changes in legislation and regulations, the land, users, etc. and the risks they pose are all considered.

2.9 Discussing risks in PFI projects

While the compendium of risks and the particular risk interests of sectoral groups can be identified, to what extent are these risks discussed and articulated in these sectors?

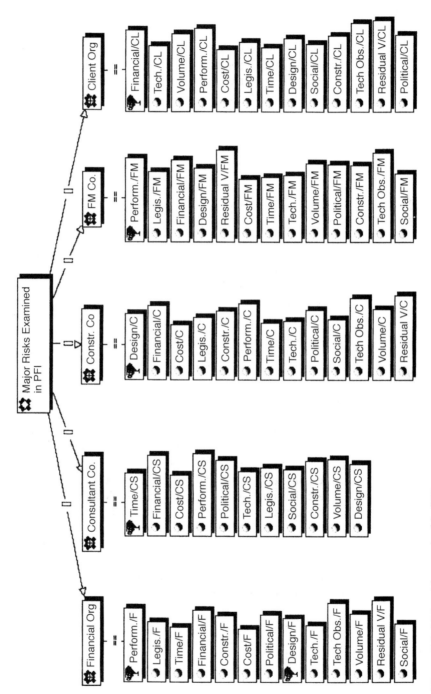

Figure 2.2 *Major risks examined in PFI.*

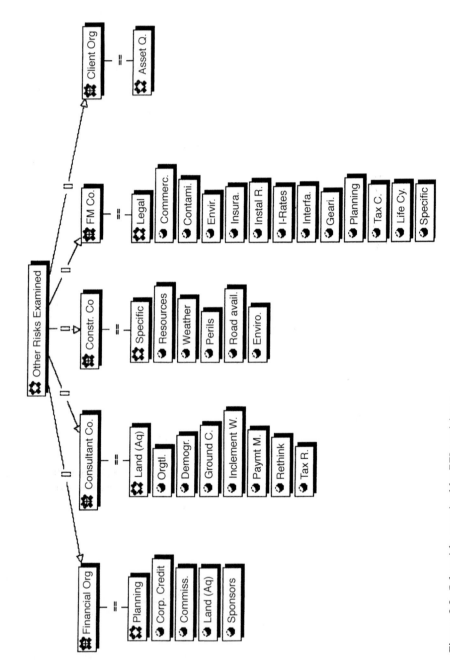

Figure 2.3 *Other risks examined by PFI participants.*

Table 2.3 *A compendium of risks prevalent in the PFI environment.*

Name	Abbreviation in Figs 2.2 or 2.3 (without suffices)	Meaning
Asset quality	Asset Q.	Quality of assets is unacceptable
Commercial	Commerc.	Commercial risks in the wider sense
Commissioning	Commiss.	On whether the project is commissioned smoothly and on time
Construction	Constr.	That construction is finished on schedule and there are no mishaps
Contamination	Contami.	Land contamination
Corporation credit	Corp. Credit	The ability of sponsors, operators and construction companies to obtain credit
Cost	Not applicable	Cost escalations
Demographic	Demogr.	Consideration of the demography around the project locality and the impact of projected changes on the project
Design	Not applicable	Incomplete designs or variations to existing designs
Environmental	Enviro./Envir.	The impact of the project on the environment and vice versa
Financial	Not applicable	Can the project can be financed in the long run
Gearing	Geari.	The funding structure (ratio of debt or bonds to equity) and its impact on profitability
Ground conditions	Ground C.	The possibility of undesirable ground conditions
Inclement weather	Inclement W./Weather	Project disruptions attributable to the weather
Installation	Instal R.	Correct installation without 'left behind' risks being passed to the service operators
Insurance	Insura.	The cost efficacy of insuring the project
Interest Rates	I-Rates	The dynamism of interest rates (inflation) and the effect on the project life cycle
Interface	Interfa.	Especially between the construction and operation phases – that inferior facilities or components are not passed to the latter
Land acquisition	Land (Aq.)	Delays arising from or inability of land acquisition
Legislative changes	Legis.	Possibility of new laws affecting the project
Legal	Not applicable	Possibility that some aspect of the project might be illegal

(Contd)

Table 2.3 *(Contd).*

Name	Abbreviation in Figs 2.2 or 2.3 (without suffices)	Meaning
	Risk	
Life cycle costs	Life Cy.	Accuracy with which life cycle costs can be projected for longer duration
Organisational	Orgtl.	The resource availability and capability of the participating organisations
Payment mechanism	Paymt M.	How any changes in the unitary payments are to be apportioned between the parties
Performance-related	Perform.	Will the service be delivered smoothly
Perils	Not applicable	*Force majeure* types of mishaps
Planning	Not applicable	Will planning permission be obtained at all, or delayed
Political	Not applicable	The possibility of a new government abandoning or changing PFI (schemes)
Residual value	Residual V	The quality of assets and standard of service at the end of the concession period
Resources	Not applicable	Risk of commitment or resources in a project that is eventually aborted
Rethink	Rethink	Clients could change their minds or radically change a project
Road availability	Road avail.	This refers to road closures and the like
Social	Not applicable	Concerns strike, disputes, picketing, etc.
Specific	Not applicable	These are risks, which are peculiar to each/certain projects, like the varying biological strength of waste and how to cope with it in a waste treatment scheme
Sponsors	Not applicable	The commitment (level) of the sponsors to the project
Tax change	Tax Ch. Tax R.	The potential effect of tax (rate) changes on the project finance and profitability
Technical	Tech.	The effective and efficient functioning of equipment, materials, processes, etc.
Technological obsolescence	Tech Obs.	The effect of technological changes (especially IT) on the standard of services at a given time
Time	Not applicable	Whether the project schedule will be achieved
Volume	Not applicable	The effect of the number of users of the service delivered on the unitary payments due the private sector

It has become clear from investigations (based on the DETR/EPSRC research) that the majority of organisations involved with PFI project delivery do discuss risks within their organisations. Figure 2.4 shows the regularity with which risks are discussed by participants. Furthermore, as PFI has made organisations more risk-conscious, it is not surprising that risks are discussed over different phases of each project.

Amongst those that had constant risk discussions, this activity can be described as 'ongoing'. These organisations are constantly weighing up the project proposal(s) against the risk of something happening.

The next category of 'frequently held discussions' refers to those being held on a (1 to 2) weekly basis. 'Informal discussions' are described as being carried out on most projects, as each is developed. For those discussions identified as 'when advising the client', there is no particular regularity attached to them. In describing such a category, a PFI participant remarked that: 'we meet regularly to discuss the risks. The more urgent the circumstance, the more regular these meetings'.

2.9.1 Staff involved in discussing risks

In seeking to identify which staff are involved there is no generic view of which staff are involved in these discussions (DETR/EPSRC). The only clear view is that discussions take place 'within a team', with the composition of the team varying within different organisations. A team meeting may be preceded by a high level preliminary meeting in which company directors and/or partners participate. The high level discussions are often limited to 3–4 core members, but there are situations when the whole project team (including line managers) is involved. Procurement specialists, banking specialists, planning and environmental and other experts may be involved in these high level meetings, if deemed necessary.

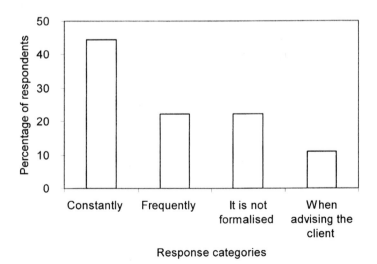

Figure 2.4 *Frequency with which risks are discussed by PFI participants.*

As a sequel to the management meeting, most organisations arrange at least one enlarged meeting where all, or most, members of the project team are involved. In organisations where there is a PFI unit, all members of that unit may attend. Instead of convening a meeting, workshops could be used, where all project advisers (financial, legal, technical, etc.) would be involved in discussing the project risks.

2.9.2 Nature of the risk discussions

The disposition of the discussions is one of a general sharing of information. If there are new guidance documents from the government, these are also examined critically. Since the private sector organisations in bidding consortia are often supplied with a tentative risk allocation table in PFI procurement, they will initially discuss the risk apportionment in those tables. They then proceed to investigate for risks that are unique to the current project and add them to the standard list. 'What if?' questions gradually permeate the discussions, wherein they assume an analytical dimension.

Proceeding further, organisations do discuss aspects of risk management. They investigate the risks that threaten the success of the current project and if there is a financial provision within to cope with these risks. Strategies for mitigating risks and their secondary consequences are then considered.

The discussions on risk concentrate on:

- Exposure to different risks
- What this project requires (i.e. resources to manage the risks)
- Pricing of the risks (adequacy of)
- Checking to see if the service is deliverable within an affordable envelope
- Where the problems and issues are likely to be in the client's risk matrix
- Identifying whether there is a risk transfer to the PFI provider (in the ITN)
- How the PFI deal affects users of the facilities
- Contract standardisation
- Value for money, and its achievement
- Key commercial issues in the contract
- The personnel who will monitor and manage specific risks
- The development of an updated risk register.

On the public sector side, clients discuss risks too. In this regard, negotiating the contractual documents, including the project agreement to be signed with the private sector (i.e. containing all the defining obligations and duties), is considered, in addition to the risk allocation that either has been or is about to be made.

Two principles underpin PFI: transfer of many risks to the private sector and achievement of value for money (VFM). As PFI participants see it, these two principles are closely interlinked, because the more unacceptable risks are transferred from the public to the private sector the more the price is piled up, and the more unattractive the cost of purchasing the services becomes. Therefore, the discussions address the balance between risk transfer and achievement of value for money.

2.9.3 Self-imposed risks in PFI

Although most risks can be foreseen, there are situations where they are not adequately covered in a PFI scheme. Such situations amount to self-imposed risks, which can arise, say, when the cost of an activity is underestimated. This is a self-imposed risk arising from inexperience, lack of information or incomprehension of the project. Therefore, the discussions will seek to identify any self-imposed risks.

Oversight too can impose risks in a scheme. For instance, some aspects of the project may not be insured. This is a case of oversight which might either be deliberate or not. Inadequate communication is also a potential source of self-imposed risks.

Discussions within PFI projects also seek to evolve ways to deal with the risks, at least those that are known. It may be suggested in this regard that the private sector should use the knowledge it has as a means of informing the process both within the short term and the long term as a way of reducing risk and the premiums associated with risk.

2.10 Conclusion

This chapter has taken the concept of risk and developed it from a theoretical perspective to practical risk considerations by parties to the PFI process. It is clear that there is a vast range of risk issues, some of which have an upside as well as a downside. When viewed from the perspective of the participants we find that many of the risk issues are common to the main parties but their importance is variable, while some risk issues are particular to a specific party. Despite the range of issues and their possible downside it is clear that the approaches used to deal with them vary widely. The message which can be taken from here is that there are many practical risk issues which are considered important by the parties to PFI but that this is not adequately reflected in the processes that address risk.

Acknowledgements

We wish to thank the EPSRC and the DETR (now the Department of Trade and Industry) for funding the research on 'Standardised framework for risk assessment and management of Private Finance Initiative projects' which informs this paper. We are also profoundly grateful to the project industrial partners who have supported the research. Many organisations have either participated in the research or supplied data. Their co-operation is fully appreciated.

References

Akintoye A. and Black C. (1999) Operational risks associated with partnering for construction. In: *Profitable Partnering in Construction Procurement: CIB W92 (Procurement Systems) and CIB TG 23 (Culture in Construction)* (ed. S. Ogunlana), pp. 25–37. E & FN Spon, London.

Akintoye A., Taylor C. & Fitzgerald E. (1998) Risk analysis and management of private finance initiative projects. *Engineering, Construction and Architectural Management,* **5**(1), 9–21.

Beenhakker H.L. (1997) *Risk Management in Project Finance and Implementation.* Quorum Books, London.

Birnie J. (1999) Private Finance Initiative (PFI) – UK construction industry response. *Journal of Construction Procurement,* **5**(1), 5–14.

Boothroyd C. & Emmett J. (1996) *Risk Management – a Practical Guide for Construction Professionals.* Witherby & Co Ltd, London.

Burchett J.F., Rao Tummala V.M. & Leung H.M. (1999) A world-wide survey of current practices in the management of risk within electrical supply projects. *Construction Management and Economics,* **17**(1), 77–90.

Chapman C.B. (1997) Project risk analysis and management – PRAM the generic process. *International Journal of Project Management,* **15**(5), 273–281.

Chicken J.C. (1994) *Managing Risks and Decisions in Major Projects.* Chapman & Hall, London.

Comerford J. & Puryear R. (1995) When outsourcing goes awry. *Harvard Business Review,* **73**(3), 24–37.

Franks J. (1998) *Building Procurement Systems: a Client's Guide,* 3rd edn. Addison Wesley Longman Ltd, Harlow.

Gallimore P., Williams W. & Woodward D. (1997) Perceptions of risk in the private finance initiative. *MCB Journal of Property Finance,* **8**(2), 164–176.

Goldsmith S. (1997) Can business really do business with government? *Harvard Business Review,* **75**(3), 110–121.

Greenspan A. (1999) Measuring financial risks in the twenty-first century. *Vital Speeches of the Day,* **66**(2), 34–36.

HM Treasury (1999) *Government Construction Procurement Guidance.* HMSO, London.

ICE and FIA (1998) *Risk Analysis and Management for Projects.* Thomas Telford, London.

Jones I. (1998) *Infrafin.* Final report of a project funded by the European Commission under the Transport RTD Programme of the 4th Framework Programme. National Economic Research Associates, Stratford Place, London.

King D. (1999) *Financial Claims and Derivatives.* International Thomson Business Press, Oxford.

Lam P.T.I. (1999) A sectoral review of risks associated with major infrastructure projects. *International Journal of Project Management,* **17**(2), 77–87.

McDermott P. (1999) Strategic and emergent issues in construction procurement. In: *Procurement Systems: a Guide to Best Practice in Construction* (eds S. Rowlinson & P. McDermott), pp. 4–26. E&FN Spon, London.

Mellish R. (2000) Greening PFI. *Private Finance Journal*, **5**(2), 20–23.

Merna A. and Smith N. (1999) Privately financed infrastructure in the 21st century. *Proceedings, Institution of Civil Engineers*, **132**(Nov.), 166–173.

Murdoch J. & Hughes W. (1996) *Construction Contracts – Law and Management* (2nd edn). E & FN Spon, London.

Mustafa A. (1999) Public-Private Partnership: an alternative institutional model for implementing the private finance initiative in the provision of transport infrastructure. *Journal of Private Finance*, **5**(2), 64–79.

Oldfield G.S. & Santomero A.M. (1997) Risk management in financial institutions. *Sloan Management Review*, **Fall**, 33–46.

Pickering C. (1999) The asset option. *Private Finance Initiative Journal*, **3**(6), p. 64.

Private Finance Panel (1995) *Private Opportunity, Public Benefit: Progressing the Private Finance Initiative*. HMSO, London.

Private Finance Panel (1996) *Risk and Reward in PFI Contracts: Practical Guidance on the Sharing of Risk and Structuring of PFI Contracts*. HMSO, London.

Raftery J. (1994) *Risk Analysis in Project Management*. E & FN Spon, London.

Salzmann A. and Mohamed S. (1999) Risk identification frameworks for international BOOT projects. In: *Profitable Partnering in Construction Procurement: CIB W92 (Procurement Systems) and CIB TG 23 (Culture in Construction)* (ed. S.O. Ogunlana), pp. 475–485. E & FN Spon, London.

Sharpe W.F. (1964) Capital asset prices: a theory of market equilibrium under conditions of risk. *Journal of Finance*, **September**, 1964.

Simon P., Hillson D. & Newland K. (1997) *Project Risk Analysis and Management (PRAM) Guide*. Association for Project Management, Ascot, UK.

Smith A. (2000) The way forward. *Private Finance Initiative Journal*, **4**(6), 10–12.

Tiffin M. & Hall P. (1999) PFI – the last chance saloon. *Proceedings, Institution of Civil Engineers*, **126**(Feb.), 12–18.

Tiong R.L.K. (1990) BOT projects: risks and securities. *Construction Management and Economics*, **8**, 315–328.

Touran A., Bolster P.J. & Thayer S.W. (1994) *Risk Assessment in Fixed Guideway Transit System Construction*, Publication No. DOT-T-95-01, Federal Transit Administration, US Department of Administration, Washington.

Wilmott P., Howison S. & Dewynnw J. (1995) *The Mathematics of Financial Derivatives*. Cambridge University Press, Cambridge.

Wynant L. (1980) Essential elements of project financing. *Harvard Business Review*, **58**(3), 165–173.

3 Value management in public-private partnership procurement

John Kelly

3.1 Introduction

This chapter addresses the management of value for money in private finance initiative (PFI) projects within the public sector through the description, and illustration by example, of a value management service. This service is applicable to establish business need, appraise the options, establish the business case and the reference project. In this context value for money gains are defined as 'improvements in the combination of whole life costs and quality that meet the user's requirements' (OGC, 2000). A value management service is defined as one that maximises the functional value of a project by managing its development from concept to use through the audit of all decisions against a value system determined by the client (Kelly & Male, 1993).

This chapter commences with a description of the background to value management. This is followed by a short discussion of the notion of projects within core service delivery, the concept of value for money, and the structuring of a value management activity for PFI stages 1–3 with reference to an example using indicative tools and techniques. It should be noted that value management has application at many stages during the development of the project (Male *et al.*, 1998), but from the perspective of the PFI client establishing the value management of the project at the early stage is vital.

3.2 The value management of projects

3.2.1 The development of value management

Value management has its foundation in the manufacturing sector of North America. The concept began in the late 1940s, when shortages of strategic materials forced manufacturers to consider alternatives that performed the same function. It was soon discovered that many alternatives provided equal or better quality at reduced cost. This led to what was then defined as value analysis whereby:

- Value analysis is an organised approach to providing the necessary functions at the lowest cost.

Value analysis was always seen to be a cost validation exercise that did not affect the quality of the product. However, it was recognised that many products had unnecessary cost incorporated by design and it was this that led to the second definition of value analysis:

- Value analysis is an organised approach to the identification and elimination of unnecessary cost where unnecessary cost is defined as a cost that provides neither use, life, quality, appearance nor customer features.

In 1954, the US Department of Defence Bureau of Ships became the first US government organisation to implement a formal programme of value analysis. It was at this time that the name changed from value analysis to value engineering, for the administrative reason that engineers were considered the most appropriate personnel to undertake the task. The term value engineering was formalised in the title of the Society of American Value Engineers, which recently changed its name to SAVE International. Value engineering in the UK began in the 1960s manufacturing sector and led to the establishment in 1966 of the Value Engineering Association. This organisation changed its name in 1972 to the Institute of Value Management. Value management is commonly used to define a value activity from the strategic to the technical/operational stages in the development of projects. Value engineering is a subset of value management, which relates to the technical and operational aspects of projects only. Value management is the term in common use throughout Europe, except France where the term value analysis is used.

3.2.2 The project within a core business activity

The *Oxford English Dictionary* defines a project as being a plan, a scheme, or a course of action. Borjeson (1976) defines a project as 'a temporary activity with defined goals and resources of its own, delimited from but highly dependent upon the regular activity'. Morris and Hough (1987) define a project as 'an undertaking to achieve a specified objective, defined usually in terms of technical performance, budget and schedule'. Accordingly a project is characterised as a unique activity with defined commencement and completion dates, generated by the core business activity, which requires the investment of resources for a return, where the investment is defined as being financial, manpower and/or material and the return is a quantifiable enhancement to the core business. The product of the project must achieve a strategic fit with the overall client objectives and the process should be so managed as to be internally efficient and externally trouble free. The temporary activity, the project, should be planned to be smoothly absorbed into the client's business plan at a point when the project is complete.

The management of a project necessarily relates to the management of information. In the early stages, this relates to the project's mission and its strategic fit with the client's business objectives. In the later stages, the information relates more to the function of space provision or more operational aspects. Value management is a team-based, project-orientated service that initially seeks to quantify the functional requirements and value criteria for project success and subsequently to monitor project delivery to ensure functional compliance and maximum value. This requires the discovery, structuring and processing of information held by a diverse number of stakeholders in a manner that permits key functions to be discovered, solutions found and actions planned.

3.2.3 Value management workshop

The most common form of a value management service is as a facilitated workshop. This implies the appointment of a facilitator skilled in value management techniques, leading an appropriate team. It should be emphasised that a value management facilitator is a team manager for the value management process only and therefore should have no other project management function. The membership of the team will comprise all of those stakeholders with an input relevant to a particular stage. The ACID test, as shown in Table 3.1, is often used to determine who should be a member of the team. Generally team membership tends to be greater in number at the strategic stage of projects when a large number of issues are being considered and smaller when the technical details of the project are being investigated.

3.2.4 Characteristic development of projects

All projects, as defined above, pass through four characteristic stages of development. These are:

- Strategic planning and business definition
- Project planning and the establishment of systems
- Service definition of the component parts of the project
- Operations and use.

Maximum value is achieved when value management services are applied pro-actively at each stage as the project develops. As demonstrated in Fig. 3.1, the lever of value, maximum value is attained when effort is applied to the lever at each stage in turn. Also illustrated is that a given amount of effort

Table 3.1 *The ACID test for team membership.*

A	Authorise – include those who have the authority to take decisions appropriate to the stage of the development of the project
C	Consult – include those who have to be consulted regarding particular aspects of the project during its evolution at the workshop
I	Inform – do not include those who have only to be informed of decisions reached during the workshop
D	Do – include those who are to carry out the tasks specified at the workshop

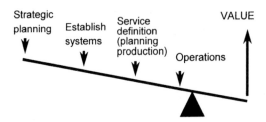

Figure 3.1 *The lever of value.*

at the strategic planning stage will give a higher value return than the equivalent amount of effort at the operations stage.

It is an unfortunate fact that value management studies undertaken re-actively at the later stages of a project (usually as a result of an overspend) often require a painful and wasteful, in respect of abortive work, climb back up the lever of value to endeavour to understand what lay behind the strategic planning of the project.

3.2.5 The structure of a value management service

Irrespective of the stage of the project at which a value management exercise is carried out, it will follow a set number of identifiable steps. These are:

- Information: all issues, information and strategies pertaining to the project are obtained through structured interviews, detailed document analysis, and the contribution of information by key stakeholders at the workshop
- The client's value system: once all the information has been obtained the client's position with regard to the project is analysed by reference to a number of key criteria namely, capital cost, operating cost, time, utility, environment, exchange or resale value, esteem, and politics
- Functional analysis: all of the information is processed by reference to the client's value system into a number of key functions. At the commencement of the project these functions tend to be of a strategic nature and structured as a function logic diagram in the form of a felled tree where the trunk is the mission of the project and the branches the functional requirements of the project separated into needs and wants. At the later stages functions are listed under each technical and operational need
- Innovation: ideas are generated, usually through a brainstorming session to satisfy all of the functional requirements identified in the previous stage
- Evaluation: the large number of ideas generated at the brainstorming session is reduced to a manageable number. At this stage a risk analysis may be developed and the result of the risk analysis may require a return to the innovation stage

- Conclusion: the end of the workshop stage is characterised by the completion of an action plan that is recorded in the project execution plan.

3.2.6 Project execution plan (PEP)

The PEP is a live, dynamic management document that records the project strategy, organisation, control procedures and responsibilities. It is updated regularly during the project's life cycle and used by all parties both as a means of communication and as a control and performance measurement tool. It begins life as an empty file with dividers indicating the documents to be included. Value management and other activities lead to the completion of various reports that are contained in the PEP.

Examples of the items that a PEP should contain are:

- The options appraisal incorporating the client's value system
- The user needs, the strategic brief
- The performance statement of all aspects of the project, the project brief
- The project execution strategy
- The project mission
- The aims and objectives of the PEP
- The procedures for updating the PEP
- The project organisation structure of the client
- A list of consultants and a description of the consultants' responsibilities
- The contractor's, management contractor's, construction manager's organisation
- The form of contract, partnering agreement, etc.
- The contract specification/drawings etc.
- The drawing register including notes of drawing development and stages reached
- A full supply chain diagram annotated with the time when the key players must be appointed
- A knowledge management structure including training required/ achieved
- Project reporting procedures and particularly the procedures for information distribution, and communication, between the client's project team, consultants and contractor
- The health and safety plan
- The quality plan
- The latest cost report and predicted cost and cash flow. This will be updated at regular intervals
- The executive summaries and action plans from value management workshop reports
- The risk management strategy and the latest risk analysis
- Copies of key permission documents such as the planning permission, the building warrant, listed building consent, tree preservation orders, etc.

- The required completion date including for any phases
- The latest contract programme with milestone activities
- Change management procedures and design freeze dates related to milestone activities
- A schedule of key meetings and workshops (including value and risk workshops)
- Procedures for PR and dealing with community and media enquiries
- Commissioning procedures
- Occupation procedures, particularly when phased
- Procedures for the closing down of the project.

3.3 Managing value in PFI projects

3.3.1 Stage 1 – establish business need

A review of current services through informal activities or formal activities such as EFQM, Balanced Scorecard, ISO 9000 quality procedures, IIP or Six Sigma, may reveal a problem or opportunity in service provision. An analysis of the problem or opportunity using Benchmarking, Goal and System Modelling, Community Survey, etc., may highlight the need for a project.

Case study

A local authority uses the balanced scorecard technique to monitor its performance in the provision of services. The balanced scorecard is output orientated and shows a performance well below target to:

- Use brownfield sites for sustainable development
- Encourage housing renewal
- Alleviate city centre traffic congestion
- Improve accessibility to city centre particularly for those without cars.

The local authority convenes a small examination and assessment panel of in-house departmental heads, under the chairmanship of an elected member, to address the above problems and suggest project solutions.

Benchmarking against a neighbouring authority it was recognised that both private and housing association housing starts were down because land was in short supply on the west of the city where the demand for housing was highest. Additionally, the benchmarking showed fewer new business opportunities. The local enterprise officer stated that it was difficult to attract new business to the derelict and contaminated industrial land east of the city, zoned for industrial development because of the poor access to the motorway east of the River Devon. A new road access and motorway interchange from the east of the city failed to obtain funding from the Highways Agency.

> The panel concluded that a latent development potential of a size uncommon in the region existed in the east of the city. The panel recommended the collection of socio-economic and other data on the east city and a facilitated value management exercise to identify the mission of the project and appraise the options.

3.3.2 Stage 2 – appraise the options

The option appraisal should present a clear picture of the current position of the client organisation, the prime project task with background information and the options open to the client to satisfy the project task. The option appraisal exercise is undertaken as a value management workshop.

VM 1 (option appraisal) – information

Pre-workshop, the value management facilitator may either undertake a document analysis through a study of data in reports, correspondence files, drawings, etc. and/or interview the key client stakeholder to become familiar with the background to the problem. Following this, the facilitator will interview a representative sample of those people regarded as being stakeholders with the following objectives:

- To identify those whose membership of the workshop team would be an advantage, an exercise undertaken by reference to the ACID test described above
- To refine the understanding of the background to the problem including any covert or hidden agenda and to assess the headings for the issues analysis
- To predetermine the tools and techniques required for the workshop and thereafter to compile the agenda of the workshop.

The workshop information stage would commence with a brainstorming of all of the issues surrounding the problem. The issues, taken in random order, are recorded on sticky notelets to be arranged by the team under headings determined by the facilitator and added to by the team. The team may vote on which of the issues are of primary importance, usually by spending a specified number of sticky dots on the notelets.

From the issues of primary importance project functions are evolved and a function logic diagram constructed. The method usually employed to construct a function logic diagram is for the facilitator to conduct a brainstorming of functions recording the team's required functions as a simple verb-noun description on sticky notelets for later sorting. The verb-noun description is a useful discipline as it forces a concise and exact description. The notelets are ordered to present a function logic diagram. This activity ends the information stage.

Case study

The local authority appoints a value management facilitator who studies all data and conducts a number of interviews to ascertain the nature of the problem. The facilitator together with the chair of the examination and assessment panel compile a team of 23 stakeholders to determine the mission of the project.

At the value management workshop the team conduct an issues analysis to understand the project. Following the issues analysis the strategic functions generated by the team and recorded on sticky notelets are as follows:

- Encourage enterprise
- Encourage new housing
- Decrease travelling time
- Bypass city
- Connect to motorway
- Enhance east city
- Establish focal point
- Establish centre
- Restrict noise/nuisance
- Span river
- Develop riverside
- Direct connection to city centre
- Comfortable living environment
- Comfortable working environment
- Create sustainable community
- Enhance environment.

These functions are re-ordered from high order needs and low order wants as follows:

- Enhance east city
- Create sustainable community
- Encourage enterprise
- Encourage new housing
- Decrease travelling time
- Connect to motorway
- Direct connection to city centre
- Comfortable living environment
- Comfortable working environment
- Restrict noise/nuisance
- Enhance environment
- Establish focal point
- Establish centre
- Bypass city

- Span river
- Develop riverside.

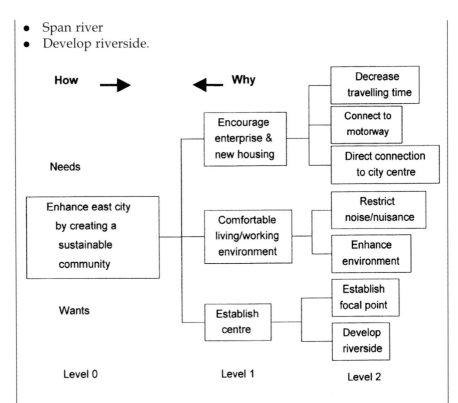

The functions are structured as a function logic diagram with functions 1 and 2 becoming the mission of the project and the remainder ordered from high order needs to low order wants, i.e. the project needs to encourage enterprise and housing otherwise it will fail. However, the community will benefit from a centre but will not fail. It is important that the strategic fit of the project is accurately recorded and often the construction of the diagram requires considerable iteration with the team until everyone is convinced that the functions have been accurately captured.

It should be noted that the functions 'bypass the city' and 'span the river' have not been included as the team decided that these were technical solutions rather than strategic functions, i.e. it may be possible to achieve all of the functions on the diagram without these two.

VM 1 (option appraisal) – innovation, evaluation and action planning

The team brainstorm ideas to answer the functions on the function logic diagram. All brainstormed ideas are evaluated and the best ideas listed for further analysis.

The purpose of the option appraisal is working from knowledge of the client and the mission of the project examining all options and recording that examination in a way that might be the subject of audit later. The value management workshop report is an auditable document which tracks the decision-making process. The document will be included in the project execution plan.

Case study

Having completed the diagram and agreed upon the mission of the project the team brainstorm solutions to the mission. During the brainstorming session 78 options were generated to meet all of the above functions. The team reduced these ideas to the following:

- Create a new east city identity of 'Devonside'
- Devonside to be a mixed development of light industrial, office park and private and housing association housing
- Service plots at Devonside
- Construct a new road bridge across the River Devon
- Construct a new LRT bridge across the River Devon
- Provide LRT from Devonside to city centre using disused railway
- Full guided bus (with overhead electric) from Devonside to city centre using disused railway
- Retro fit existing buses to use guided busway using disused railway
- Convert disused railway to new road
- Sell service ducting across new bridge
- Build new bus station to west of city centre (a new offices and shops development is planned for this site).

An action plan was created and individuals tasked with ascertaining the technical and economic viability of the options. On completion of the action plan the value management workshop was adjourned.

Four weeks following the value management workshop the team met for a facilitated implementation workshop. Having considered all ideas in detail the following were carried forward as projects.

- The 3000 acres of largely derelict industrial land to be compulsorily purchased and named and marketed as Devonvale, an offices, shops, light industrial and housing development
- A new road bridge to be constructed together with a north city link road to connect with an existing road leading to the motorway interchange
- New bridge to carry new water main and foul sewer
- A new guided busway over new bridge and via a disused railway through the city centre to a new bus station west of city will reduce travel times from east city to city centre from 1 hour to 15 minutes. A partnership agreement with a local bus company is anticipated to retrofit existing buses with guided busway steering.

It is anticipated that the 3000 acre development together with the bridge, north city link road and the guided busway will form a single PFI project.

3.3.3 Stage 3 – business case and the reference project

The business case is defined as 'a clear definition in terms of service delivery of what is being sought: the output specification. This should not be a description of a particular asset, but a prescription of a particular service' (OGC, 1999). This is a useful definition from a value management perspective as it implies the understanding of function, a factor that lies at the core of value management.

The reference project demonstrates that an investment option exists which should be affordable. A full value management study of the reference project, which could become the Public Sector Comparator, will not only demonstrate the existence of an affordable option but will provide a clear and auditable statement of this fact.

VM 2 (business case and the reference project) – information

The same facilitator will assemble a different team from that in VM 1 to consider the business case and reference project. Some of those from the VM 1 team will be invited to join the VM 2 team but only in recognition of their expertise in connection with the project in hand. The facilitator will, in a similar manner to that in VM 1, identify new stakeholders that may include consultant designers. At this stage the facilitator will be liaising with the project sponsor. The workshop will follow the same format as described in 'the structure of a value management service' above.

At the workshop the information stage is a structured gathering of facts and opinions surrounding the embryo project. This may be undertaken by brainstorming issues, or by presentations from knowledgeable stakeholders, or by interrogation by the facilitator. Whichever method is used it is important that information on at least the following is obtained:

- The context and definition of the project: a short description of the project and how it is to achieve a strategic fit with the overall objectives of the organisation
- The location of the project: a short description of the geographical location of the project including any restrictions with respect to noise, vibration, working hours, offloading restrictions, storage of materials, etc.
- Near-neighbours and local community: an overview of the nature and likely attitude to the project of near-neighbours and the community as a whole
- Political policies: a view on whether central and local government are for, against, or neutral towards the project
- Financial planning: any restrictions with regard to cash flow and/or expenditure profiles
- Timing of the project: any issues with regard to the project commencement and completion date
- The form and type of project procurement: is there a preferred contractual arrangement?
- Environment: does the project impact global or local environments and/

or does the client have an environmental policy with which the project must comply

- Specific client policies with regards to: internal organisation and executive control of the project, communication routes; policies with regard to technological solutions which may have an impact on safety and security.

Case study

This part of the case study considers for illustrative purposes the construction of a new road bridge to be constructed together with a dual carriageway north city link road to connect with an existing dual carriageway road leading to the existing motorway interchange.

The bridge will be 300 m long and span the River Devon at a height of 8 m above the river. The bridge will carry a dual carriageway road, a guided busway, a footpath, a cycle track, a water main and a foul sewer.

The bridge and link road will form a strategic link and allow the development of the east city. The bridge and link road are not anticipated to attract opposition except from those living on the north city estate. It is important that the bridge is completed at the same time as the guided busway. The link road must be completed before the first house, office or unit completion at Devonvale. It is anticipated that the link road will involve a deep cutting in the area of the north city estate; transportation of excess excavated material should be kept to a minimum for environmental reasons. Construction traffic must be kept off city roads as much as possible to avoid increasing congestion.

VM 2 (business case and the reference project) – information – the client's value system

The client's value system is at the heart of the mission of any project and remains central to any audit process. Exploring the following seven elements and ranking them using pairs comparison exposes the client's value system:

- Time: the time from the present until the completion of the project
- Annual costs: the cost of the PFI to the client (note: in a traditional non-PFI contract the value system would account for capital cost and operating cost separately)
- Environment: the extent to which the project is to be sympathetic to the environment measured by its local and global impact, its embodied energy, the energy consumed through use and other 'green' issues
- Exchange or resale: the monetary value of the project at the end of the concession period. In a PFI project this may be considered to be the amount the client may need to spend in the years following the end of the concession
- Aesthetic/esteem: the extent to which the client wishes to commit

resources to an aesthetic statement or to the portrayal of the esteem of the local authority
- Politics: the extent to which it is important for the project to be popular with the voters
- Utility: the extent to which the design can go beyond the legislated minimum to for example reduce noise, dust, etc. and generally make living in or next to the project more comfortable.

Case study

Through discussion an overview of the client's value criteria is obtained and a paired comparison diagram completed.

- Time: there is a commercial requirement for the road and bridge to be completed within 30 months of today's date. Not to be operational within three years would be a disaster
- Annual costs: annual costs are to be kept to the minimum
- Environment: environment is seen to be important to the average voter
- Exchange: the road and bridge will be taken over at the end of the concession on the basis that their condition will require no maintenance for three years
- Aesthetic/esteem: planting and landscaping are important
- Politics: the whole scheme must be popular with the voters as it will add £5 per household per annum to the council tax
- Utility: living next to the road must remain comfortable.

The diagram below demonstrates that living next to the bridge and road has to be comfortable, this is the most important factor with cost, time and environment equal second.

A. Time

A	B. Annual Cost						
A	C	C Politics					
D	B	D	D. Environmental impact				
A	B	C	D	E. Exchange			
A	B	C	D	F	F. Esteem		
G	B	G	G	G	G	G. Utility	
A	B	C	D	E	F	G	
4	4	3	4	0	1	5	Totals

VM 2 (business case and the reference project) – information – function analysis

The function logic diagram from VM 1 becomes a key piece of linking information for VM 2. The logic of the diagram is explained to the VM 2 team. This strategic understanding is important before the team begin to construct a technical/operational functions list. To construct a technical/operational functions list it is first necessary to focus on an element or component of the project and then ask 'What does that element do?' Often the element will perform more than one function.

Case study

The strategic function diagram is examined and the team note that the bridge and link road solution to the option appraisal is reflected by level 2 functions. The team quickly conclude that a bridge is the only feasible means to 'decrease travelling time', 'connect to motorway' and 'direct connection to city centre' but to check this the functions of a bridge are listed.

- Span river
- Support traffic/people
- Support services
- Focal point for transport routes
- Minimise maintenance downtime
- Minimise bad weather downtime
- Facilitate flooding of river
- Segregate people, bicycles, busway and vehicles
- Meet demand.

VM 2 (business case and the reference project) – innovation

The facilitator normally conducts an innovation exercise as a brainstorming session focusing on each of the functions in turn. Brainstorming is carried to a standard set of rules including record every idea no matter how stupid, no judgement or discussion during the brainstorming session, no criticism of contributors and no negative comments.

Case study

Brainstorm the function – span river

- Single-span bridge
- Multi-span bridge
- Tunnel
- Ferry
- Floating bridge
- Suspension bridge
- Architectural wonder
- Over precast culverts
- Embankment
- Dam
- Air lift.

VM 2 (business case and the reference project) – evaluation

The brainstorming session will end with many suggestions for each function, often amounting to over a hundred ideas in total. At the evaluation stage these are reduced and grouped to form a suitable number for subsequent development (and often for risk analysis). The first exercise is to quickly reduce the list of ideas to a manageable number of good ideas. This can be achieved through voting, asking the team for a show of hands in support of the idea. Alternatively, and more efficiently, the facilitator will read out the ideas and wait for no more than two seconds for a champion amongst the team to shout out 'keep it'. This may result in over a hundred ideas being reduced to say 70.

The next step is to determine whether the ideas that remain are:

- Functionally suitable (FS): that is they exactly meet the functional requirement that gave rise to their suggestion
- Economically viable (EV): based upon a first impression, the idea appears to be one that can be afforded
- Technically feasible (TF): again based upon the knowledge of the group the idea can be achieved technically without resorting to innovation in manufacturing or installation. Obviously some projects by their nature call for this but the majority of projects do not
- Client acceptable (CA): the idea is conducive with the client's value system.

Ideas that meet all of the above criteria are carried forward to the next stage, those that do not are reconsidered to ensure that there was a good reason for their rejection. The next stage involves grouping and combining ideas to generate an entire solution that can be subject to further and more accurate analysis.

Case study

In the case study a number of ideas were generated for the function 'span river'. These ideas are subject to a two-stage evaluation, selection by a champion and the four-stage evaluation, FS, EV, TF and CA. These are demonstrated in the table below where the first stage results in four ideas being crossed through.

The ideas are then grouped for further analysis.

In the group discussion following the brainstorming, the discussion centred on the options being a bridge of some form or to culvert the river through an embankment. To drop the road to the level where it could pass over the top of 2 m high culverts was not considered technically feasible. During the discussion the idea of constructing a dam and forming an ornamental lake began to gain momentum. In this case the road would run across the crest with a short bridge over a deep spillway. Following the discussion the embankment idea was to be linked with the dam idea.

Idea	FS	EV	TF	CA
Single-span bridge	✔	✔	✔	✔
Multi-span bridge	✔	✔	✔	✔
Tunnel	✔	✘	?	?
~~Ferry~~				
~~Floating bridge~~				
Suspension bridge	✔	✔	✔	✔
~~Architectural wonder~~				
Over precast culverts	✔	✔	✘	✔
Embankment	✔	✔	✔	✔
Dam	✔	✔	✔	✔
~~Airlift~~				

VM 2 (business case and the reference project) – action planning

Action planning requires the team to select the best value for money solution from the three options from evaluation stage. At the end of the workshop the team decide:

- Who is responsible for taking action
- By when is the action to be taken.

At the end of the workshop all stakeholders should have a task commensurate with their ACID test responsibilities i.e.:

- Those responsible for authorisation will investigate authorisation in accordance with the agreed action plan
- Those who were attending the workshop to be consulted have been consulted and will have no active part in the action plan unless another team member's action activity requires further consultation.

Those who were to 'do', will take away the actions on the action plan for further detailed investigation.

VM 2 (business case and the reference project) – implementation

At an agreed time, usually between two and four weeks following the workshop, the team meet to consider the options which have been worked up in sufficient detail to fully understand the technical and economic viability and the risks associated with each option. Because those who are to authorise the decision are a part of the implementation workshop a decision can be taken.

Case study

Project	Cost	Comments
Single-span bridge	£8.5m	Relatively risk-free simple solution could result in elegant design. Site investigation shows foundation on rock at 2 m below riverbed. Estimated construction time 24 months
Multi-span bridge	£6.7m	As above with four support piers, two of which need protection from flooding river otherwise comments above apply
Suspension bridge	£12m	Span too short and deck too wide to make this a viable option
Embankment with culverts	£4.5m	Simple solution, accommodates sufficient culvert capacity for maximum envisaged flood water. Makes use of excess excavated material from link road. Accommodates gravity sewer as opposed to pumped pressure main of other options. Construction time anticipated to be 21 months
Dam with bridge over spillway and lake	£8m	Constructed as an earth and rock fill dam using excess excavated material from link road, a viable solution in location anticipated. Will result in the flooding of an area in excess of that anticipated to be in council ownership. Construction time anticipated to be 30 months

The facilitator takes the team back through the client value system, in which utility, i.e. comfort of those affected by the scheme, was of highest importance followed by environment, time and cost. The strategic function diagram is studied and the team noted that development of the riverside is a low order want. Although the dam is an exciting option the team decide to set this to one side. The embankment with culverts best meets the value system and function diagram and is accepted as the reference project and public sector comparator.

3.4 Conclusions

This chapter has examined in some detail the first three stages of the PFI process, namely, establish business need, appraise the options, and the business case and the reference project. The chapter demonstrates that use of the value management method provides a logical, progressive, pro-active and inclusive approach in ascertaining a cost-efficient reference project. The case study illustrates that the value management approach can lead to:

- A strong case for investment which is expected to be cost-effective
- The construction of a business case which suports the investment and possibly the PFI approach
- A realistic assessment of what is possible and is not a listing of hypo-thetical impractical proposals
- A possible solution to the current below target performance areas which is capable of being studied in terms of capital investment, operations, maintenance and ancillary services
- The development of an affordable reference project and public sector comparator
- A full value management study report which is an auditable record of decision making.

The primary question is, however, how far should the local authority team progress the project before handing the initiative to the private sector? The Treasury Green Book (1997) states that the PFI should emphasise:

- The output required to achieve policy goals, as opposed to the inputs needed
- The public sector as a procurer of services as opposed to a demander of capital assets
- The specification of output requirements to provide maximum flexi-bility for competing private sector bidders.

This may be interpreted in the context of the case study example as indi-cating that the output goals should be in terms of traffic flow and environ-mental requirements. The question of how these should be technically achieved should be left to the private sector. There is, therefore, an apparent presumption that the private sector is more innovative in seeking risk manageable solutions to a performance-specified service than the public sector. However, the Green Book also states that the problem and assessment of options should include the costing of options in sufficient detail to pro-vide a broad check of the likely affordability of the project before procure-ment begins.

The clue to this apparent dilemma comes in the later statement that privately financed projects offering the greatest potential for value for money gains are likely to be those:

- With the most scope for transferring manageable risk
- Which maximise the service to be provided by the private sector supplier in association with the asset
- That have the greatest scope for generating additional revenue by sales to a third party.

This last statement indicates a primary criterion for a successful PFI in that the local authority should define the policy and the infrastructure in performance and technical terms and allow the private sector maximum scope for commercial innovation. In the case study example, the local authority is in the best position to determine the overall strategy for the development of the new district of Devonvale together with a technical statement of the supporting infrastructure. However, the private sector is undoubtedly in the best position to undertake the commercial development of the private housing and office, shops and factory units. The question is one then of packaging the project to maximise the commercial innovation offered by the development. This might mean that the river crossing, the new link motorway link road, the guided busway and the development of 3000 acres of industrially contaminated land might be better handled by the private sector as a package. The private sector is certainly in a better position to take the risk involved in decontaminating the site and its commercial letting.

It is therefore the conclusion of this chapter that the development to PFI stage three, business case and reference project, should be undertaken by the public sector as a value management study to produce a well-referenced, researched and validated public sector comparator of infrastructure assets. Private sector expertise can be used to overlay the infrastructure in such a manner that it offers best value for money.

References

Borjeson L. (1976) *Management of Project Work*. The Swedish Agency for Administrative Development, Satskontotet, Gotab, Stockholm.

HM Treasury (1997) *Appraisal and Evaluation in Central Government: Treasury Guidance*, 2nd edn. HMSO, London.

Kelly J. & Male S. (1993) *Value Management in Design and Construction: the Economic Management of Projects*. E&FN Spon, London.

Male S., Kelly J., Fernie S., Gronqvist M. & Bowles G. (1998) *The Value Management Benchmark: a Good Practice Framework for Clients and Practitioners*. Thomas Telford, London.

Morris P. & Hough G. (1987) *The Anatony of Major Projects: a Study of the Reality of Project Management*. Wiley, New York.

OGC (1999) *How to Achieve Design Quality in PFI Projects*. HMSO, London.

Office of Government Commerce (2000) *Value for Money Measurement*. OGC, London.

4 | Risk perception and communication in public-private partnerships

Peter Edwards and Paul Bowen

4.1 Introduction

This chapter commences with a description of the communication environment of project stakeholders. This is followed by an exploration of risk perception and risk communication in public-private partnerships.

The viewpoint assumed is predominantly *normative*, i.e. the approach is about what *should be*. A descriptive approach (*what is*) is adopted only where existant research allows comparisons to be drawn between the two approaches.

4.2 The communication environment

Applying different perspectives specifically to the process of human communication raises issues which focus on the nature of communication. More specifically, it provides an indication of potential points of analysis. Fisher (1978) characterised different approaches to human communication as mechanistic, psychological, interactional, and pragmatic. The mechanistic perspective is adopted here for illustrative purposes.

Viewing communication within a mechanistic perspective implies a form of conveyance or transportation across space. The components of this perspective consist of the *message* (travelling across space from one point to another), the *channel* (the mode of conveyance of the message), the *source* and *receiver*, *encoding* and *decoding* (the process of transforming a message from one form to another at the point of transmission and destination), *noise* (the extent to which the fidelity of the message is reduced), and *feedback* (a message that is a response to another message).

Central to the mechanistic perspective of human communication is the element of transmission – the movement of the message by means of an appropriate channel. This channel, linking source and receiver, is clearly directional; the directional aspect implying impact on the receiving end, fostering the notion of the source influencing the receiver, i.e. *if* (specified message variables), *then* (receiver effects). The transformation of the message via the encoding/decoding process is highly complex, involving linguistic codes, paralinguistic cues, learned behaviours, cognitions and sociocultural norms (Fisher, 1978). Barriers to communication are seen mainly as existing within the individual's limited capacity to process information received from multiple sources, as opposed to the perception that barriers stem mainly from psychological barriers inherent in the individual's cognitive capacity for encoding and decoding messages.

Within this perspective of human communication, the meaning of a message is seen as being a function of the location of the message at some specific point along the process of communication. In other words, meaning prior to encoding, meaning encoded and transmitted, meaning received and meaning decoded. The attractiveness of the mechanistic perspective lies in its simplicity and its emphasis on the *physical components* of communication.

4.2.1 Models of human communication

Various models have been developed in attempts to explain the communication process. Examples of such models include those developed by Lasswell (1948), Shannon and Weaver (1949), Johnson (1951), Newcomb (1953), Schramm (1955), Gerbner (1956), Westley and MacLean (1957), Berlo (1960), Ross (1965), Barnlund (1971), Feldberg (1975) and Tubbs and Moss (1981). These models all view communication from different standpoints and, consequently, can be *loosely* categorised into one or another of the perspectives of human communication mentioned above. The inability of the various writers to agree on a universal model of communication is reflected in the plethora of models contained in the literature (Bowen, 1993).

It is not within the scope of this chapter to present an overview of the various communication models in existence. Rather, the intention here is to provide a brief description of just one such model, to facilitate a normative approach to risk communication associated with public-private partnerships (PPP). This particular model is chosen not for its superiority over any of the other models presented in the literature, but rather by virtue of the fact that it encapsulates the majority of the concepts required for the analysis of risk communication. Indeed, one measure of the effectiveness of a model is the extent to which the model permits one to organize data and then make successful predictions about such data (Tubbs & Moss, 1981). The *mechanistic* model developed by Feldberg (1975), will be used to describe and analyse the *components* of the communication process surrounding the theory of risk communication.

4.2.2 Feldberg's model

The Feldberg (1975) model is presented diagrammatically in Fig. 4.1. Although the model is presented by Feldberg (1975) in several phases, for the sake of brevity this approach is not adopted here.

The model, linear and mechanistic in nature, may be described in terms of four distinct stages. The first stage is one in which a simple two-person communication process is assumed. The relevant components at this stage are the original sender, the final receiver, the message, the medium, the

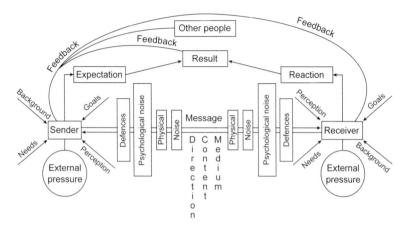

Figure 4.1 *Feldberg's model of human communication (Feldberg, 1975).*

expectations, the reactions, the result, the direction of the message and the content of the message.

The sender communicates the message to the receiver with some objective in mind. This objective may be conscious or unconscious, structured or unstructured, and is translated into the sender's expectations. The message is transmitted to and received by the final receiver, who reacts in some way to the message. The sender's expectation (anticipated result) and receiver's reaction (actual result) are translated into a result. Whether the receiver's reaction conforms to the sender's expectation depends on many factors, the aim of the communication process being to ensure that the sender's expectation and receiver's reaction are as congruent as possible. Prior to transmission, the sender should know the identity of the receiver if the correct expectations are to be established.

The degree of congruency is affected by the direction, medium and content of the message. The direction of the message refers to the route taken by the sender in transmitting the message, whilst the medium refers to the means used to transmit the message. The sender should endeavour to select the most effective method for transmission, the choice of medium depending on the availability of media, the content of the message, the nature of the receiver, and the distance between sender and receiver. The message transmitted by the sender will possess content or information. The exact content will be influenced by the direction of the message, the nature of the receiver, the nature of the medium, and by the objectives of the sender.

The second stage involves examining some of the major reasons why the sender's expectations and receiver's reactions are incongruent. The principal concepts here are personal factors, external pressures, physical noise, psychological noise, and personal defence mechanisms. Personal factors include such influences as age, gender, status, profession and value system. Differences in these personal factors cause different perceptions of reality. External pressure refers to those pressures emanating from other individuals or groups that may cause incongruent expectations and reaction notwithstanding similar perceptions of the message (e.g. lobby groups reacting to a controversial project). Physical noise is self-explanatory. Psychological noise and personal defence mechanisms are similar in nature. The former refers to noise caused by fear, anxiety and insecurity on the part of the receiver (and here stakeholder risk perceptions are seen as influential). Psychological noise can stem from fear of the sender and/or the contents of the message, and serves to distort the communication process. The latter form of noise refers to the fact that many of the receiver's personal defences are based on deep-seated experiences and values, many of which may be based on the relationship with the sender or on prejudices. It is these experiences and values which influence stakeholder attitudes towards, and responses to, project risk.

Stage three constitutes the mechanism for evaluating the relative success of the communication. Evaluation takes place on the part of the sender, the evaluation being in the form of feedback. Feedback on the relative congruence between expectation/reaction and anticipated result/actual result derives from three main sources. Firstly, from the receiver. The receiver may be required to, or may of the individual's own volition, inform the sender of the result of the reaction to the message. Secondly, it may derive from the sender. The sender may check on the receiver's reaction, and the result of this reaction. Lastly, it may derive from other people or groups. Other people may have an interest in the receiver's reaction, reporting their evaluation of this result to the sender. In all three cases, the senders evaluate the validity of the feedback in terms of its source and in terms of their own expectations.

The final stage of communication revolves around the notion of two-way communication. Immediately the receiver responds to the sender's message with a message, the communication process is again initiated, but in reverse. In essence, all the factors mentioned in the previous three stages are brought into play, being now applicable to the receiver as sender and the sender as receiver. In all likelihood the original sender will respond to the original receiver's message, and the entire cycle of communication will begin afresh and continue until the communication process is terminated. An example here might be the allocation of a risk to a project stakeholder; that risk taker's response in the form of a bid price; and a subsequent process of price negotiation.

In its simplest form, the mechanistic model assumes direct communication between sender and receiver. In any organization a substantial amount of communication passes through, and is interpreted by, several intermediate receivers. In a project, this might be exemplified by the nature of client – contractor – sub-contractor relationships. This diminishes the likelihood of the final receiver's reaction being congruent with the original sender's expectation. According to Feldberg (1975), the main responsibility for ensuring that expectations and reactions are congruent lies with the sender or originator of the message. To facilitate the attainment of congruence, it is the sender's responsibility to ensure that the appropriate receiver is selected, the most appropriate medium is chosen, the message contains the correct content, and that there is an absence of psychological barriers inhibiting the ability of the receiver to accept the message.

Tubbs and Moss (1981) developed a model of human communication which addresses the static inadequacies of the Feldberg model, introducing the concept that communication transactions are dynamic. Their model portrays the movement of communication in time, and emphasises the mutually influential nature of the communication event. Dance (1967) observed that communication, while moving forward, is at the same moment coming back on itself and being affected by past behaviour. Communication is seen as ever-changing (dynamic), requiring the active participation of both sender and receiver. It is perceived as convergent in that the source and receiver work together over time to create and share meaning – they converge on shared meaning.

An interesting feature of the mechanistic perspective is the concept of the *gatekeeping* function (Lewin, 1951), illustrated in Fig. 4.2. The gatekeeping function exists on the channel between the source and receiver, and is performed by a person serving as an intermediary between the two. The gatekeeper should be seen as a receiver and re-transmitter of messages rather than in the context of the traditional role of source and receiver. The gatekeeper regulates (filters) the flow of messages, and may function to modify messages so that the initiated message is dissimilar to the message that is ultimately received by the receiver. For example, an architect could act as a (potentially harmful) gatekeeper in a situation where the client and the architect were the only participants in the project briefing process; with all other stakeholders being excluded from the process and receiving briefing information through the communication filters of the architect.

4.2.3 The communication of risk

We have seen that information, taken in its broader sense and pertaining more to knowledge than to data *per se*, can only be transferred between

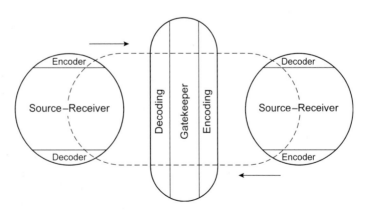

Figure 4.2 *The gatekeeping function (Westley & MacLean, 1957).*

parties if that information is transmitted (and received) in a manner which is meaningful to both. Clearly, the more closely the models of reality of the transmitter and the receiver correspond, the more effective becomes the resultant communication. Thus, the effectiveness of the PPP can be a function of the effectiveness of the communication between the stakeholders. In building procurement, communication constitutes a major problem (Griffith, 1985). This problem is compounded by the increasing partition of the building procurement process into more and more specialised disciplines, resulting in little mutual appreciation of each other's knowledge base. The purpose of communication between project stakeholders, and the context within which that communication occurs, are as important as the communicators themselves.

A little-acknowledged aspect of the risk communication context is that, for every project stakeholder, perception (i.e. knowledge and understanding) of risk and risk management is rarely acquired through first-hand experience. We rely on a variety of media – and thus the communicating power of others – for our information: e.g. through lectures, videos, audio tapes, case studies, journals, conferences and text books. We learn about risk largely at second or third hand, and then use this learning to inform our first-hand communication and decision making about risk in the real-life project environment. If the risk-learning messages distort risks, e.g. exaggerating some risks or diminishing others, or use inconsistent descriptive framing for risks, then our subsequent decision-making judgement is likely to be at least clouded, or even skewed.

4.3 Risk

According to the Royal Society (1991), risk is 'the probability that a particular adverse event occurs during a stated period of time'. This definition is preferred over many others as it incorporates concisely the three essential elements of risk: chance of occurrence; unfavourable or harmful impact; and duration of exposure.

Risk arises out of individual or organisational decision making. At its widest, this might include whole societies. A nation, for example, through democratic processes, might decide to adopt nuclear power generation for its main source of electrical energy and thus accept the technical risk involved. The decision making will be associated with activities relating to undertaking tasks or commitments, or achieving objectives. Any exploration of risks in PPPs, therefore, needs to consider not only the tasks, commitments and objectives of the partnership, but also the decision-making processes associated with the organisational structures (the stakeholders and their relationships to each other) in such partnerships. Since these partnerships are inevitably concerned with the procurement of projects, it has become fashionable to talk about 'project risks', but this should not disguise the fact that these are actually the risks of the stakeholders (partners) in such projects.

Risk is associated with, but distinguishable from, uncertainty. Since, by definition, uncertainty is some state short of certainty, it implies a lack of complete information, and it is therefore associated with risk in terms of the inherent variability of one or more of the input variables relating to the probability, impact or time elements of risk. Modelling risk is more often than not a question of investigating and treating this uncertainty, and may require setting acceptable boundaries on the range of expected values for each of the affected variables. The question which may arise is 'Acceptable to whom?'

4.3.1 Risk management

Risk management is 'a systematic approach to dealing with risk' (Edwards, 2001). A risk management system should typically include processes to deal with risk identification, classification and allocation, i.e.:

- Risk analysis
- Risk response
- Risk monitoring and control (including risk recovery)
- Risk outcome recording and evaluation.

Many techniques are available for each of these processes; the choice of technique being influenced by the nature of the risks involved and the amount of information available. Clearly, effective communication is a requirement for each of these processes.

Importantly, an effective stakeholder project risk management system will span the whole procurement process, incorporating pre-project planning and resource allocation; monitoring and control during operational activities; and post-project evaluation and feedback. A guiding principle to effective risk management is that the more serious the risk, the higher should be the level of management involved in the decision making which spawned it and in the subsequent processes for dealing with it. Clearly, effective communication between the stakeholders in a risk-rich environment is central to the processes inherent in any risk management system.

4.4 Risk perceptions

Since risk is a social construct, in that people decide what constitutes a risk (Royal Society, 1991), it follows that any risk must be perceived by human beings. Such perceptions are founded within the intra-personal and inter-personal communication networks that exist between stakeholders. These perceptions are diverse because they are influenced by value systems, and hence by attitudes, judgements, emotions and beliefs – all components of the process of human communication. Different risks will mean different things to different people. Some risks will even mean different things to the same people at different times in their lives or in different circumstances. Risk perception is therefore a field of considerable interest in cognitive and behavioural psychology. More importantly, perhaps, this social perspective of risk cautions against an over-reliance on exclusively mathematical approaches to assessing and modelling risk, and against overly mechanistic 'hard systems' methods of managing risk. It also places a greater emphasis on risk communication, particularly in the context of PPP in private finance initiative (PFI) project procurement.

Against the difficulties arising from the risk perception diversity perspective stands the countervailing view of the role of professional judgement in assessing and managing risk. Faced with a diversity of views about risk, we seek the advice of experts we can trust. We assume that they will understand, and be capable of dealing with, perceptual dissonances. However, we must recognise that experts too are susceptible to perceptual heuristics and biases (Mak, 1992; Birnie, 1993). We also assume that they are capable of adequately communicating (to the appropriate receiver(s), utilising the appropriate channel and medium of transmission, and conveying the desired message content), without imposing their own risk value system. This may be an overly optimistic expectation.

4.5 PPP project procurement context

Within this social perspective of risk, with its perceptual diversity and its professional judgement, must be considered the procurement approach of PPP in PFI projects. Since risk is personal and not abstract, the risks in PFI projects are those of the parties participating in them – the project stakeholders. In most instances, project stakeholders will comprise organisations or groups, rather than individuals. This gives rise to two issues, namely, who are the perceivers? and whose risk management system is it?

The risk perceivers are the project stakeholders, and a stakeholder is any entity which has the power to influence project decision making directly. The concept of influence is important, as a key component of communication is the exertion of influence towards some desired outcome. The closer the expectation and the result, the more effective the communication process. Risk arises out of the decision making associated with the achievement of project objectives. Risk perceivers will therefore be involved with project decision making at some level. To some extent, the level of decision making will reflect the hierarchy of project objectives (Fig. 4.3) and the stakeholder identity. Key to project success is the effective communication (and adoption) of project objectives by stakeholders, who may themselves have conflicting goals. Indeed, the identity of stakeholders, not least of all the identity of the 'client' in large public organizations, together with their associated (often concealed) objectives, can act as not inconsiderable barriers to effective stakeholder communication (Bowen, 1993).

Project objectives may relate to project procurement, i.e. bringing the project to fruition. For the most part, these objectives include completing the

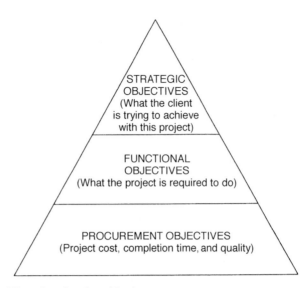

Figure 4.3 *Hierarchy of project objectives.*

project on time, within an approved budget, and to an appropriate standard of quality or fitness for purpose. Project procurement stakeholders will comprise the client organisation, design team, finance provider, contractors and sub-contractors.

Functional objectives for a project relate to what the project is required to do after it has been procured. PPP projects now innovatively bring the finance provider and contractor organisations (and to some extent the design team) into what was formerly the exclusive preserve of the client. Whilst innovative in terms of organisational structure, the potential for barriers to effective communication caused by the introduction of additional stakeholders should not be underestimated.

Research in South Africa into the effectiveness of building procurement systems in the attainment of client objectives highlights the inadequacy of the briefing process as a communication environment and provides useful insight into the (often haphazard) manner by which procurement teams are constituted. Professional advisors, whilst acknowledging the need for clients to be guided through the procurement process, admit to not infrequently failing to advise clients of procurement path alternatives, as well as to failing to associate themselves with client objectives (Bowen *et al.*, 1999). There is no reason to believe that these communication difficulties will apply in any lesser degree to PPPs.

At the strategic level of project objectives, project decision making will be concerned with what the client is trying to achieve with the project. This is different to the functional objectives. At a functional level, for example, a new high school building will provide the appropriate physical amenities for a given number of teachers, administrators and pupils to engage in educational activities. Strategically, however, the project reflects a contribution to the political outcome of a particular education policy in a given region. Traditionally, decision making at the strategic level has been entirely *intra*-client based, but the PPP approach introduces other stakeholders to the strategic process.

Perceptions of project risks, therefore, will relate to how particular stakeholders are engaged in project decision making at levels concerned with achieving particular project objectives. In PPP projects, these perceptions may be occurring in environments which are unfamiliar (in terms of previous experience) to many, if not all, of the project stakeholders; and perhaps unfamiliar too to the professional advisors on whom they rely for advice. Indeed, research in South Africa has established that many clients, professional advisors, and contractors possess limited knowledge of procurement method alternatives, tending to favour methods with which they are familiar, e.g. the traditional procurement route (Bowen *et al.*, 1999). Moreover, formal feedback from projects as a vehicle for learning and open communication proved to be the exception rather than the rule. No greater justification is needed for encouraging clear and open communication in such environments. Smith (1999) notes that stakeholder inexperience is an important factor to consider in assessing the 'riskiness' of projects.

The second issue is concerned with perceptions of project risk management. It is a conveniently tempting trap to see risk management operating as a single one-off system, perhaps based upon a standard design, and implemented for each project. The reality is different. We have already noted that different stakeholders will perceive different risks on the same project, that even the same risks will be perceived differently and that stakeholders' risk perceptions may be influenced by the levels of decision making in which they are engaged and the nature of the objectives they are seeking to achieve. Such potentially incongruent perceptions need to be acknowledged and addressed as part of the process of risk communication and risk management system.

A single system, 'one size fits all' concept of project risk management is impractical. It would have to be so multi-dimensional (in terms of stakeholders, objectives, decisions, risks, responses, treatment plans and outcomes) as to be incapable of effective implementation, particularly over the notional life span of a PPP project. A single system view also presupposes a single point of responsibility for creating and managing such a system – a clearly unworkable arrangement – and is based on a false assumption that the risks being dealt with are all exclusively *project* risks, whereas they are actually the risks arising for each individual stakeholder from their involvement in the project. A stakeholder, for example, may face an ongoing financial risk of insufficient operating capital which is exacerbated by involvement in a particular project. In essence, risk management is a 'human' problem rather than a project problem and, as such, human communication should play a pivotal role.

The practical inadequacies of the single system concept of project risk management lead naturally to consideration of a plural systems approach, and here project *quality* provides a useful analogy. As an important objective in project procurement, project quality is not nowadays perceived as achievable through the overlaying of a single quality management system on a project, but rather as the expected outcome of the sum of the quality assured processes contributed by each project participant. Each contributor's assurance of quality is backed by certification of conformance with an appropriate standard such as the ISO 9000 series.

Interestingly, many construction-related organizations (contractors, professional consultants) see their quality assurance systems as a substitute for risk management – a means of limiting their liabilities to third parties (Edwards, 2001). However, while it is possible to regard quality and risk as different points along a continuum of project management, it should be noted that they really address different objectives. Quality assurance is concerned with 'getting things right'; while risk management is 'dealing with things that might go wrong.'

Conceptually, a plural systems view of project risk management might be represented in Fig. 4.4. Even this 'flower petal' view might be simplistic in reality, particularly where organisations engage in consortia arrangements to participate in PPP projects, as Fig. 4.5. shows. The point is that each stakeholder organisation must bring its own organisational risk management system to bear upon the project risks to which it is exposed. Effective communication should be seen as the means of linking these (potentially different) risk management systems.

While the plural systems approach addresses the issue of whose risk management system it is, the analogy with quality assurance presently fails at the point of project integration. To date there is no evidence that project

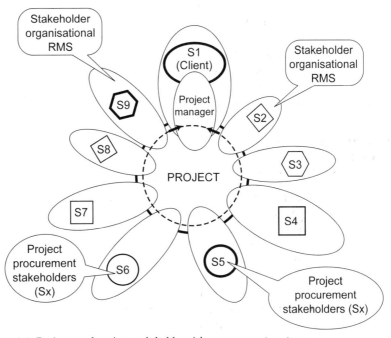

Figure 4.4 *Projects and project stakeholder risk management systems.*

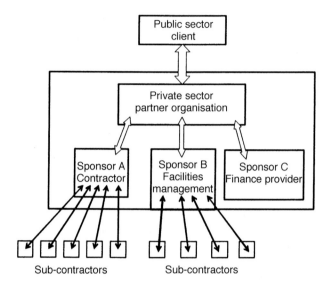

Figure 4.5 *Organisational structure for PPP projects (after Akintoye et al., 2001).*

managers call for evidence of the existence of systematic organisational risk management – to an appropriate and universally recognised standard – from each of the project stakeholders. What evidence we do have suggests that any risk management which does take place tends to be *ad hoc*, unsystematic and applied inconsistently by the organisations involved. Stakeholders prefer to use their positions of dominance and power to transfer as much risk as possible down the supply chain, either at best using obscure legal language in largely untested contract clauses, or at worst relying on common interpretations of unwritten practice, to do so.

Such practice really stands in contradiction to all that we should be seeking to achieve in effective risk communication. This is the challenge facing clients and consultants alike.

4.6 Conclusions

In PPP projects, risks arise from the decision making of project stakeholders engaged in pursuing project objectives at procurement, functional or strategic levels. Perceptions of these risks will vary among the different stakeholders, thereby creating the need for effective risk communication.

Risk management is not practicable at a single-system, project-based level, but should rather be implemented as a plural systems approach through which the organisational risk management systems of each participant are overtly brought to bear upon the project risks. In order to achieve this, effective communication about risk is of paramount importance. Effective risk communication requires not only placing appropriate emphasis on the risk messages themselves, but also giving attention to the message media, the message senders and receivers, and the provision of feedback to ensure that mutual understanding of project risks has been achieved.

References

Akintoye A., Beck M., Hardcastle C., Chinyio E. & Asenova, D. (2001) *Standardised Framework for Risk Assessment and Management of Private Finance Initiative Projects*, Report No. 5. Department of Building and Surveying, Glasgow Caledonian University.

Barnlund D. (1971) A transactional model of communication. In: *Language Behaviour: a Book of Readings* (ed. J. Atkin). Mouton, The Hague.

Berlo D. (1960 *The Process of Communication*. Holt, Rinehart and Winston, New York.

Birnie J. (1993) *A Behavioural Study using Decision Analysis of Building Cost Prediction by Chartered Quantity Surveyors*. Unpublished DPhil. Thesis, University of Ulster, Jordanstown, Northern Ireland.

Bowen P. (1993) *A Communication-based Approach to Price Modelling and Price Forecasting in the Design Phase of the Traditional Building Procurement Process in South Africa*. Unpublished PhD. Thesis, University of Pretoria, South Africa.

Bowen P., Pearl R. & Edwards P. (1999) Client briefing processes and procurement method selection: a South African study. *Engineering Construction and Architectural Management*, **6**(2), 91–104.

Dance F. (1967) Toward a theory of human communication. In: *Human Communication Theory* (ed. F. Dance), pp. 289–309. Holt, Rinehart and Winston, New York.

Edwards P.J. (2001) *A Study of Risk Perceptions and Communication in Risk Management for Construction Projects*. Unpublished PhD. Thesis, Department of Construction Economics and Management, University of Cape Town, South Africa.

Feldberg M. (1975) *Organizational Behaviour: Text and Cases*. Juta and Company, Cape Town.

Fisher B. (1978) *Perspectives on Human Communication*. Macmillan, New York.

Gerbner G. (1956) Towards a general model of communication. *Audio-Visual Communication Review*, **3**, 3–11.

Griffith A. (1985) *Buildability: the Effect of Design and Management of Construction*. Report, Department of Building, Heriot-Watt University, Edinburgh.

Johnson W. (1951) The spoken word and the great unsaid. *Quarterly Journal of Speech*, **32**, 421.

Lasswell H. (1948) The structure and function of communication in society. In: *The Communication of Ideas* (ed. L. Bryson), pp. 37–51. Institute of Religious and Social Studies, New York.

Lewin K. (1951) *Field Theory in Social Science*. Harper and Row, New York.

Mak S. (1992) *Risk Management in Construction: a Study of Subjective Judgements.* Unpublished PhD. Thesis, University College, London.

Newcomb T. (1953) An approach to the study of communicative acts. *Psychological Review,* **60**, 393–404.

Ross R. (1965) *Speech Communication: Fundamentals and Practice.* Prentice Hall, Englewood Cliffs, New Jersey.

Royal Society (1991) *Report of the Study Group on Risk: Analysis, Perception, Management* (Chairman: Professor Sir Frederick Warner). The Royal Society, London.

Schramm W. (1955) *The Process and Effects of Mass Communication.* University of Illinois Press, Urbana.

Shannon C. & Weaver W. (1949) *The Mathematical Theory of Communication.* University of Illinois Press, Urbana.

Smith N. (1999) *Managing Risk in Construction Projects.* Blackwell Science, Oxford.

Tubbs S. & Moss S. (1981) *Interpersonal Communication.* Random House, New York.

Westley B. & MacLean M. (1957) A conceptual model for communications research. *Journalism Quarterly,* **34**, 31–38.

PART TWO

Stakeholders' perspectives on public-private partnership risks and opportunities

5 A construction perspective on risk management in public-private partnership

Ezekiel Chinyio and Alasdair Fergusson

5.1 Introduction

Construction is a risky endeavour. Many things can go wrong in a construction project, especially if it is very complex. Sometimes too, the client's needs are not precisely defined *ab initio*. Occasionally, some client's objectives conflict with each other, with the implication that many subsequent activities cannot be undertaken on time.

It would be naïve for participants to pretend that nothing will ever go wrong in a particular construction project. It is rather better to assume that all that can go wrong might go wrong. From such a premise, one can plan fallback options, should something(s) go wrong in the course of the project. This is not being pessimistic but accountable, because, if one does not plan for potential downsides, events might catch such a person unawares and expose that person to negative consequences.

Construction risks impact on time and require money to be redressed. If potential risks can be foreseen upfront and addressed beforehand, then the potential delays and economic implication of such risks can be minimised or eliminated. Risk management from a construction perspective therefore should involve an examination of what can go wrong and what to do about it.

The wherewithal of construction risk management in private finance initiative (PFI) projects is discussed in this chapter. PFI is a type of public-private partnership (PPP) that gained an entry and wide usage in the UK market in the 1990s. This form of procurement is not discussed in detail in this chapter; as a modest understanding of PFI is useful for the comprehension of this chapter, readers are referred to Chapter 2 for a detailed discussion of PFI. The Private Finance Panel (1995, 1996) also explain the concepts of PFI.

An investigation into PFI risk management practice informs the considerations of this chapter. The investigation was sponsored by the Engineering and Physical Science Research Council and the Department of the Environment, Transport and the Regions, and conducted by personnel from Glasgow Caledonian University. The risk management of PFI projects examined in this investigation spanned two years, i.e. December 1999 to December 2001. Further details on this research are discussed by Akintoye *et al*. (2000).

5.2 Construction project delivery and the PFI form of procurement

Construction facilities play a significant role in the delivery of public services, with or without PFI. For example, a road in part serves as a basis for people to move between two or more locations. Similarly, hospital premises serve as a basis for people to obtain health care; an airport enhances transportation; a classroom facilitates the delivery of education; etc. However, construction project delivery fits into PFI perfectly as a jigsaw puzzle. Although the essence of PFI is to deliver public services, facilities are almost always used in this quest. Thus as the PFI market matures in the UK, diverse types of construction facilities are used as a means to an end in public service delivery. In this regard, police colleges, schools, hospitals, prisons, fire stations, housing estates, etc. have been built as aspects of different PFI projects.

The actual way in which construction works are managed may not change significantly in a PFI setting. Given a particular design and method statement, two constructors can evolve two construction projects with relatively similar outcomes. It is therefore the procurement procedure that primarily distinguishes a PFI construction product from a non-PFI product. However, the risks in a particular PFI project are higher than if that same project were not done via PFI. The reason being that the scope of responsibilities of the participants in a PFI project is higher than in non-PFI projects. The next section describes PFI project delivery to establish how it amplifies its construction and other risks.

5.2.1 PFI procurement

In the PFI form of procurement, private sector organisations are directly involved in delivering public services through a consortium (Goldsmith, 1997). Since the public sector can contract with several private sector organisations for the delivery of a given service, the formation of a consortium by the private sector participants in a PFI project is a distinguishing feature of this form of procurement. In addition to the partnership between the private sector participants, private finance is also almost always utilised for PFI projects. In essence, the private sector partners in a project undertake to design, build, finance and operate facilitates to achieve the objectives of a client, i.e. concerning service delivery. Sometimes, the tasks of land acquisition, demolition of old facilities, obtaining planning permission, etc. are also transferred to the private sector participants.

By transferring many project tasks and their associated risks to the private sector, the public sector is inadvertently transferring away many risks that would have been borne by it (Kangari, 1995). That way, one principal objective of PFI is inadvertently achieved. The 'value for money' criterion, which is also expected to be achieved fundamentally through PFI, is attained partly through negotiations between the public and private sectors.

Since its inception in 1992 in the UK, several hundred PFI projects have been started (Robinson, 2000), examples of which are shown in the PFI map of Great Britain (Hillgate Communications Limited, 2001). Having passed through the learning phase, the PFI market is ever maturing. Also, given that the present UK government prefers the PFI form of procurement, many more projects will be procured through PFI.

5.2.2 Construction facilities and PFI

Many PFI projects involve an element of construction (Pickering, 1999). Hospitals, schools, prisons, roads, courts and many other types of facilities have been constructed by means of PFI. The huge transaction costs that attend PFI schemes make this form of procurement much more suitable for larger projects. In its earlier years, some smaller schemes were procured through PFI, but the transaction costs were relatively high. Therefore, some clients now combine their smaller schemes into one cluster PFI project. The tendency is for PFI schemes to be characterised by large-scale facilities, which somehow amplifies some of the risks.

Although most PFI projects are large, the consortia that deliver these projects adopt a very lean structure, employing very few employees. The several responsibilities of a consortium are usually outsourced to tertiary organisations, except on a few occasions when a conglomerate might opt to undertake some or most tasks. The construction sub-contractor, often engaged through a 'design and build' contract, may in turn choose to sublet some aspects of its work to other organisations (Fig. 5.1). Therefore, the construction sub-contractor must manage both the construction risks and those of the supply chain in the PFI contract.

5.2.3 Construction-related risks in PFI

Risk pertains to 'uncertainty of an outcome' (Boothroyd & Emmett, 1996). It is a chance that some event may or may not occur. In a PFI project, there would be a whole range of risks, e.g. (Akintoye *et al.*, 2001):

- Planning permission may be denied
- Some design features may conflict with each other
- The proposed technology may become obsolete
- Legislative changes may render illegal some aspects of a PFI scheme
- The services to be provided may not be in full demand
- The facilities may not have any residual value at the end of the concession period.

Effectively one can consider the risks of PFI projects during any of its phases, i.e. pre-construction, construction, operation and post-operation.

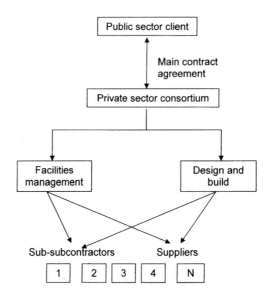

Figure 5.1 *Organisational structure of a PFI contract.*

The land selected for a particular PFI project may pose its own peculiar risks, like being amenable to subsidence due to coal seams. Heavy traffic around the site may interfere with access and slow down the works. The design process may fail to address the risks pertaining to health and safety, layout of different buildings (or facilities), structural fitness, etc. A host of other risks could face the construction team. The term risk is defined elaborately in Chapter 6 of this book, and different types of risks are explained there in detail. This chapter will therefore not cover this; suffice it to say that construction-related risks prevail in PFI projects and that a construction company involved in a PFI project should be able to identify the risks associated with the construction aspects.

PFI is underpinned by two related principles: transfer of many risks from the public to the private sector and achievement of value for money. The transfer of risks must be done optimally otherwise the value for money objective will be eroded. Therefore some form of risk quantification should underpin the transfer of risks. The risk management process and how quantification forms part of it is discussed in this chapter. The discussion is from a construction perspective. The principles discussed in this chapter are also drawn from leading edge organisations in the PFI market.

5.3 Risk management

Risk management involves the identification, mitigation and evaluation of risks (Steele, 1992). Risk management should be approached as an iterative process, and not in the discrete phases of identification, evaluation and control. As shown in Fig. 5.2, the identification of risks should be followed by a search for solutions that can ameliorate or eliminate these risks.

If the risks are successfully eliminated, then there is nothing to evaluate. If, however, solutions are derived to counter these risks, then the cost implications of the mitigating solutions should be evaluated. The outcome of the evaluation should be fed back to the identification task to re-appraise the new risk profile of the project. Sometimes the mitigation of risks gives rise to secondary risks, which must in turn be addressed through identification-mitigation-evaluation. An organisation should try to deal with the realities of a situation.

There would always be the issue of probabilities to be considered, but if, for example, a construction company were assigned to build where there has been an existing building, then it should carry out a full site investigation to understand the peculiarities of the site. If the project concerns a greenfield site and archaeological risks are also anticipated, then a historical survey should be conducted. This may involve searching relevant maps thoroughly to assess what could be on the site, or an independent professional organisation could be assigned the task. Either way, such an investigation will cost some money, but it may prevent a worse outcome. Whatever the site investigation establishes would provide a basis for making more informed decisions.

The iteration process of risk identification-mitigation-evaluation continues until a satisfactory position is reached. Figure 5.2 shows that risks are still monitored and controlled, after they have been finally evaluated. The main activities that encompass the risk management process are examined more closely below.

5.3.1 Risk identification

The greater burden of risk identification in PFI lies with the client. This is so because of the nature of PFI procurement where the client's project documentation includes a risk matrix. Usually, the public sector client organisation will give an indication of those risks which it is prepared to take. The public sector will also identify those risks it feels the private sector should

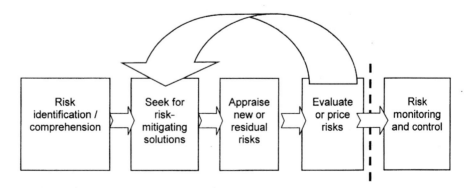

Figure 5.2 *Illustration of the risk management process.*

bear and those to be shared between the two sectors. Ideally, the task of the private sector is to either accept the client's proposition or negotiate on the re-allocation of some risks. However, the spirit of co-operation prevails in PFI, and so some private sector consortia find themselves identifying risks, especially those that had been omitted in the client's risk matrix.

PFI procurement also encourages innovative inputs from the private sector. As this sector thus envisages the design solution in a particular PFI scheme, it endeavours to identify and assess the risks associated with it. Such assessment enables a consortium to price its bid more competitively. Therefore, private sector participants are heavily involved with risk identification in PFI.

Some organisations use checklists or risk prompts to help them identify the current project risks. Given the hindsight of experience, there will be headings that people readily use as a checklist. However, some organisations develop elaborate checklists as a guide to risk identification. Checklists are a very valuable resource, however it is noteworthy that new risks cannot be found in a checklist and certain risks may not be documented in the checklist. Therefore, checklists should be used with caution.

Some organisations tend not to start with a preconceived framework, because there is a danger that key parts of the risk assessment could be missed. For such organisations, a tick in the box approach is not the best way of starting. It is rather better for them to start on a blank sheet and then go back and check it against a framework because a checklist would not be able to identify some project-specific risks. Whichever way the task of risk identification is approached, the essence is to make sure that the risks are identified, at least the significant risks.

Many risks could plague a project and identifying all of them and assessing their cumulative effect could take a long time or be cost-ineffective. At a point in time in the bidding process, an organisation would have to draw the curtain on risk assessment. Therefore, the main objective is to see that the major risks that could impact on the project most adversely are not left unidentified.

Where an organisation or a consortium is running out of bidding time, the 80–20 principle might be applied, in which the key risks, about 20%, must be identified and addressed. Such key risks normally account for 80% of the cost and time implications of the entire risks, hence the 80–20 principle. However, risk management should not be restricted to the key risks only. The minor risks should be addressed as much as possible, time permitting.

There are different avenues through which an organisation can identify risks (Edwards, 1995). Given that the risks facing a PFI project are many, different ways should be used to identify them. Useful ways of risk identification are discussed below, though real-time practice should not be limited to only these avenues.

Risk identification through use of personal and corporate experience

If an organisation or consortium embarks on a PFI scheme without having done one before, it will face so many types of risks that it might be difficult to understand them all. Even after some risks have been identified, it is sometimes difficult to know what to do with them. However, experience acquired over time makes it relatively easier to identify the key risks facing a PFI project. In this regard, a policy of dwelling in one or a few sectors is inadvertently helpful to risk assessment. For example, some utility organisations have been involved in water treatment for a long time. Such organisations then find it easier to identify the risks facing a PFI waste management scheme because the core functions involved are not significantly different from what they are used to doing. To such an organisation, it is really a matter of sitting down and looking through the project and identifying where the risks are. Likewise, some construction firms undertake PFI projects concerning housing and schools, where they are experienced, but not prisons or hospitals, where they are inexperienced.

Apart from familiarity, specialisation and experience also enable organisations to build databases, which can be used to draw insight for risk management. Databases serve as an information resource for diverse things, like construction methodology, method statements, defect analysis, patterns of traffic, costs and prices, etc. Databases can thus aid in risk identification. From a cost point of view, a number of databases are used to filter information that will cast a light on the current project. The 'Building Cost Information Service' (BCIS) and some of the journals on PFI produce databases of projects. There are also various project intelligence databases, which provide varying and useful information.

Checklists and risk matrices can also be developed from databases for future use. Also, through reviews of previous projects, there are some elements such as experience gained from a previous project, that are useful to determine how to deal with current risks. It is thus not surprising that organisations involved in PFI project delivery are keen to employ experienced personnel. Also, while forming consortia, the constituent organisations scrutinise their potential partners to check for, *inter alia*, their level of experience and how it combines with theirs.

Sub-contracting is often inadvertently utilised to augment experience. Where a consortium lacks experience in a particular aspect, it should use its sub-contractors as an avenue to supply the missing link. Given the long-term nature of PFI projects, a partnering arrangement between a consortium and sub-contractors could be worthwhile, as it amplifies the experience of a consortium and minimises bureaucracy, transaction costs and conflicts.

Risk identification through safety reviews

The routine review of previous projects can give an insight into what can go wrong in the forthcoming projects. Reviews can be done on in-house projects that have been undertaken by an organisation. They could also be carried out on other public projects where information can be accessed. Such reviews are targeted at risk and safety issues, and involve a retrospective evaluation of how safety standards were achieved in the projects under study. It is easier to criticise looking back than to plan effectively looking forward. It is therefore possible to make mistakes in the planning and execution of a project, and expose the operatives to safety mishaps. Reviews of projects can identify where such mistakes had been made, and take pre-emptive correction against future schemes.

Risk identification through intuitive insights

Intuition plays a part in risk identification. When you have got a risk like a potential change in law, there is nothing that will tell you definitively whether the law will change, or not. You have to take a view of the future, and make provisions for it at the onset, so to some extent intuition is helpful to risk identification.

Innovation is encouraged in PFI schemes, and experience may fail to identify what could go wrong with an innovative product. Intuition is readily available for pinpointing the long-term implications of innovative solutions.

Risk identification through brainstorming

Workshops and other forms of meeting are often held, sometimes over several weeks, to identify where risks lie in a project. When puzzling issues are encountered, brainstorming is often used as a tool to resolve them. Therefore, brainstorming is used to identify risks, and think of solutions on how to deal with them. Organisations responsible for delivering the project should have brainstorming sessions at various stages throughout the project if particular issues are difficult to resolve.

Risk identification through site visits

Site visits are not new to construction. One of their functions is to identify risks, and this has been put to use in PFI. For example, a site visit was used to identify and assess the danger posed by asbestos in a PFI project in Scotland. This school project involved the demolition of old buildings which contained asbestos. Site accessibility, wayleaves requirement and ground hazards are some risks that could be spotted through a site visit. Procurers will thus find it useful to visit the sites of their projects, albeit to identify some risks facing their schemes.

Risk identification through the use of organisational charts

Organisational charts are very useful for assessing the quality and competence of personnel available within an organisation or set of organisations for a project. The organisational structure of an establishment can also give information on the nature and efficacy with which the entire firm is managed.

In PFI project delivery, construction companies coalesce with other types of organisations to form consortia. At times two or more construction companies could team together in a particular consortium. So, as a general prelude to forming consortia, construction firms should scrutinise the personnel resources of their potential partners, especially if those potential partners are not well known. Even when the partners are known, a re-appraisal of their employees will cast a light on their new employees and their effect on the company profile. If, for instance, a new manager were employed in a partner organisation, a construction company would be assessing the ease with which they can relate with him/her and if established procedures of dealing with predecessors would have to change. Organisational charts thus serve as a useful tool for identifying potential bottlenecks to a forthcoming project.

Risk identification through the use of flow charts

Flow charts are particularly useful but not limited to depicting flow processes. For example, flow charts can be used to show the flow of materials in a factory, the flow of crude oil as it is processed in a refinery, etc. A flow chart can equally be used to show the movement of cement, sand and aggregates until they are placed as a concrete floor or component in a building. By depicting flows this way, it makes it easy to spot the processors (concrete mixer or other machinery) that can go wrong. The number and capability of personnel needed to be at various locations along the line of flow can equally be spotted. That way, the risk posed by employees, like absenteeism, mistakes, etc. can be quickly identified.

Flow charts have been used by some construction companies that are involved in the delivery of PFI projects. In this regard, flow processes in waste management projects have been used as a source of risk identification. In one such project in the North of Scotland, the risk posed by a specified treatment plant in terms of number of possible breakdowns was assessed to be high. Given that defaults attract severe penalties in PFI, the risk posed by this plant informed the consortium's decision to choose another type of plant. In this project also, the construction company involved was able to use one of the flow charts to compare the estimated amount of waste that would flow into the drain pipes with the specification of the pipes. This helped the consortium to refine the specification of the pipes to cope effectively with the waste load.

Risk identification through research, interviews and surveys

Sometimes, the foregoing techniques may not offer sufficient insight into some risks. Researches, surveys and interviews come in handy for making enquiries into some risks. Take an issue like 'planning' on a new job where previous information or experience may be unavailable. A construction organisation that faced such a situation went out to talk to local authorities in the area, to assess if obtaining planning permission for the project would be a major problem. On a road project, another construction firm interviewed local resident associations to find out what they thought about the project, assessing the risk of strike actions that could disrupt the project. Dangers posed by a PFI project to the environment, and vice versa, could be checked through research and interviews.

Research is vital where refurbishment works are involved, or where existing facilities will be utilised. Latent defects may be present in such existing facilities. Any organisation going into such a contract will have to price the risks on the basis that rectifying latent defects will cost them more money. They may have to drill holes in the walls, put telescopes into the cavities, look at the foundations, and going into the infrastructure thoroughly to ascertain any latent defects that are pre-existing in the property. Before such an organisation signs the contract it should conduct a full intrusive survey of the property, where the opinion of surveyors and other consultants should be obtained.

Risk identification through analysis of assumptions

A whole PFI project will be based on a series of assumptions. The design especially is often based on assumptions, when information on the client's requirements is not fully certain. An organisation will have to continually go back to check these assumptions in the light of emerging information. It will also need to check whether the assumptions pose a real threat to the project. It is thus worthwhile to catalogue the assumptions that have been made in the course of a project's development, and to revisit them regularly.

Risk identification through consultation of experts

Sometimes, the in-house expertise needed to assess certain risks may be unavailable. In such circumstances, outside assistance should be sought. PFI consortia normally consist of construction, finance and facilities management organisations. Although this structure is not rigid, and is not utilised on all occasions, many consortia employ this formation. In any case, it is sometimes impossible for a consortium to have all the needed expertise for a given PFI project in-house.

Consulting experts in the course of risk assessment is like an extension of the use of experience. Different experts specialise in different subjects and have the built-up experience that goes with their vocation. So, by using consultants, a consortium is acknowledging that they are not experienced in a particular subject, and that consultants know better and will be engaged to assist. For example, a traffic forecasting company was hired by one consortium, to help evaluate and explain the traffic risks facing a motorway project in the north of England. Likewise, an environmental company might be used to assess the environmental risks in a project.

If an organisation or consortium wishes to assess risks more accurately, then it should not hesitate to engage the services of requisite consultants when needed. One trait with PFI practice is that consortia always have to hire a lot of experts, i.e. legal, financial, design, environmental, planning, etc.

Such consultation ensures that many risks are inadvertently identified and mitigated before the construction is embarked upon.

Overview of risk identification

Risk identification is very crucial because if the risks facing a project cannot be determined *a priori*, then any or some of them can materialise at any time in the life of a project and interfere with the achievement of the project's objectives.

Construction practitioners have found the foregoing risk identification approaches useful in PFI practice. By and large, practice leads to perfection. Thus, experienced practitioners tend to know what can easily and significantly go wrong in the projects within their normal scope of work. If many personnel who have experience of diverse project aspects are involved in the process, then the risks facing a particular scheme can be identified almost exhaustively.

However, risk identification is not a task that should be approached casually, even by the most experienced organisations. The reason being that new risks can often emerge, and risks which had previously been minor could suddenly become key issues in a project. Risk identification should thus not be taken for granted.

5.3.2 Risk evaluation

A risk may materialise in the course of a project. There is no guarantee that it would, but if it did, there would be a consequence. There are two features that characterise risks (Carter *et al.*, 1994; Simon *et al.*, 1997):

- The probability (chance) by which they can happen
- Their ultimate impact on the project, if they do materialise.

An accurate assessment of these two aspects will enable an organisation or consortium to decide on a course or courses of action. In one particular PFI scheme, the engineers employed by the consortium established that the strength of the ground was not certain. This scenario led the consortium and its design consultants to opt for a raft instead of a rift foundation. Risk evaluation influences decision making, especially in the course of project formulation.

The probability of a risk occurring and its impact on a project are used in tandem as decision aids. For example, if the chance of a risk happening is assessed to be high and its potential impact is equally high, then such risk is accorded high priority. Figure 5.3 shows a macro prioritisation of risks, where a risk designated with five stars is accorded utmost priority, given that its impact is high and its chance of happening is high too. The effect of adverse weather on the progress of a construction project may be rated by, say, five stars if significant delays are expected.

A one-star risk in Fig. 5.3 has a low chance of occurring, and its impact is not significant. For example, a one-star risk can concern the discovery of

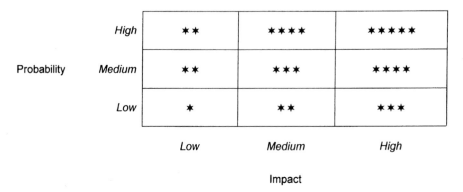

Figure 5.3 *A prioritisation of risks.*

brackets which are undersized for a particular sanitary fitting, as the time it takes to effect a replacement may not delay the project. During the planning phase, lesser attention may be paid to such minor risks.

Paying lesser attention to a one-star risk is not tantamount to ignoring it. Management needs to be sure that everything is in place for the project, however management time is better served if their energy and capabilities are directed towards the crucial issues. Most organisations that have operated for some time would normally have a system that sorts out minor risks and problems. This then frees their time to face the new and complex issues.

However, risks cannot be mapped on a permanent basis, given their dynamism. What may be a major risk today may turn out to be a minor issue tomorrow. Risk management growth and experience embellishes itself. And so with time, some ways of mitigating some known high risks are found, and such risks then become minor issues. Therefore, the profile of risks is always changing and many risks are managed unconsciously. As each scheme is approached, the risks that impact heavily on the project should be established. This reinforces the triadic management principles of risk identification-mitigation-evaluation. If a high risk is established, means of mitigating it should be sought, upon which the ultimate scenario is evaluated. The process could weave between mitigation and evaluation several times before a comfortable position is reached hence the iterative nature of risk management.

Types of risk evaluations

A prioritisation of risks as in Fig. 5.3 warrants some form of assessment. There are different ways in which the two features of risks can be assessed. These different ways are classified in Table 5.1. An assessment could be qualitative, quantitative or somewhere in between. The amount of information and time available and the need for the assessment determine the type of evaluation to be utilised.

Table 5.1 *Risk assessment classification.*

Type	Outlook
Qualitative	Both probability and impact are assessed subjectively
Semi-quantitative	Probability assessed subjectively but impact assessed objectively
Full quantitative	Both probability and impact assessed objectively

A qualitative assessment is employed when uncertainty is prevalent. In the absence of information, subjectivity prevails. A semi-quantitative assessment can be employed where the impact of risks can be established fairly accurately. On the risk concerning a general change of law, a semi-quantitative evaluation can sometimes be utilised because the impact can be assessed numerically, while the likelihood is often difficult to establish that way. It is difficult to define when government is going to change the law.

A full quantitative approach is adopted when information is available. Through safety reviews, use of databases and other sources, information can be generated to facilitate a quantitative assessment of risks. However, information is scarcely available fully. Therefore only a few risks get to be evaluated quantitatively while the numerous other risks cannot be assessed that way.

Assessing the probability of risks occurring

Statistical analysis is ideally employed to assess the chance of a risk happening. For instance, it can be estimated that the probability of the mechanical works delaying the project is 0.17%. Such a figure can be derived from past records of similar delays. 'Bayes theorem' can be used to combine the statistics of previous events to derive the probability of a complex scenario (Lindley, 1965; Steele, 1992). The assessment of probabilities can then facilitate the accurate mapping of contemporary risks.

However, most risks are difficult to quantify in terms of measuring a real probability because the underpinning information is usually unavailable or insufficient. In the UK, PFI was introduced in 1992, and virtually no project has run its full course. There is thus no hard data to say how things will work in PFI. Projections have had to be made from other types of schemes. In the absence of reliable information, a subjective estimation of the probabilities might suffice.

Many types of projects have now been procured by means of PFI and data on this type of procurement is being consolidated. In the near future, participants will be able to generate sufficient data to support the statistical computation of probabilities concerning PFI-related risks. For now though, many of these PFI participants are satisfied with the way they have assessed the probabilities of the risks of their projects.

Assessing the impact of risks

Organisations are basically involved in PFI to make a profit. So the impact of a risk on a project is what happens to the return on that project. For example, if an organisation planned to make a 20% return on a project, if a risk materialised and the company had not made any provision for it, the impact of that risk on the project might be that their return will go down to, say, 13%. Now if they are not prepared for a return of 13% on their investment, then they should increase the percentage of their expected return. So the impact of risks is usually assessed in terms of how it affects an organisation financially.

Initially, risks are assessed on several dimensions, like potential delays to the project, embarrassment to be faced, effect on function or quality of product, etc. However, all these considerations are subsequently translated into financial terms. So, monetary units are ultimately used to assess the impact of risks.

Risk assessment strategies

Not only has risk assessment matured as a project development practice, it has become a tool for winning bids. Organisations and consortia manage and assess risks such that they underbid other competitors. In this regard different organisations use different strategies while assessing risks, including the following:

- Assess every risk as it is
- Assess every risk but model the price via probabilities
- Assess only the main risks
- Benchmarking
- Adjudication in risk evaluation
- Reactive risk assessment
- Pro-active risk assessment
- Sensitivity analysis in risk assessment.

Assessing every risk
Under this strategy, every risk besetting the project is assessed, and probably priced into the bid. When the chance of a risk occurring is very high, or if it is ascertained that the risk will definitely materialise, then it is probably safer to price its full impact into a bid. Sub-contractors, who may be facing fewer but higher-impact risks, may find it easy to employ this strategy. When the risks are few, it is also viable to assess their impact in detail.

Assess every risk but model the price via probabilities
An alternative strategy, which takes account of most or all risks, is to price all the identifiable risks but control their cost consequences through probabilistic considerations. A simple example will illustrate the point.

The chance of failure of a component is assessed to be 0.15% and its financial impact is estimated to be £100 000. Since the probability value of 0.15% is relatively low, an organisation might decide that nothing will happen, and so feel unjustified in adding £100 000 into their bid. However, if they add nothing at all, and the risk did occur, then they would loose £100 000. So the company should strike a balance on how much to cover for each risk, or the whole combination of risks in the project. In simple terms, one way in which this risk will be priced by some organisations, is:

Risk cover = 0.15 × £100 000 = £15 000.

Instead of pricing £100 000 into the bid for this risk, the analyst would add £15 000, thus minimising the extent to which the bid is beefed-up. The effect of each risk is considered in the foregoing manner and added into the bid. The cumulative effect of all risks is either obtained by summation, or through an integrative formula (Sawczuk, 1996).

The foregoing approach adopts the view that not all risks will materialise in the project, and that the amount allowed as risk cover will be sufficient to take care of those risks that eventually materialise. This form of analysis also aims to balance the losses of some projects (i.e. many risks occurring) against the gains of others (where fewer risks had materialised). Striking the right balance can be very difficult though. An organisation can introduce an excessive buffer for the risks to guard against making a loss. However, that may render the bid uncompetitive. On the other hand, an organisation might go for a lean buffer and expose itself to losses, should many or all risks materialise. Different organisations adopt different policies in this regard, and some of the approaches are discussed in the paragraphs below.

Assessing the main risks only

For an organisation involved at the top tier of a PFI project, the risks of a project are numerous. Risk evaluation at that level is partly used as a bid-winning tool. Therefore, many organisations find it worthwhile not to price every risk but to concentrate on the key issues. In a simplified form, Fig. 5.4 shows nine different scenarios of a risk. Many more scenarios can be evolved, as well as a numerical disposition.

At the top-right cell of Fig. 5.4 are the risks with a high chance of occurrence and having a high impact on the project. Given that most PFI contracts last for about 25 years, it would not be hard to find some risks that fall into this category. Examples include boiler units and decorations, which may have to be replaced at some time in the life of a PFI scheme. The top end risks are usually priced in the bids, because their impact is high and they will almost always materialise in the course of a project. Some organisations tend to focus on the top risks, and only price such into their bids.

At the lower-left cell of Fig. 5.4, are the low-risk items, such as an electrical socket needing replacement. In between the two ends are risks with an impact and probability of occurrence that range from moderate to high. The private sector PFI participants are usually big organisations that are not affected significantly by lower end risks. Therefore, the tendency is to acknowledge such risks but not to price for them in a bid, more so given that their chance of occurrence is low.

In between the two extremes of top and lower end risks are a host of risks

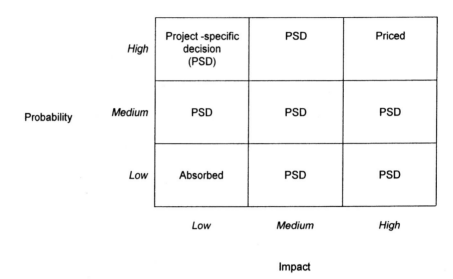

Figure 5.4 *Probability-impact matrix of risks.*

that are treated differently by different organisations. What is eventually priced into a bid depends on how each organisation can manage those risks, and how much it is willing to forfeit in profit should many risks materialise in the life of a contract. The decision to be reached should be judged on the basis that risks present opportunities for both gains and losses. If a financial provision is made for risks, and they failed to materialise, then the company will end up with excess profit. If, on the other hand, more downsides are experienced than the provision, then a loss will be incurred. So a balanced view has to be made as to what each organisation is comfortable with, and striking the right balance is difficult.

Benchmarking
In addition to the foregoing three types of assessment, some organisations use a template as a starting point for assessing risks. When a current project is being assessed, its risks are compared with the template to see how their profile deviates from the template. This approach is applicable where data is available. A construction firm with a head office in London, and a long experience of building schools projects has developed a risk assessment template on school projects. This organisation uses this template to assess the risks of a new scheme. Obviously, templates can be developed for other types of schemes, which will serve a useful basis for subsequent use. However, the information on which a template is developed should be known to the users to enable them to account for the specificity of individual projects.

Adjudication in risk evaluation
There is a fair degree of subjectivity in risk evaluation. Most organisations look at risks, assess their probabilities and values, put them together to see how comfortable they feel with the balance of the outcome. It is a weighing

of how comfortable they feel in terms of how many of those risks might occur and how many would not occur. There is no definitive measure for predicting such outcomes, and judgement is often based on intuition. Such a decision is often made on a collective basis involving the key personnel of an organisation, most often the board of directors.

Adjudication is largely a commercial decision that depends on what an organisation is comfortable with. The performance of other existing projects is usually taken into consideration in the adjudication of (risk) pricing. In this context, a risk owner or manager may see his/her risk assessment values undercut by management. However, management is in a position to see where the potential losses from one project can be balanced with the gains from other schemes, some of which may not yet be in the pipeline.

Reactive risk assessment

Reactive risk assessment entails waiting for risks to manifest before they are assessed and/or addressed. Risks that are known are assessed initially, but those that are not known are not assessed until a negative event has occurred. Given the high downside of some risks, it is worthwhile that as many risks as possible be identified and strategies mapped for addressing them up-front.

In very complex or novel projects, it may be impossible to identify all the things that could go wrong, as some dynamic interactions might be unknown. Also, the numerous risks in some projects may not render wholesome risk assessment cost effective. In such cases, the concentration should be on risks with high impact or frequent occurrence.

Pro-active risk assessment

Pro-active risk assessment is being alert and not leaving anything to chance. All potential risks are identified in the pro-active approach, and solutions sought for them up-front. In one PFI hospital scheme in Scotland, for instance, the ironmongery was identified as a major concern, given that the many doors in the wards will be in use constantly. The building contractor pro-actively endeavoured to ensure that a durable ironmongery product was used. In doing so, several products were compared, and instances where they had been used and proven to be durable were checked. In addition, warranties on the products were ascertained to be in place, in case the ironmongery failed. The building firm that constructed this hospital was seeking to ensure that the ironmongery will last, as well as protecting itself from liability, should the product fail too soon.

Financiers of PFI schemes often adopt a pro-active approach, especially since this form of procurement is based on non-recourse financing. They (financiers) always question issues that can impede the progress of the project. They always want to be sure that a project will be completed on time and within budget. They also want to be sure that the PFI project will generate enough revenue to repay the loans.

However, a balance must be struck while assessing risks pro-actively.

Given that thousands of risks and combinations of risks can be associated with a particular project, time may not be available for implementing a first-principles approach. The benefits of a thorough risk assessment must thus be weighed against the cost. An organisation should consider if it should commit its costly expertise to the unending assessment of what might happen if a light bulb burnt out. The importance of light is not being undermined here, but the point is that some risks are so basic that they are dealt with unconsciously, so that valuable expertise is not misapplied in addressing such issues. Along this line, risk allocation and risk ownership is important. Management can assign different risks to different personnel. Senior personnel can deal with the high risks while junior officers are empowered to address the routine and minor issues. The various activities can then be aggregated in a co-ordinated manner.

Sensitivity analysis in risk evaluation

Whichever way the risks are evaluated, some form of sensitivity analysis should be conducted to identify the most volatile risks, i.e. those that have a high knock-on effect on the achievement of the project's objectives. In sensitivity analysis, therefore, the cumulative influence of the risks on the project's objectives is assessed. It is viable to conduct sensitivity analysis, after the myriad of project risks have been individually assessed. The sensitivity analysis will then assess the impact of one or more risks on the overall project outcome. The impact of the risks on project price and time can be assessed, as well as running a check on other project features.

If an organisation has got a 3% capital cost risk on 'the change of law', which is capped at 3%, it can perform sensitivity analysis on this risk over five-year periodic intervals. It could, for example, be assumed there will be no changes in law for the first five years, but as the assessment stretches to 20–25 years, some effect on the project outcome is registered, albeit minor.

Sensitivity analysis is more often a numerical exercise in which risks are priced or assessed in other numbers and put into a model to determine their effect(s) on different project features, for example, affordability. If, for instance, some risks impacted on the affordability of the project, then means of mitigating such risks should be sought. If, however, they cannot be mitigated to make the project affordable then the organisation may wish to consider walking away from the deal. Apparently, therefore, risk evaluation is a means to an end, i.e. it informs decision making.

5.3.3 Risk mitigation

Risk mitigation involves finding solutions to counter risks. Although risk analysis is important, it is ultimately aimed at facilitating risk management. Instead of simply pricing for risks, a way of getting round them is better. If a construction site is suspected to pose archaeological risks, then an organisation might find it worthwhile to conduct a full site investigation to establish what is underground. That will inform a better choice on courses of action that are open to this organisation. Risk mitigation is therefore an important stage in risk management (McKim, 1992).

Risk mitigation should last continuously throughout the life of a project, as new solutions can emerge that will change previous actions. For example, a new and less risky system of heating can come into the market, by means of innovation, and a building facility may have to be translated from the old to the new system. Each time a risk is controlled, the overall risk profile of the project is altered. For the sake of accountability, therefore, a process of risk monitoring in the course of project delivery is necessary.

There could be endless ways in which risks could be mitigated. Thus the decision that is taken on each risk or a set of risks is circumspect, depending on the peculiarity of an organisation, and the surrounding circumstances. For instance, if an organisation anticipates that its project might be delayed by strike actions, the unions could be involved in some aspects of the procurement process to address their concerns up-front and curtail the potential delays. For each identified risk, a construction organisation should seek for a solution that addresses it effectively and efficiently.

When the consequence of a risk is very high, or when the risk falls outside the expertise of the firm, outsourcing of such risks to another party should be explored. The guiding objective of PFI is to allocate risks to the parties best able to manage them. Therefore, construction organisations involved in PFI project delivery should not hesitate to involve outsiders in their risk assessment, if their in-house expertise is wanting. In this regard, an engineering firm that was undertaking a PFI road project (motorway) found it worthwhile to consult traffic and planning experts on relevant issues they could not cope with. For this organisation, it was not expedient for them to maintain such experts on their payroll, as their services were not needed on a daily basis.

Having undertaken many projects in the past, some organisations have developed standard procedures for mitigating certain known risks. These procedures, which are used in-house, may be documented in brochures, and may be reflected in the policies of some firms. Apart from serving as a guide to standard practice, the documentation is also useful for inducting new employees on the practices and procedures of the company. Obviously, such brochures are updated and changed with time. Examples of issues that have been documented by some construction organisations include threshold values concerning energy consumption, degradation of buildings, liquidity ratios, etc.

The big facilities management organisations in the UK have established their own individual routines by which they conduct major re-appraisals of facilities under their care. Some of these organisations inspect and test utility services and equipment every five years, in addition to non-routine checks. If a change in technology warrants the replacement of some components, then the effect of such a change on the services is assessed without waiting for the five-year interval to elapse. That way, potential problems are spotted and solved before they materialise into a hazard. Risk mitigation is thus an on-going function.

Competition too, drives many organisations to endeavour to excel at risk mitigation because, if a major risk with adverse consequences materialised in the course of a project, the event could generate bad publicity for the construction organisation involved. In future competitions then, negative publicity could discredit that organisation, as clients may be hesitant to engage the services of an organisation that has received recent bad publicity. Therefore, organisations endeavour to mitigate the risks facing their projects, especially the major risks.

Risk mitigation strategies

There are four general risk mitigation strategies, namely:

- Risk elimination
- Risk reduction
- Risk transference
- Risk retention.

These strategies are discussed in the sections below. The discussions are based on practical interactions with PFI participants, and reviews of literature (i.e. Cost Engineer, 1993; Institution of Civil Engineers and Faculty and Institute of Actuaries, 1998; Baker *et al.*, 1999).

Risk elimination

Risk elimination is also often referred to as risk avoidance or risk aborting. Actions to avoid the risk can involve the complete elimination of risk. These actions can be drastic, as in a client refusing to proceed with a very risky project. A contractor could refuse to bid for a very risky project, thus avoiding the risks that would have been faced.

If there is ground water on site, it is worthwhile spending a considerable amount of money on investigations. A construction company experienced this problem and designed the project so that the building was lifted above the water table. This company knew the ground water problem was going to be a major problem. After much consideration, following the site investigations, it ultimately decided it had to physically lift the ground floor of the building by about half a metre. Risk elimination may not come freely. For this organisation, an additional cost was incurred in raising the building. However, the initial amount it spent in raising the building and avoiding the risk of ground water was relatively small, compared with the frequent maintenance costs of a saturated building.

Risk reduction

If not eliminated, risk or uncertainty can be reduced, by acquiring more information. In view of their adverse consequences, and given that risks are inevitable, attempts should be made to minimise their effects. Actions that could be taken to minimise some risks concern the redesign of facilities to minimise health and safety risks, interacting with unions to minimise disruptions to work, etc.

Risk transfer

Responsibilities for some risks can be transferred to other parties whenever it is possible and sensible to do so. For instance, some risks can be transferred through the use of insurance and performance bonds.

In PFI, consortia do not retain many risks. They usually transfer the construction tasks and thus risks to a distinct construction outfit. Similarly, the 'facilities management' (FM) functions and risks will be transferred to a specialist FM services provider. Figure 5.1, in part, shows how responsibilities flow from the client, through the private sector consortium to subcontractors. Having transferred most of their tasks and risks, PFI consortia can afford to maintain lean structures, with very few key staff.

There is usually a flow down of many risks from the SPV, because lenders do not want the SPV to bear significant risks. If the project site were suspected to be contaminated, the banks would want to see that an expert has tested and certified the site for the project to proceed. If inflation were to rise sharply, the lenders would want to be satisfied that the project would not be aborted.

It is usually more effective and efficient to transfer the risks to specialists who can handle them better. Therefore, PFI consortia transfer many tasks and risks to different experts, depending on the requirements of each project. In hospital schemes, some tasks are outsourced to organisations that specialise in catering, pottering, laundry, security, etc. Similarly, outsourcing is employed in other types of projects, as and when appropriate.

Risk retention

Risk retention is also known as risk absorption and risk pooling. After reducing the potential impact of risks, those that cannot be eliminated or transferred away are absorbed by the organisation. The risks that are suitable for retention by any organisation are those with minimal consequences. Another criterion that influences organisations to accept risks is their ability to control the risks in question.

Different organisations retain different sets of risks, endeavouring to limit their exposure. The banks will definitely have a view on a company's decision, but the ultimate decision rests with the organisation. The bank will have a concern whether the organisation will have the ability to repay its debt at all times.

Risk mitigation tools

In controlling risks through one of the aforementioned strategies, a number of tools can be applied. These tools are, generally, more prominent in the financial sector. An overview of some risk mitigation tools now follows.

- **Guarantees** – these are issued on behalf of contractors by banks, governments, or their agencies to ensure that the client has recourse to compensation, in case of the contractor's default
- **A 'letter of credit' (LOC)** – is a form of guarantee, issued by a bank on behalf of a contractor that is operating overseas. The LOC entitles the

client to withdraw cash on production of certain documents or upon fulfilling certain conditions. Usually the exercise of such right is associated with the non-performance of the contractor

- **Bid bonds** – are issued to safeguard the client, such that if and when a contractor's bid were accepted by the client, that contractor would not renege on entering into a contract with the client
- **Performance bonds** – are issued by a surety company to cover the aspect of non-performance on the part of a contractor
- **Surety bonds** – are a form of guarantee that other forms of resolution would be sought, in the face of non-performance, before the cash-withdrawal penalty is applied
- **Insurance** – can be used to mitigate risks that cannot be managed in any other way. Insurance is usually used to protect an organisation from the consequences of disasters
- **Risk premium** – the equivalent of this term in construction is the contingency sum, which is usually added to an estimate to account for unforeseen eventualities that cannot be fully priced when an estimate is prepared
- **Risk-adjusted discount rate** – is mostly used in banking and business to adjust a risk-free discount rate by accounting for future inflation and extraordinary risks.

In addition to the foregoing instruments, financiers maintain a tight scrutiny on PFI projects in a bid to recover the debt owed them by private sector consortia. Issues they often examine include:

- The type and capability of employees who will run the project
- Assessing that the proceeds from the business will be channelled to an account that is tightly regulated
- Ensuring that sponsors are forced to contribute equity into the project
- Checking that independent expert opinions have been sought on different aspects of the project.

The role of attitudes in risk mitigation

The risk attitude of a project participant will determine the courses of action taken in the face of risks (Smith *et al.*, 1999). An individual or organisation can be risk neutral, risk seeking or risk averse. The disposition towards risks is flexible and depends on the type and nature of risks being faced, and the magnitudes of the risks. In general, people are risk averse when the downside consequences are high, however attitudes can change with time and circumstances. When the impact of risk(s) is small, construction organisations tend to be risk seeking. However, as the aggregate value of risks increases, they increasingly become risk averse. The progression from risk seeking to risk aversion may be slow or fast. Risk analysts need to be wary that peoples' attitudes towards risk do influence their decisions and opinions.

Each organisation should know the level of risks they are comfortable with, and act accordingly. Being comfortable with risks depends in part on an organisation's competence. For example, while a construction organisation may be comfortable with the buildability of a complex design, it may not be very conversant with the efficient utilisation of energy in the same facility.

5.3.4 Risk allocation in PFI schemes

Risk allocation is a prevalent feature of PFI, which starts with the initial risk matrix prepared by the client. A public sector body procuring a PFI project would state its preference as to how the project risks should be shared between it and the private sector participant. Individual bidders would assess the client's proposition and either concur or disagree. The iterative assessment of risks by the private sector bidders enables them to reach an ultimate decision on whether they should bear certain risks or not. If a certain risk that had been allocated to the private sector by the client was assessed to have a high impact, or knock-on effect on other risks, the negotiation process will be used to either shift the responsibility for such risk back to the client or steer the course of the project away from such risk.

The public and private sector participants involved in a PFI project negotiate with each other over several months to sort out the ownership of risks. As decisions evolve, the new risk profile of the project is continuously assessed. The negotiations continue until all risks have been priced and allocated to one of the parties. Accounting Standards drive this process, in part, in a bid for the risks to achieve 'off balance sheet' status. The quest of optimum risk transfer by means of the absolutist approach to best value is thus sometimes restricted through observance of the Accounting Standards (Fig. 5.5).

In the past, each project was negotiated on an *ad hoc* basis, but now there is a much more realistic view of what risks can be carried by the private sector. Gone are the days in which the public sector sought to transfer every risk to the private sector. However, different projects are underpinned on different circumstances, and so the impact of risks varies with different projects. On some projects, it will be more cost-effective for the client to acquire the site, demolish any existing old buildings and make the site ready for the private sector participants. On some other projects, these functions will be better transferred to the private sector as well. Therefore, risk allocation cannot be standardised on a permanent basis as individual circumstances determine what is best. However, a template can be established which will inform risk allocation in the current project.

5.3.5 Risk documentation and reporting

As risks are identified and mitigated or assessed, the eventual decisions reached in respect of each risk are documented, and the records are used in monitoring the risks throughout the life of a project. Right from the risk identification phase, working sheets, as exemplified in Fig. 5.6, are used to portray information on different risks and to inform the preparation of

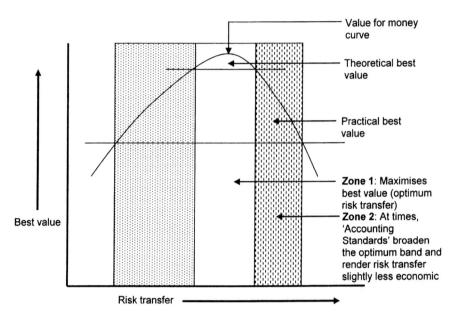

Fig. 5.5 *Accounting Standards.*

risk registers. By means of records, which are now computerised, information on any risk can be culled easily, and updated as the profile of each risk changes.

Within organisations or SPVs, reports by different officers to their superiors often have sections dealing with risks. The ownership of risks and the mitigating strategies adopted would be documented somewhere. Minutes of meetings also show decisions reached in respect of certain risks. Having mitigated and assessed all risks, the contract documentation would ultimately reflect the final allocation of risks between the public and private sectors.

5.4 Knotty issues in risk management – a practical perspective

In the course of our investigation at Glasgow Caledonian University, several interviewees with involvement in PFI project delivery were asked to relay any difficulties they had had with risk analysis and management. The intention of the research was to ensure that impediments to efficient risk management in PFI were ameliorated. The problems divulged to the investigators are highlighted below, with frequent reference to the interviews that were conducted:

(1) **The absence of a risk management culture**: It was explained that getting people to imbibe a formal working culture of risk assessment was difficult. The commitment to manage risks for long is yet to

A Risk Note Pad

Risk Reference No.: ....   Brief Title ....

Champion ....

	High		
Probability of occurence	Medium		
	Low		
	Low	Medium	High

Impact/cost

Concise definition of risk identified
....
....

Subsidiary risks
....
....

Linkage to other risks
....
....

Management of risk (statement)
....
....

Mitigation of risk (statement)
....
....

Cost estimate (details) ....
.... (£....................)

Ownership of risk: Client .... Provider ....

Shared (define relationship) ....

Figure 5.6 *A risk documentation proforma.*

properly take root. For instance, it was argued that some builders do not want to hang around in PFI after the construction phase. They usually want to finish the construction aspect, get their money, leave the consortium and go away.

(2) **The efficacy of risk assessments cannot be ascertained**: In view of the foregoing issue, a logical problem was unveiled concerning the efficacy of risk assessment. According to a respondent in the research under discussion, it is easy to add to a bid a premium of 22% for cost over-runs, while in reality it may be 10%. Another interviewee remarked that 'although projects are being completed and are running success-fully, the client may not be getting optimal value for money. In most cases the client is probably getting quite good value but not necessarily the best'. This problem is an apparent echo of one of the issues iden-tified in literature.

(3) **Lack of historic data to support risk assessment**: There is a lack of relevant historical data to support risk evaluation in some aspects of PFI. For instance, how many times will the private sector fail to keep a hospital ward in an acceptable clean condition, is a question they cannot answer. With PFI being relatively new, there is often no experience or track record to revolve round the participants. In such cases, there is nothing to measure against. Risk assessment in such scenarios was described as a speculative guessing game.

The other problem with statistical analysis of PFI projects concerns the different characteristics of the projects. With road schemes, for example, there are totally different types of roads with different characteristics. There is no relevant database now that captures the varieties of projects.

(4) **Recourse to subjective assessments**: When there is a dearth of data, the risks concerned are assessed on the basis of subjective judgement. That is a factor in the process that is uncertain, because a subjective opinion is based on personal perception. There may be no way to tell if a subjective opinion is right or wrong.

(5) **Differing perceptions on the magnitudes of risks**: Following on the previous issue, risk assessment in PFI is a collective judgement on what might (never) happen. The different members of a 'Special Purpose Vehicle' (SPV), i.e. the consortium delivering the project, should reach a collective decision on the magnitudes of risks. However, these dif-ferent parties sometimes have different views concerning the severity of risks, i.e. the likelihood of risks materialising and the amount of money needed to cover for them. One reason advanced for such dif-ferences is the vested interests of the PFI participants.

(6) **Laid-back clients**: Clients are sometimes not forthcoming with their precise requirements until a preferred bidder has been appointed. It often takes several months before a preferred bidder is selected, and waiting that long to obtain some information is frustrating to private sector participants.

The lack of a clear strategy and unwillingness to make committed decisions are other perturbing issues attributable to clients' passive-ness. This is a high source of uncertainty, which is counted as a client-induced risk. Clients are laid-back, partly because they do not thoroughly understand the PFI process.

(7) **Occasional lack of requisite expertise**: In complex projects, particularly information technology (IT) schemes, some PFI participants have found it difficult to identify the technical experts who are qualified to provide the kind of risk assessment that was required. It takes some time before each type of scheme passes through its learning curve, within which the relevant expertise may be insufficient. However, the types of schemes on which PFI was utilised, only recently, are fast catching up with the pace-setters, not having to start from scratch.

(8) **Long duration of PFI schemes**: PFI schemes entail long-term contracts, sometimes in excess of 25 years. The length of such contracts gives rise to its own risks. For example, facilities managers may not remain in the contract for 25 years, the technology on which a particular IT scheme was based will almost surely change, etc. It may not be exactly clear (i.e. uncertainty), what will happen when such changes occur.

(9) **The late start of risk assessment**: Although the PFI form of procurement has fairly matured, its risk assessment, which is not a one-stop task, is allowed to evolve over time. The efficacy of the assessment thus improves over time and with the acquisition of more information. A late start of the assessment in the course of the procurement process incurs rushing, and the missing out of vital details. The process should be embarked upon quite early and not later, as has been the case in certain PFI projects.

(10) **The dynamic nature of PFI risks**: Risks pertaining to changes in law, traffic volumes, and the like, are very difficult to assess, as risk analysts may not often know how much these will change. Environmental laws and the social acceptability of some schemes are also dynamic, and difficult to assess. In such situations, putting a value on risks becomes difficult, especially as risks have to be priced up-front in PFI, for periods of 25–30 years.

(11) **The unstructured nature of risk assessment**: In the private sector the understanding of some risks is sometimes based on the experiences and gut feelings of the assessors. The risk assessment process is not completely structured or documented. Other personnel within an organisation may not be able to reproduce an assessment, as the expertise for doing so is not passed-on in a co-ordinated and structured manner.

(12) **Transient expertise**: Most of the organisations that form a PFI consortium are big international organisations with diverse professionals. However, their employees sometimes move on, leaving a vacuum to be replaced. When that happens in the course of risk assessment, a company could be caught off-guard without sufficient or requisite personnel to analyse their project risks. Although it is always possible to employ new hands, it is sometimes very difficult to get the right risk assessors at the right time, as there are not many out there.

5.4.1 Alleviating the difficulties

In view of their enormity, the foregoing problems could not be set aside by an investigation which was seeking to develop a framework for best PFI risk management practice. Solutions were thus sought on how these knotty issues could be redressed. The notable solutions proffered from the ensuing discussions are discussed below:

(1) **The consolidation of databases for PFI risk assessment purposes**: Relevant information is really the antidote to inaccurate risk assessment. After running PFI contracts for a couple of years participants will be able to develop more reliable information on this type of procurement and also understand better which resources are most unreliable. Although some existing databases are currently being utilised, it is conceivable that a consolidated databank that is specifically structured to support risk analysis in PFI undertakings would be developed.

(2) **The training of risk analysts**: Training programmes would enable those organisations or personnel who are not conversant with detailed or current risk analysis to update their skills. Attendance at conferences or seminars can also improve the skills of risk assessment, as participants can learn new techniques in the market. Training need not be expensive or drawn out. Some big organisations have acknowledged the importance of training, and now run training programmes on risk management for some of their employees.

(3) **Adequate time to be devoted for risk analysis**: In view of the fact that PFI procurement takes so long, an interviewee commented that 'the time allocation for considering risks should be planned into the PFI programme so that analysts have sufficient time for a thorough risk assessment. It should not be a rushed process'. Sometimes, investigations or inquiries are necessary for the understanding of some risks. The more time an analyst has, the more thorough an evaluation can be performed. Therefore, considerable time should be accorded the risk analysis function.

(4) **Facilitating a better understanding of risks**: Some organisations, especially from the public sector, felt that a thorough identification of the risks they were going to bear in PFI projects was their biggest problem. Avenues that could facilitate a better understanding of risks were thought to include brainstorming, holding of workshops and simplification of the issues.

(5) **Motivating clients**: Given that public sector clients are sometimes laid-back, the need for motivating them to be more forthcoming was expressed. In this regard, the payment mechanism could be structured to reward clients that are readily forthcoming with information or decisions.

(6) **Seeking advice from experts**: There are many types of risks in PFI, and one person may not be versatile enough to understand all of them

thoroughly. Therefore risk analysts in PFI should not hesitate to seek external advice where necessary.

(7) **Detailed planning**: Some companies do not conduct rigorous risk reviews before proceeding to sign deals. This laxity has been extended to PFI undertakings, which obviously is a different domain. More detailed planning would help such organisations identify a myriad of risks that can easily go unnoticed.

(8) **Adopting a more structured approach**: Some organisations conceded that their risk assessment procedures were not as sophisticated as they wished them to be. It was felt that a more structured approach, especially for PFI procurement, would be better. This structured approach was described by one organisation as 'a formal framework, a process that says that it will be done this particular way'. Although a common sense approach to risk assessment might be effective, a structured approach would at least introduce transparency in the process.

(9) **Risk assessment should not be over engineered**: Although the accurate assessment of risks is worthwhile, it was cautioned that the process should not be over engineered or mechanised without discretion. According to a respondent, 'just to stick money on risks is not the ideal solution. It is a reliable identification of the risks that pose the greatest threats and the alternative solutions for ameliorating them that matters most'. Preliminary efforts towards circumventing risks should be exerted, before pricing those that cannot be avoided.

(10) **Acquiring and retaining experience**: The requisite personnel for risk assessment in PFI are yet to saturate the market. One way of speeding this development up is to train potential and existing professionals to be able to cope with PFI in general and risk assessment in particular. While a moderation of educational curricula could address the former aspect, conferences, seminars, workshops and the like would address the latter, at least in part. If the market can be saturated with knowledgeable personnel, then needed expertise can be accessed more quickly.

(11) **Standardisation of risks**: One way of alleviating the risk assessment difficulties is to have a greater standardisation of these risks through standard contracts. Such standardisation would circumvent the risk identification phase and enable participants to devote more time to finding risk management solutions. However, standardisation will not get rid of some project-specific risks that will have to be identified in each scheme.

5.5 Conclusions

PFI projects are expected to offer the client value for money (VFM). Therefore, a client organisation should weigh its risk allocation decision against the effect on VFM. The achievement of VFM should be optimised. So, if the transfer to the private sector of many risks that cannot be mitigated by the client undermines the achievement of VFM, then a client should consider bearing those risks.

Similarly, after bidding has commenced, the private sector bidders should use the VFM criterion to decide which risks are sensible for them to accept from the client. It is sometimes not cost-effective for the private sector to bear responsibility for some risks that have been allocated to them. For instance, where the 'planning permission' is facing objections from members of the public, it may be worthwhile for the client to handle this aspect.

Sometimes, it happens that a public sector client would allocate a certain risk to the private sector, but that sector might feel otherwise. If the client then insists and prevails on the private sector to accept such risk, a higher premium will be charged, and VFM will be eroded. Given the dynamic nature of the prominence of risks over different projects, the decision to be made on each type of risk for each project should depend, in part, on how the private sector feels the particular allocation would be cheaper for both sides. If the private sector is not enthusiastic about accepting a particular risk, it may be worthwhile for the public sector to think it over again.

Although standardisation is gaining ground, PFI procurement is still characterised by lengthy and complex negotiations. One reason for this is that different risks impact on projects differently. It behoves the participants to assess the magnitude of the risks fairly accurately, and to determine the best allocation that would yield optimum VFM. If a confrontational disposition is adopted, the negotiations will be lengthened further. On the other hand, a sharing of resources, particularly in terms of risk assessment expertise, can help the two sides to quickly reach a common understanding on the implication risk allocation. Openness in terms of some organisational objectives is also a plus for the two sides to strike a deal faster.

The PFI market in the UK has matured, and so vying for its projects has become very competitive. Prudent risk management is fast becoming a bid winner. Given that risk transfer is a fundamental requirement of PFI, organisations that undertake projects via PFI must be able to identify and especially manage the risks that could be encountered. Experience helps in both tasks. Risk and safety reviews, reference to databases, site visits, consultation with experts, brainstorming, intuition and other techniques can be used to identify risks.

Having identified risks, it is advantageous that solutions to mitigate them are found. There will be different ways in which to mitigate different risks, and there may be no perfect solution. Each organisation is free to select the risk mitigants that best suit its circumstances. What may be a best risk mitigation strategy for one organisation might not be for another.

The evaluation of risks involves an assessment of the consequences of the mitigating solutions that have been evolved to counter risks. While risk evaluation may initially be subjective, the task is eventually made numerical, and expressed in financial terms. Experience, gut feelings and policies have a bearing on risk evaluation, as management often moderates the pricing of risks and/or bids. At the end of the day, it is a commercial decision that depends on what amount of profit an organisation is willing to forfeit to gain a competitive edge.

Risk management does not stop when a bid is won. An organisation should continue to review the risk profile of its projects to see if something can be done to gain time or save costs. New or innovative ways of managing facilities can bring unexpected gains in risk management. Therefore, risk management should last the lifetime of a project.

Acknowledgements

We wish to thank the Engineering and Physical Science Research Council (EPSRC) and the Department of Environment, Transport and the Regions (DETR, now Department of Trade and Industry) for funding the research on 'Standardised framework for risk assessment and management of Private Finance Initiative projects' which informs this paper. We are also profoundly grateful to the project industrial partners who have supported the research. Many organisations have either participated in the research or supplied data. Their co-operation is fully appreciated.

References

Akintoye A., Beck M., Hardcastle C., Chinyio E. & Asenova D. (2000) Management of risks in the PFI project environment. In: *Proceedings, Sixteenth Annual Conference of ARCOM*, Association of Researchers in Construction Management, **Vol. 1**, pp. 261–270. University of Reading, Reading.

Akintoye A., Beck M., Hardcastle C., Chinyio E. & Asenova D. (2001) Risks in private finance initiative projects. In: *Proceedings of International Conference on 'Public and Private Sector Partnerships: the Enterprise Governance'* (eds L. Montanheiro & M. Spiering), pp. 1–15. Sheffield Hallam University, Sheffield.

Baker S., Ponniah D. & Smith S. (1999) Risk response techniques employed currently for major projects. *Construction Management and Economics*, **17**(2), 205–213.

Boothroyd C. & Emmett J. (1996) *Risk Management – a Practical Guide for Construction Professionals*. Witherby & Co Ltd, London.

Carter B., Hancock T., Morin J. & Robins N. (1994) *Introducing Riskman Methodology: the European Project Risk Management Methodology*. NCC Blackwell, Oxford.

Cost Engineer (1993) RISKMAM – the new European methodology for project risk management. *Cost Engineer*, **31**(1), 16–17.

Edwards L. (1995) *Practical Risk Management in the Construction Industry*. Thomas Telford Publications, London.

Goldsmith S. (1997) Can business really do business with government? *Harvard Business Review*, **75**(3), 110–121.

Hillgate Communications Limited (2001) *PFI Map of Great Britain 2001*. Hillgate Communications Limited, London.

Institution of Civil Engineers and Faculty and Institute of Actuaries (1998) *Risk Analysis and Management for Projects*. Thomas Telford, London.

Kangari R. (1995) Risk management perceptions and trends of U.S. construction. *Journal of Construction Engineering and Management*, **121**(4), 422–429.

Lindley D.V. (1965) *Introduction to Probability and Statistics* (Part 1). Cambridge University Press, London.

McKim R.A. (1992) Risk management – back to basics. *Cost Engineering*, **34**(12), 7–12.

Pickering C. (1999) The asset option. *The Private Finance Initiative Journal*, **3**(6), 64.

Private Finance Panel (1995) *Private Opportunity, Public Benefit: Progressing the Private Finance Initiative*. HMSO, London.

Private Finance Panel (1996) *Risk and Reward in PFI Contracts: Practical Guidance on the Sharing of Risk and Structuring of PFI Contracts*. HMSO, London.

Robinson P. (2000) The private finance initiative: the real story. *Consumer Policy Review*, **10**(3), 83–85.

Sawczuk B. (1996) *Risk Avoidance for the Building Team*. E & FN Spon, London.

Simon P., Hillson D. & Newland K. (1997) *Project Risk Analysis and Management (PRAM) Guide*. Association for Project Management, Ascot.

Smith N.J., Merna T. & Jobling P. (1999) *Managing Risks in Construction Projects*. Blackwell Science, London.

Steele A. (1992) *Audit Risk and Audit Evidence: the Bayesian Approach to Statistical Auditing*. Academic Press, London.

6 A financial perspective on risk management in public-private partnership

Darinka Asenova and Matthias Beck

6.1 Introduction

The private finance initiative (PFI) was officially launched in the UK in 1992, as a means for provision of services to the public in areas such as healthcare and education, which have been traditionally governed by the public sector. The need for large-scale capital investments and the pressure from different sources on government resources have contributed to an increase in the use of PFI over recent years. As a procurement method, PFI appeals to some public sector clients, because it allows them to retain overall control over assets and core activities, while the private sector is responsible for the provision of supporting services. In addition, it has been suggested that PFIs can bring wider benefits to the community in terms of job creation, the promotion of a commercial culture within the public sector, and other types of efficiency gains. Practically, procurement through PFI has presented advantages to government departments in terms of accounting rules, which allow expenses towards PFIs to be put off current balance sheets, and advantages in the accounting treatment of PFI have primarily stood behind the signing of 300 projects with a total capital value over £15 billion since 1992.

Under the PFI regime, clients are required to ensure value for money (VFM). VFM is usually demonstrated by comparing private sector bids with an independent public sector comparator (PSC), which describes in detail all costs if the project was procured directly by the public sector (TTF, 1998). Given that the public sector can usually borrow at lower rates than the private sector, it has been argued that the efficient transfer, allocation and management of project risks is key to achieving VFM in PFIs. This risk transfer involves the utilisation of two essential elements: firstly, the use of a specific payment mechanism which is subject to availability and performance of the facility; and, secondly, unambiguous contractual terms which clarify which party is responsible if a particular risk materialises.

Some of the most significant risks transferred to the private sector in a PFI transaction relate to project finance. In a PFI transaction, the private companies which take on the obligation to build and manage the facility usually provide only a small part of the project's capital. Most of the capital is borrowed from banks and other financial institutions. The project loan is later recovered when the client starts paying for the service provision. This process effectively removes most duties relating to the raising of capital and the payment for debt from the public sector. A recent investigation conducted by the National Audit Office (NAO) has reported that the scope for innovations relating to PFI finance was not exhausted. Among other innovative setups, the NAO examined the PFI contract for refurbishment of the Treasury building in Whitehall, where for the first time the client ran a separate funding competition which resulted in 7% savings on the project costs over the life of the project (Timmins, 2001). Today, the NAO recommends that, despite the possibility of additional risks of delays and cost increases, such approaches can be appropriate for improving the overall VFM in future PFI projects.

This chapter commences with an investigation of the risk management practices UK financial services providers utilise in the context of PFI projects. Section two contextualises PFI finance in terms of its relationship to project finance, while Section three investigates the main financing options available to PFI participants together with the risks associated with them. Section four gives a general overview of financial risk management instruments. Sections five and six, lastly, provide an empirical analysis of the financial risk management measures utilised by 14 financial companies in the UK, all of them with substantial involvement in PFIs. (The data was collected as a part of a wider research project conducted from 1999 to 2001 at Glasgow Caledonian University on 'Standardised framework for risk management in PFI projects'.)

6.2 Project finance and the PPP/PFI environment

The term project finance (PF) refers to situations where a loan for the capital costs of a project is repaid on the basis of cash-flows associated with the operation of that project. Most PF (and PFI) loans are financed on a limited (or non-) recourse basis (Zakrzewski, 1999; Carrick, 2000) with the lenders' recourse being restricted to project assets and/or cash flows. In the past, PF has been widely used for financing infrastructure and public sector facilities like hospitals, power stations, prisons, etc. The key characteristic of PF is that long-term assets are being funded by long-term capital (Carrick, 2000). In most PF projects, the credit risk associated with the borrower is of relatively low importance. As a consequence, financial risk analysis centres on those risks which threaten the project's completion or operation. Another important criteria for PF projects is whether the project can provide an adequate return on the investment (Sarmet, 1980). Strict non-recourse PF is now rare, as financiers insist on some risks being borne by the sponsors (Sarmet, 1980). With regard to equity investors, the recourse is usually restricted to their equity investment and commitment, completion guarantees in respect of the construction, and performance guarantees in respect of facilities operator. Due to the non-recourse nature of PF, financial institutions have developed thorough approaches for the scrutiny of risk which are aimed at ensuring that no relevant risk has been left unchecked. As a subgroup of PF projects, PFIs essentially require a similar focus on potential risk, which has given risk identification, allocation, evaluation, and management paramount importance in PFI procurement.

In the UK, relatively few banks have established themselves as principal financiers of PFI projects. As a rule, these banks combine financial strength with the ability to analyse large projects with diverse risks and complicated financial structures. Currently the main form of capital raised for PFI projects is senior bank debt. In most cases, the private sector partners create an independent company known as Special Purpose Vehicle (SPV) that takes on the responsibility for building and operating the facility. Although an SPV is normally a partnership between well-known reputable companies, the lender must understand how the project works in order to assess the risks attached to future revenue forecasts. Typically, one of the main functions of financial companies involved in the PFI is to ensure that a proper financial structure is in place, which will guarantee that the project's financial requirements are properly met. Therefore, the financial company usually acts both as a project's loan arranger and a financial advisor to the SPV. Other financial companies are often involved in PFIs as equity providers, senior debt providers, insurers, or bond underwriters.

In order to ensure the soundness of a PFI project, financiers need to examine the proposed risk allocation among project participants within the specific political and economic environment of the project (Leeper, 1979). Stein (1995) noted that this involves a number of issues. For example, financiers must ensure that all costs for project completion are covered without further recourse to the lender and that, once completed, the project will be able to generate enough revenue to service the debt. In case of anticipated difficulties, it is essential that the project is backed by creditworthy parties. Finally, in order to ensure that the project is not abandoned before completion, it must be ascertained that it is of sufficient national or strategic interest to elicit the required commitment from relevant parties.

Risk management in PFI, however, is an ongoing process. Before committing funds, financiers typically investigate relevant potential risk factors in some detail. In order to protect their investment after the financial close, financial institutions continue to monitor the project accounts while being able to apply certain restrictions to the borrower (Carrick, 2000). For example, a financier might set a certain debt-to-equity ratio which is required to ensure the availability of debt service reserves. If this ratio is approaching critical values, the financiers can prohibit any additional borrowings. In order to ensure that the payments to the loan provider are met, the bank can also restrict the dividend payments and can exercise controls over the cash flows. In addition, as the project's assets constitute the main source of repayment, some banks impose restrictions on the selling of assets (Ex-Im BUS, 1999).

6.3 Methods for financing PFI schemes

Finnerty (1996) noted that the specific features of the PFIs allow SPV members to finance their projects by reference to the ultimate service purchaser's credit rating. This possibility positively impacts on the financing structure of PFI as well as the degree of leverage involved in the respective financial setups. However, the possibility for low cost finance can be offset if there are high contractual costs and, in particular, high legal expenses. In order to satisfy the needs and requirements of its sponsors, a PFI project usually has to rely on several sources of finance. Thus, while the main part of the capital requirement is usually met by external financiers, the SPV members are often expected to provide a certain amount of equity capital in order to demonstrate their commitment to the project.

As concerns sponsor's equity, it is traditionally assumed that this includes any subscription for share capital in the SPV made by the SPV members or other sponsors. However, the term is now commonly used to describe any form of investment akin to equity (Sapte, 1997). Equity is the lowest ranking capital layer of a PFI project and, in case of project failure, the equity investor is therefore likely to bear the highest risk of loss. Since financial rewards in terms of expected rates of return typically correspond to risk exposure, equity carries the highest rate of return (Finnerty, 1996), making this type of investment attractive to some financiers. Sapte (1997) has investigated a number of factors which determine the ability of SPVs to limit their equity contributions to the total capital requirement. The main factors include economic considerations, costs attached to the equity, requirements of the jurisdiction of the SPV, government requirements, and lender requirements.

As concerns external capital provision, Horne and Wachowicz (1998) have argued that two main principles dictate the relationship between profitability, liquidity and risk. The first principle states that profitability and liquidity are inversely correlated. The second principle asserts that profitability and risk are positively correlated, i.e. that there is a trade-off between the two variables. This principle explains the motivation behind the approaches of different parties which provide capital for the PFI projects. Initially, at the emergence of the PFI market, the financing options were very limited and the major capital requirements were met by the traditional PF methods, i.e. through tranches of bank debt. However, the market demand and the affordability issue required the creation of alternative approaches. In recent years, more sophisticated capital markets products have been developed to provide long-term debt at competitive conditions in respect of credit term and margins (Morrison 1998; Middleton & Richardson 1999). Thus, the popularity of fixed income products such as bonds has grown significantly. Banks, meanwhile, have responded to the new challenge by extending the maturities beyond the traditional 20–23 years and lowering the margins (Ellis, 1999). (The term maturity refers to the length of time to the expiry of a loan/debt.) Different financial institutions such as pension funds and life insurance companies provide fixed rate financing for PFIs, while commercial banks provide floating rate financing. As a result, in addition to the small amount of sponsors' equity (about 10–15%) required for PFIs, a number of different financing options are now available such as the bond market, commercial lending through bank debt, leasing, mezzanine debt and mortgage finance, etc. (Sapte, 1997; Ellis, 1999; Pickering, 1999).

Large projects can be financed by a mixture of bank loan, fixed rate or index-linked bonds, sometimes provided with the participation of international project finance banks. The choice of financing method for a particular project depends on its specific requirements, the project risks, the amount of equity available, and the perceived quality of the consortia. Mutual understanding is essential and the borrower needs to appreciate the bank's objectives and the operation of the restrictive covenants. In the past it has been argued that no single financing option is ideal for all projects, as each new project carries unique risks, has a different risk profile and is accessible to different funding sources (Wynant, 1980). However, there is a possibility that as the number of PFIs increases further, financing approaches will converge towards a few proven and workable models.

With regard to the provision of bank loans, a small group of commercial banks have quickly adopted PF and PFI financing and developed skills for analysing large-scale projects with complex risk profiles. Their loans are provided with a floating interest rate or margin over LIBOR rates. (LIBOR is the London inter-bank offer rate, i.e. the rate at which banks are willing to lend funds in the London inter-bank market.) Finnerty (1996) distinguishes between four alternative types of bank credit, which include revolving credit, term loan, standby letter of credit or performance bond and bridge loan. Revolving credit provides the sponsors with the opportunity to use a line of credit repeatedly within certain limits as project needs arise. The term loan is a fixed time loan over more than one year, which is used to cover the capital costs during the project's construction period. The standby letter of credit or performance bond impose an obligation on the issuing bank to make payments to the commercial papers holder, if the bank's customer is unable to do so. Letters of credit and performance bonds are often used as a credit enhancement. Finally, a bridge loan is a form of interim financing covering the time lag before the expenditures are covered. Bridge loans are commonly used to replace short-term financing with longer term financing. Term loans are most common in PFs and PFIs, as their repayment profile is structured from future profits. In large transactions borrowing requirements are often financed by several banks through a syndication agreement. In this context, the underlying financial structure is usually arranged by a single bank, which subsequently sells parts of the loan to other banks. Syndicate banks are often long-term partners, which lend at the same conditions and have the same priority for repayment. The syndicator, meanwhile, is usually a lead banker and investment manager who keeps a small amount of the whole loan (10%).

The highest project expenses usually occur during the construction phase. In most cases, the senior debt providers start supplying a loan after the SPV members have made their shareholders contribution. Following this cash input from the senior debt provider, future payments become dependent on the completion of certain construction phases or milestones. During the entire construction stage, parts of the project's overall loan are continuously drawn, so that the interest and the actual amount of debt are increasing up to the operational phase. When the client starts repaying, the debt decreases until it is eventually repaid a couple of years before the end of the concessional period.

Mezzanine debt is relied upon in cases where there is a gap between senior debt and sponsor equity. This situation typically arises when senior debt providers are not prepared to increase the level of debt and the sponsors cannot invest more equity. This can be due to the small size of equity provided by the sponsors or specific project circumstances (Morrison, 1998). In such cases, mezzanine finance provided by other parties outside the SPV can bridge the shortfall by providing a third layer of capital in the range of up to 20%, in a form of subordinated debt (Morrison, 1998). Normally, mezzanine finance is exposed to greater risk and higher returns compared to the senior debt. It also ranks ahead of equity in terms of payment distributions, and is therefore more akin to equity than to senior debt. However, due to the higher risk being carried by equity, equity typically provides for the highest potential returns. The attractiveness of mezzanine finance to the investors thus arises from the fact that it provides for the possibility of achieving good commercial returns with excessive risks being taken. Over recent years, some commercial banks have started to provide financial packages to suit projects' funding requirements at different stages of a project. These packages, called comprehensive credit facility, are gaining popularity because they offer two main advantages. Firstly, they ensure a single point of reference for the SPV (and the client), and secondly, they can increase flexibility by allowing switches between different financing options and types of loans (Finnerty, 1996).

A PF lease involves fixed term lease contracts between large financial institutions (lessors), which own the assets for tax purposes, and the SPV (lessee), which pays the agreed series of payments. The lessor is entitled to depreciation allowances, which are normally not available to the SPV due to the lack of trading income (Sapte 1997). Lease finance has been used in some earlier water treatment and power projects, but now is becoming relatively less important.

Mortgage finance refers to situations were the asset is owned conditionally by the SPV. The borrower has the right to use the property while the mortgage is in effect and agrees to pay on a regular basis towards the principle and the interest.

In addition to these sources of funds for PF, there are other capital providers, which by the nature of their core activities collect and manage capital. These include life insurance companies, public pension funds and private pension funds. Life insurance companies operate with relatively steady annual cash flows and are prepared to invest in projects that are economically viable and offer competitive rates of return. Project debt is usually sold to them in the form of privately placed securities. Public placement funds utilise funds which are collected from the retirement funds of public sector employees in projects which meet set standards in terms of risk and return. Private pension funds, by contrast, invest funds from corporate pension funds, and therefore can be flexibly invested with few restrictions.

The European Investment Bank (EIB) has also been involved in some PFI transactions in the UK. It acts as the lending agency of the European Union (EU) and provides long-term, low-cost loans at a fixed rate. The loans are usually available in some strategic areas of development such as education, health, etc. The EIB does not absorb construction risks and, as a consequence, only provides letters of credit during the construction phase. EIB's actual involvement starts at the completion of the construction when the project is fully operational.

The relationship between the risk profile and the rate of return for the main financial components used in PFI financing are illustrated in Fig. 6.1.

Bonds are long-term interest-bearing documents of debt, issued by public as well as private sector organisations, which oblige themselves to pay the principal amount after a specified period of time called maturity (Fitch, 1997). Banks, insurance companies and individual investors comprise the main market for bonds, while the bond issuer can raise funds sufficient for large-scale business projects. In order to reduce the risk of adverse interest rate changes, some bond issuers attach the bond's interest to the value of interest rates through some type of index. These index-linked bonds can be beneficial to purchasers where there is considerable uncertainty about interest rate movements. In November 1997, the first bond-financed PFI transaction in the UK was concluded with a capital value of £88 million

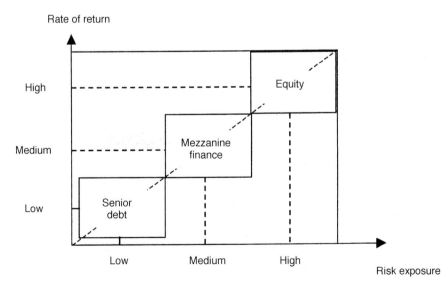

Figure 6.1 *Relationship between the risk profile and the rate of return for the main financial components used in PFI financing.*

(Euroweek on-line, 1999). This transaction has been used as a template for similar contracts in the UK and in Europe. Another form of innovation in PFI financing involved the financing of a hospital project (£91.2 million capital value), through an index-linked bond in July 1998 (Euroweek on-line, 1999).

Bonds have different credit ratings, which reflect the risk of default. These ratings are assigned by specialised credit agencies. The main advantages of bond financing are the length of the credit terms and lower margins. With the completion of a bond-based transaction taking as long as two years of negotiations, same financiers have argued that the margins do not correspond to the time and effort involved (Oliver, 1998). Increased competition in bond markets has resulted in an extension of maturity periods to above the traditional 20–25 years (Dixon, 1999; Ellis, 1999). This affects overall risk profiles, because longer tenure is often associated with high uncertainty and is considered a risk by itself. Meanwhile, long lending periods have opened new business opportunities for institutions with long-term lending traditions (e.g. building societies and ex building societies) who are to able to utilise their expertise and enter a new market (International Projects 500, 2000). Financial experts predict that some projects with long duration, low performance and low technology risks such as hospitals, university and other accommodations, etc. will attract increased capital market activity. Other, more dynamic, sectors like IT, meanwhile, are not considered as suitable by the financial market and will therefore have to be either self-financed or externally financed through short-term bank loans (Euroweek on-line, 1999).

6.4 Instruments and tools for financial risk management

Although basic types of financial derivatives have been known for decades, they have become a major feature of financial markets since the 1970s (Fig. 6.2). At that time, the development of derivatives was accelerated by increased instability in financial markets in terms of exchange rates, interest rates and price fluctuations. The main types of derivatives include futures, options and swaps. In addition to these, there are a number of new, more complex *off-balance sheet* financial instruments. In most cases, the latter can be de-composed to a basket of the main instruments.

Although derivatives, as contracts, do not possess initial value, they can have a substantial impact on the risk exposure of different *on-balance sheet* assets and liabilities, and therefore on the assets of the institutional investors. The value of the financial derivatives (or derivative securities) is based on cash market instruments such as stocks, bonds, currencies and commodities. The basic idea underlying the use of derivatives is to provide protection against adverse price movements and rates by fixing their future transactional values (Blommestein, 2000).

Most widely used amongst derivatives are conventional fixed interest rate swaps. In addition, Cuny and Gethin (1999) emphasise the growing

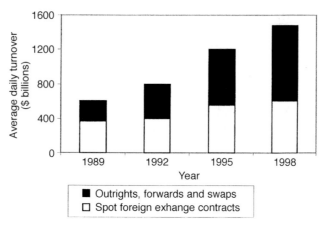

Figure 6.2 *Global activities in foreign exchange markets (1989–1998).*

importance of the Retail Price Index (RPI) hedging market and the event-driven hedging in PFI projects. The use of these instruments is based on the observation that, while the basic revenue stream paid by the client will be certain to be within some limits, the inflation-linked revenue cannot be estimated exactly. The argument for RPI hedging is that inflation-linked revenues cannot be forecasted exactly, but that uncertainty can be removed through the use of the hedging tool, which would fix the RPI at project inception.

Other types of derivatives include forwards and futures, options, and swaps, which are, with the exception of swaps, used to a lesser degree in PFI finance. A forward is an obligation to buy or sell an asset at a specified forward price on a known date (Neftci, 1996). Swaps open the possibility of obtaining a fixed price for an asset despite changes in the cash price and are often used for currency contracts or interest rate markets (Eales, 1995). Futures are a standardised form of forward contracts and, although they are conceptually identical, differ from them in operational terms. Thus, futures involve a commitment to buy or sell a fixed amount of specified underlying asset at a fixed future day (Das, 1997). While forwards are not exchange-regulated and are traded over-the-counter, futures are traded in formalised exchanges (Neftci, 1996), where access is restricted only to the circle of members. The main advantage of using options is that the holder (the purchaser) is in a position either to use or to disregard them. Options contracts avoid the effect of adverse changes in the price of the underlying instrument.

According to Neftci (1996) 'a swap is the simultaneous selling and pur-chasing of cash flows involving various currencies, interest rates, and a number of other financial assets'. Swaps are considered to be fairly com-plicated instruments, which may stretch over long time periods. Banks are often involved in swaps as an intermediary. The basic interest rate swap, called 'vanilla swap' (Winstone, 1995), between fixed rate and floating rate is commonly used to illustrate the principle of swap operation. In this process any losses of interest rates from one customer are compensated with the corresponding gains from another customer.

Risk adjusted discount rates are used in banking and business activities. The underlying principle of this financial instrument is to adjust a risk free discount rate by accounting for future inflation and extraordinary risks (Flanagan & Norman, 1993). The risk premium, which is often used in construction projects, constitutes a contingency sum added to an estimate to account for unforeseen eventualities that cannot be fully priced when the estimate is prepared (Raftery, 1994). Risk premiums in construction range from 5% to 15% of the project price (Ruster, 1996). Similar allowances for contingencies are sometimes made in financial transactions. Another type of risk premium involves the allowances which are added to a risk-free discount rate, by investors, to reflect the perceived risk in an estimate (Flanagan & Norman, 1993). Ranasinghe (1998) has argued that a differential adjustment of the estimates of individual bill item costs or activity durations is better than adjusting the project cost or duration through the general contingency allowance.

Escrow or trust accounts are established in order to ensure that revenues from a project are appropriately used to finance the operation of the project and to service its debt (Stein, 1995).

6.5 Financial risk management in PFIs: an empirical analysis

In order to identify which risk management techniques were most frequently used by finanical services providers involved in PFIs, the project team conducted 14 interviews with senior financial experts across the UK. These interviews were conducted by two researchers between April and December 2000, whereby each interview lasted between an hour and an hour and a half. Although the researchers encountered initial difficulties with respondents changing and postponing prearranged interviews, in the end a sufficient sample of detailed expert interviews was obtained.

Interviewees were selected both on account of their past involvement in PFIs and their general expertise in matters of project finance. The starting point for the selection of respondents was the *PFI Journal* in which companies advertised their services, and subsequently a database of PFI participants was constructed in conjunction with the team's research contacts. To avoid potential biases which could arise from respondents having been involved in certain types of projects only, interviewees were selected according to their involvement in a broad spectrum of PFI project types, e.g. schools, hospitals, water treatment, courts, MOD projects, prisons, roads, etc. and respondents were selected from a couple of sources. Given the commercial sensitivity of some questions and in order to ensure honest and comprehensive responses, the research team assured respondents upfront of its commitment to commercial confidentiality and the anonymity of the companies, individuals and projects involved. In order to improve the quality of responses, respondents were briefed over the phone on the main discussion points one or two days before the interview.

The questions bank for the interviews was based on an in-depth literature review phase which focused on the risk management practices of PFI participants during different stages of project planning, development and implementation. The interviews themselves were semi-structured and were tape recorded for convenience, and in order to speed up the process of analysis. Pilot tests were conducted with industry contacts who had previously worked with the research team. Following the data collection all interviews were transcribed, with further analysis being conducted with the help of dedicated software for textual analysis.

The sample of the financial companies for the survey included 14 leading financial organisations, based or operating in the UK, all of them with substantial records in PFI transactions. Senior representatives from all areas of financial engagement in PFI were interviewed, that is staff who were associated with senior debt providers, financial consultants and equity providers. Table 6.1 presents a breakdown of respondents by area of involvement. Accordingly, most respondents played multiple roles in PFI transactions, including involvements as debt arrangers to the SPV and as providers of senior debt and equity. (Debt arrangers are usually banks that have acted as consultants to the client, but for legal reasons cannot be named as consultants.) One manager from a financial company defined his role as a 'principal', which is a staff member employed by the organisation which puts the SPV together.

6.5.1 Risk identification by financial institutions

Based on respondents' statements, three risks can be identified as being of principal importance to the financial risk management effort associated with PFIs (Fig. 6.3). These include, in order of importance, performance-related risks (92.9%), general financial risks (85.7%), legal risks (78.6%), time and cost escalations and other construction risks. In addition to these key risks, respondents made reference to other risk factors, such as political, social, commissioning risk, risks arising from the relationship between the SPV members, environmental risk, protestor's actions, reputational risk, etc.

As concerns the management of these risks, several respondents stated

Table 6.1 *Breakdown of the financial companies by area of activity.*

Debt arrangers	5
Senior debt providers	10
Equity providers	8
Consultants	4
Principal	1

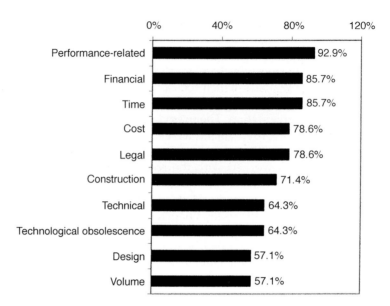

Figure 6.3 *Top ten risks investigated by the financiers in the PFI schemes.*

that they started identifying risks from the project's documentation, focusing on the terms and conditions of the concession agreement. In most they then focused their attention on other agreements such as fuel supply, service provision, etc. Several respondents, moreover, noted that risk categories were not clear cut and that their investigation of different types of risks often overlapped.

In addition to identifying three core risks which had to be examined in every project, respondents noted that there were a number of risk categories which could be eliminated from the list at the very early project stage in the context of specific projects. Moreover, several respondents suggested that, depending on the sector specifics, it could take a very short time to clarify the relevance of certain risks. Nonetheless, there was consensus among project participants that, when assessing a new project, a complete list of possible risks had to be used as a starting point. Thus it was suggested that, for example, in an IT transaction technological obsolescence would have been a priority, while in school or housing projects this issue would be of no or little relevance.

6.5.2 Risk identification techniques

During the interviews, most of the respondents reported that their financial organisation had a standard format or scheme for the identification of relevant risks. This scheme is typically based on knowledge from previous projects and includes a checklist (or a risk matrix) for the main risk categories such as construction risks, operational risks, etc. Usually, an examination is conducted to identify sector-specific risks, risks related to the financial structure, risks associated with the sponsors, environmental considerations and other factors influencing the risk profile. The main risk identification methods used by the practitioners are illustrated in Fig. 6.4. All respondents noted that they relied heavily on their previous experience in forming a broad initial judgement about the feasibility of a project. Furthermore, respondents suggested that later on in the PFI process the employment of consultants to investigate the risk details became more relevant. All respondents noted that a sensitivity aimed at analysing the impact of different risk factors on the project's performance was only conducted once a financial model had been constructed.

As concerns the initial stages of project planning, respondents noted that they used checklists, databases and experience/intuition to a varying degree (Fig. 6.4). When queried further on the nature of experience/intuition, respondents suggested that this involved an expectation of potential problems, which was based on past experience, and did not necessarily involve strong empirical evidence. A comparatively small number of respondents noted that they used workshops or brainstorming sessions in order to bring risk assessors together. Likewise, a small number of respondents suggested that site visits could assist in identifying problems and clarifying ideas.

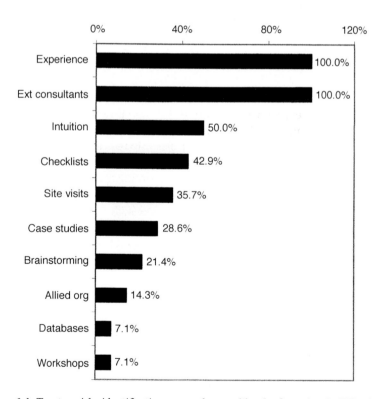

Figure 6.4 *Top ten risks identification approaches used by the financiers in PFI schemes.*

6.5.3 Information aids for risk identification

As a part of the interviews, respondents were presented with a checklist of potential sources of information and asked to identify their main risk identification aids. The top five sources, which the respondents utilised either 'always' or 'frequently', are illustrated in Fig. 6.5.

With very few exceptions, all companies employ external consultants (93%) (Fig. 6.5). Other important sources included government publications, PFI agencies and DETR information, and PFI panel documents. However, as one respondent noted, apart from external consultants these other sources of information were merely used 'as a starting point to establish the ground rules'. Moreover, it was noted that text-based information sources were more appropriate for public sector representatives who lacked sufficient experience in PFIs, than for financial service providers.

6.5.4 Risk evaluation conducted by financial institutions

The risk evaluation practices of the financiers are geared towards ensuring that all project risks are thoroughly investigated, properly allocated and fully understood by the parties involved in a PFI project. Normally, construction companies and sub-contractors provide fixed price contracts which evaluate in-depth the risks that fall to them. The role of the technical advisors employed by banks includes the performance of independent checks on the accuracy of the risk evaluation. All respondents noted that financial companies use at least three types of external consultants to support their project risk evaluation, notably legal, technical, and insurance experts. In addition,

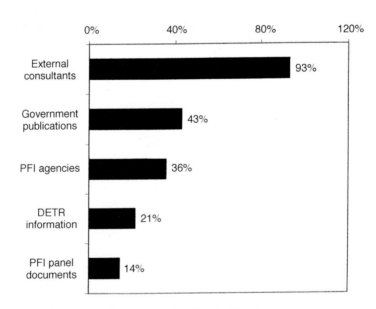

Figure 6.5 *Main sources of information for risk identification.*

some respondents noted that their risk management process involved external financial experts. As concerns the function of these advisers, the repondents noted that experts provide specialised knowledge and support for the decision-making process. One respondent explained the interaction with the external consultants as follows:

> 'At the start of a deal we internally develop a very broad overview of the key risks. As we progress with the transaction we bring the external advisors to conduct risk assessment to overlay our initial view. Subsequently, we work on the key issues that they outline, which would hopefully mirror the main ones that we had identified. It is initially 100% internal assessment but ultimately the balance moves down the scale as we bring in the external advisors and eventually may change to 80% external input.'

According to respondents, the project team of the debt provider conducts a detailed analysis as part of the credit application, which is then submitted to the bank's credit committee for approval. The credit application describes the project and includes appendices such as the financial model, the base case, and the results from the sensitivity analysis. It also includes a descriptive analysis including the risk matrix. This matrix includes risk categories such as construction risk, operation risk, shareholder risk, financial risk, regulatory risk, etc. Some institutions assign a subjective number from 1 to 10 to each risk, which allows an overall rating to be produced from that matrix. This number can then be used as a final credit risk rating for a particular transaction.

All respondents emphasised that the financial model was their key risk assessment tool. When queried as to the nature of the financial model, respondents stated that this was a mathematical expression of the project's future cash flows, which included all project revenues and costs, as well as taking account of the risk factors affecting them. One of the principal goals of the model was to estimate the financial impact of different risks through semi-quantitative analysis. According to respondents, the starting point for all such financial models was the base case scenario which included the main financial ratios together with an analysis of the borrower's case, i.e. the borrower's expectations for the project development. Taking into account the suggestions of technical advisers, the financial model was then used to investigate different 'what if' scenarios, which could include anything from an increase in operational costs, rising inflation, construction delays, and pessimistic life cycle scenarios. In this context, much of the analysis focuses on the identification of different parameter variations which could threaten the repayment. In some cases this analysis is extended further to assess the combined impact of several randomly selected negative scenarios on the repayment stream. A number of respondents noted that, based on the analysis of the financial model, experienced project teams are able to allocate and focus on a few 'weak spots', so that there is no need to investigate hundreds of potential scenarios.

As concerns the concrete elements of their analysis, respondents noted that they also investigated the base case in the model for the construction price of the asset and run a sensitivity that assumes price inflation by 10%, 20%, or 30%, according to the advice of their technical consultant. Similarly, it was noted that all other financially important variables are incorporated in the analysis. When analysing a number of scenarios, a common sense view usually determines the analyst's estimation of the likelihood of different risk scenarios and, consecutively, their worst case scenario. Ideally, many financiers strive for a situation where even in a worst case scenario, the payment of the debt service is not affected because of the gearing of the project. If the worst case indicates that the senior debt service could be affected, funders will usually reduce the amount of debt requiring the SPV to increase its equity contribution.

6.5.5 The role of external advisers in risk evaluation

Our previous analysis has indicated that financial risk analysis in PFIs relies heavily on the contribution of external advisers. In order to further analyse the relationship between financial service providers and external advisers, the research team queried respondents' views on the the quality and reliability of the external advice. As Fig. 6.6 indicates, nearly half of the respondents (49%) considered that the information received from the advisers was on average 'Very accurate'. By contrast, 38% of respondents note that they had not always been satisfied with the information provided by external advisers. Respondents who were dissatisfied with the service of advisers attributed this to a lack of experience, especially among junior advisers, or communication problems which arose when the advisers used strictly technical approaches in their analysis of project risks. Despite some criticism, 13% of the respondents noted that they currently receive better quality advice compared with the earlier PFI stages. When queried about the availability of qualified advisers, some respon-

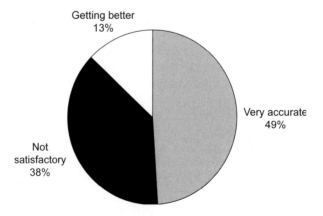

Figure 6.6 *Rating for the quality of the service provided by the external consultations.*

dents reported skill shortages in the area of technical consultancy. Accordingly, this problem was particularly severe in the IT area where there was a lack of experts who could provide advice in the context of complex projects.

When queried about the software used in connection with their risk evaluation activities, all respondents noted that they relied on Excel spreadsheets models, with 28% having used Lotus at some point of the time. Meanwhile, none of the respondents stated that they had made any use of dedicated risk analysis software.

6.5.6 Problems with risk analysis

When queried about problems arising in the context of PFI risk assessment, there was some consensus that differences in the risk perception of different parties were a principle source of difficulty. These differences, it was further explained, typically involved both estimates of likelihood of occurrence of some risk events and of the amount of money needed to cover for them.

As part of this analysis, respondents' statements about problems encountered in PFI risk evaluation were subjected to a textual network analysis which relied on the software package ATLAS.ti. This network analysis highlighted two thematic categories, notably those relating to the communication with the client, and those relating to the nature of the PFI risks in general (Fig. 6.7). More specifically, 35% of respondents reported that they had encountered communication problems with clients, whereby some respondents suggested that, in some instances, the public sector imposed unreasonable requirements in terms of the level of detail in private sector's risk assessment. In addition, it was noted that there was some uncertainty over particular projects relating to affordability and the level of support from central government institutions. One interviewee explained that 'the Government's commitment to a particular sector is probably the hardest thing to understand, i.e. how committed the politicians are to making something happen'.

Another 35% of the respondents noted that they encountered problems due to the complex nature of PFIs. PFI transactions involve multiple participants which, depending on the size of the project, can mean that more than 20 to 30 companies are involved. In some cases this complexity can, as one respondent noted, make risk evaluation 'a little bit of a guessing game and more or less a speculative process'. In addition, some respondents highlighted that the lack of an agreed definition for different risks can lead to misunderstanding during the negotiation process.

Some respondents reported that they had encountered cases where some project participants had failed to fully understand their obligations and risks before signing an agreement. This created difficulties for them later on, since, once they recognised their problem, they were already legally bound under the fixed price contract. It was felt that many of these problems could be avoided through the use of external experts.

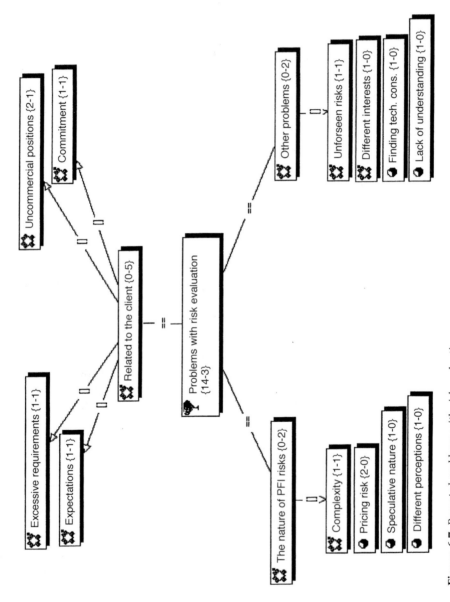

Figure 6.7 *Reported problems with risk evaluation.*

6.5.7 The reliability of risk assessments

As part of this analysis, the interviewees were asked to indicate how they felt about the reliability of their risk evaluations. Most of the respondents noted that they felt fairly confident with their risk evaluation practices and indicated that they were able to properly account for the majority of risks. In this context, one respondent pointed out that risk assessment was 'not a precise science' but required a fair amount of guesswork and intuition. This view was fairly widely supported by respondents in terms of statements such as 'we hope what we do is reliable', 'our approach is reasonably robust, and gives an approximation of what the impact of risks may be', 'we are confident that 75% of risks are adequately evaluated' and, lastly, 'the reliability is quite high, within 5 and 10% of actual cost'.

Regarding the overall quality of risk assessments, there was broad agreement that the quality of such assessments depended very much on a good understanding of the project details or, in other words, reliable and detailed inputs. In this context, one respondent explained that 'it is ultimately down to the accuracy of the information that goes into the model'.

As regards the ability of SPV partners to come to abroad agreement on risk assessments, the overall view was mixed. Amongst nine respondents who discussed their views on this issue, five felt that this was hardly ever possible, while three stated that they almost always agreed. One interviewee, meanwhile, stated that this very much depended on the specific case. Specifically, one respondent noted that 'there is too much conflict' in PFI negotiations, while another suggested that 'the bank is bound to disagree with the contractor' in their views on risk. One of the respondents who suggested that agreement on risks assessment was usually possible, attributed this to the fact that his company had a close collaborative relationship with a contractor which allowed them to focus on the project instead of organisational and management issues. As concerns levels of disagreement over the project span, several respondents noted that risk assessments typically diverged the most at the beginning stages of a project and then converged as the project moved further along.

6.6 Risk mitigation practices

The second batch of questions of this survey queried which risk mitigation practices respondents thought to be most useful. In this context, specific reference was made to techniques of risk transfer, retention, reduction and/or elimination. In response to this question, respondents noted that, regardless of their precise function (as debt or equity providers), their risk mitigation practices were very similar, with the exception of situations where they acted as financial advisers only.

As an overall rule, respondents suggested that, due to the tight cover ratios, they tended to manage the risk by seeking first of all to mitigate them contractually. In this context, the decision to enter into the PFI contract was actually conditioned on the fact that most important risks were mitigated upfront.

6.6.1 Risk transfer

Respondents noted that their primary risk mitigation aim was to structure the PFI transaction contractually in such a way that most of the risks were transferred to the appropriate parties. This concern was based on the fact that the SPV company itself has no financial resources to cover substantial project risks, so that leaving risks with the SPV would necessitate higher cover ratios or a higher degree of equity investment. In this context respondents emphasised that risks remaining with the SPV needed to be supported by equity rather than by debt, so that they could be hedged.

According to respondents, a bank can deal effectively with financial risks but, from a senior lender's point of view, the margin in this type of projects is in the range of 1%, i.e. the bank can afford to lose money in one of a 100 projects. As a consequence most banks seek to be fairly certain that most relevant risks have been passed on to other parties. In line with the idea for allocating the risks to other parties, financiers transfer all major construction risks to the construction companies, e.g. construction time and cost over-runs, design, etc. (Fig. 6.8). All operational risks such as escalating life cycle costs or technological changes, meanwhile, are usually transferred to the operational companies, while the political and some legislation risks are transferred to the public sector.

6.6.2 Risk retention

Despite their desire not to bear significant risks, financial organisations usually have to accept some risks which are related to their core activities as well as some residual project risks. Risks relating their core activities include a range of financial risks such as interest rates and inflation, which they seek to hedge through an agreed hedging policy. They also have to retain counter-party risks such as the credit risk, which is normally mitigated through a thorough investigation of the borrower's financial standing (and the parent companies). In this context respondents noted that the PFI regime has promoted a high degree of responsibility among private sector parties. This means that companies that are heavily involved in PFI usually are willing to solve problems, in order to protect their reputation, even where they are not directly liable for them.

Despite their focus on risk mitigation in PFI contracts, financial organisations are inevitably exposed to some of the project risks that have been passed through commercially. For example, if the sub-contractor fails to perform or becomes insolvent this may affect the funder. In most cases, PFI contracts specify caps on the liability of the construction company and the FM provider. Respondents noted that if the construction company has a reliable track record with the type of schemes involved, the bank would be

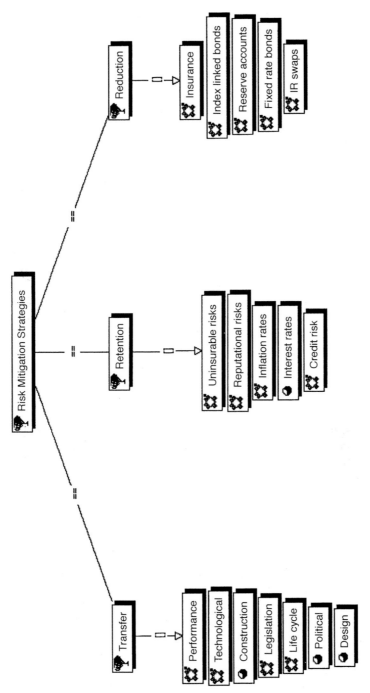

Figure 6.8 *Risk mitigation strategies used by the financiers in the PFI projects.*

prepared to accept a lower capping level on the liquidated damages than otherwise. According to respondents, examples of risks that are frequently capped include liquidated damages for construction, or risks arising in the context of new technology. *Force majeure* risks (beyond one's control) and/or uninsurable risks are also often left with the financiers, as well as some degree of political risk, taxation risk or change in law.

6.6.3 Risk reduction

As concerns risk reduction, respondents noted that they would almost always seek to reduce the magnitude of virtually every risk retained. According to respondents this could involve the use of financial instruments like interest rate swaps, which provide a fixed rate of loan repayment, as well as other hedging instruments which provide index-linked rates or interest rate caps.

Amongst respondents in this study, two bank representatives reported the usage of fixed-rate bonds, which guaranteed a fixed interest rate for the length of the bond. Moreover, it was noted that indexed-linked bonds that tied the interest payment to the rate of inflation had also become popular. Indexed-linked bonds provide for a greater level of risk reduction as they link project revenues to inflation. All respondents noted that, as senior debt providers, they had used debt syndication, which involved selling parts of the debt to other banks in order to balance their overall portfolio risk exposure.

As concerns the possibility of risk elimination, respondents noted that on many occasions it was not possible to eliminate any project risks. The only possibility for the private sector to eliminate particular risk was not to accept it from the public sector in the first place. Thus, on the whole, there is clear evidence that the risk management practices of financial service providers involved in PFIs focus on maximising the degree to which risks are mitigated or hedged, whereby the usage of financial instruments is seen as a preferred risk mitigation strategy.

6.7 Conclusions

Over the last decade, a number of banks and other financial institutions have become specialised in the provision of capital and financial advice in the context of PFIs. Based on the principle of project finance, PFI finance entails the repayment of debt from a revenue stream which is generated by a capital project. As part of this setup, there is typically no recourse to the assets of any of the companies involved. If SPV members do not deliver the project according to the specifications, they incur penalties and revenue deductions which can threaten the loan repayment. Delayed project delivery can also create cash flow problems because the interest on the loan rolls up, thus increasing the amount of debt. From the perspective of financiers, all of these factors conspire to give prime importance to the early and comprehensive assessment and management of risks in PFIs.

Our research indicates that the main risk categories examined by the financiers include performance-related risk, financial risk, time and cost overruns and other construction risks, etc. In terms of their risk assessment methodology, financial companies place heavy reliance on previous experience, advice from external consultants, checklists, site visits, case studies, etc. Moreover, there is a suggestion that most financial companies utilise fairly standardised assessment approaches.

Given the growth in public expectations with regard to the conduct of the private sector, be it in terms of social responsibility or environmental protection, there is, in our view, a strong possibility that current risk management practices will become insufficient. As the significance of PFI in terms of public sector service provision increases, so does the reputational risk associated with a company's involvement in PFI. Previous research has highlighted the importance of reputational risk and the need for companies to assess the full range of possible public responses to a project (Drennan *et al.*, 2001). While financial companies may be able to transfer many of the operational and financial risks to other companies, it is necessary for them to understand that reputational risk can hardly ever be evaded.

Acknowledgements

We wish to thank the Engineering and Physical Science Research Council (EPSRC) and the Department of Environment, Transport and the Regions (DETR, now the Department of Trade and Industry) for funding the research on 'Standardised framework for risk assessment and management of Private Finance Initiative projects' which informs this paper. We are also profoundly grateful to the project industrial partners who have supported the research. Many organisations have either participated in the research or supplied data. Their co-operation is fully appreciated.

References

Blommestein H.J. (2000) The changing nature of risk and the challenges to sound risk management in the global financial landscape. *Financial Market Trends*, **75**, 171–194.

Carrick M. (2000) *Commercial Debt Raising for PFI Projects*. Ernst & Young UK, Corporate Finance, Internet article (www.budgetnews.co.uk/Template1.nsf/HomePages), accessed on 19 June 2000.

Cuny C. & Gethin I. (1999) Hedging risk in PFI deals. *The Private Finance Initiative Journal*, **4**(3), 76–78.

Das S. (1997) *Risk Management and Financial Derivatives*. Macmillan Press Ltd, London.

Dixon I. (1999) Life on the margin. *The Private Finance Initiative Journal*, **4**(6), 66–67.

Drennan L., Beck M. & Henry R. (2001) From Cadbury to Turnbull: finding a place for risk management. *Insurance Research and Practice*, **16**(1), 27–34.

Eales B. (1995) *Financial Risk Management*. McGraw-Hill, Maidenhead.

Ellis S. (1999) Financing options. *The Private Finance Initiative Journal*, **3**(6), 60–61.

Euroweek on-line (1999) UK capital markets – March 1999. *PFI: Capital Markets Deals Swell*, Internet article (www.euroweek.com/public/supplements/ukmarch99/6.html), accessed on 11 July 2000.

Ex-Im BUS (1999) *EX-IM Bank Project Finance*, Internet article (www.exim.gov/mpfprogs.html), accessed on 15 March 2000.

Finnerty J. (1996) *Project Financing, Asset-based Financial Engineering*. John Wiley & Sons Inc., USA.

Fitch T. (1997) *Dictionary of Banking Terms*. Barron Educational Services Inc., Hauppauge.

Flanagan R. & Norman G. (1993) *Risk Management and Construction*. Blackwell Scientific Publications, London.

Horne J. & Wachowich J. (1998) *Fundamentals of Financial Management*. Prentice-Hall International Inc., London.

International Projects 500 (2000) *Banks PFI, the International Projects 500 Comments* (ed. J. Pritchard). The International Projects 500 in association with 500 Publishing/Legalease Ltd., Internet article (www.projects500.com/projects/edit/edit/eu139.htm), accessed on 29 June 2000.

Leeper R. (1979) Perspective on project financing. *The Banker*, **129**(643), 77–83.

Middleton N. & Richardson M. (1999) Surveying the scene. *The Private Finance Initiative Journal*, **4**(3), 24–26.

Morrison N. (1998) United Kingdom, *International Financial Law Review*, March, 47–51.

Neftci S.N. (1996) *An Introduction to the Mathematics of Financial Derivatives*. Academic Press, San Diego.

Oliver C. (1998) UK project financiers turn to PFI as government seeks increased targets. *Euroweek*, July, 30–35.

Pickering C. (1999) The asset option. *The Private Finance Initiative Journal*, **3**(6), 64.

Raftery J. (1994) *Risk Analysis in Project Management*. E & FN Spon, London.

Ranasinghe M. (1998) Risk management in the insurance industry: insights for the engineering construction industry. *Construction Management and Economics*, **16**(1), 31–39.

Ruster J. (1996) Mitigating commercial risks in project finance. *World Bank Note* No. 69. World Bank, New York.

Sapte W. (1997) *Project Finance, the Guide to Financing Built-Operate-Transfer Projects*. Euromoney Publications, Essex.

Sarmet M. (1980) International project financing – the European approach. *The Banker*, **130**(654), 89–95.

Stein S.W. (1995) Construction financing and BOT projects. *International Business Lawyer*, **32**(4), 173–180.

Timmins N. (2001) Call for separate finance tenders in PFI deals. *Financial Times*, Nov 9.

TTF (1998) *Treasury Taskforce Policy Statement No. 2, Public Sector Comparators and Value for Money*. HMSO, London.

Winstone D. (1995) *Financial Derivatives – Hedging with Futures, Forewords, Options and Swaps*. Chapman & Hall, London.

Wynant L. (1980) Essential elements of project financing. *Harvard Business Review*, **58**(3), 165–173.

Zakrzewski R.A. (1999) Risk minimisation in project finance. *The Poland Library*. <www.masterpage.com.pl/outlook/risk.html>, accessed on 15 March 2000.

7 A legal perspective on risk management in public-private partnership

Andrew Walsh

7.1 Introduction

'Appropriate risk allocation between the public and private sectors is the key to achieving value for money on PFI projects. If the private sector are asked to accept responsibility for a risk that is within their control, they will be able to charge a price for this part of the deal which is economically appropriate. However, if the Department seeks to transfer a risk which the private sector cannot manage, then the private sector will seek to charge a premium for accepting such a risk, thereby reducing value for money. The Department should therefore have sought to achieve not the maximum but rather the optimum transfer of risk, which allocated individual risks to those best placed to manage them'. (National Audit Office (NAO) report, *Examining the Value for Money of Deals under the Private Finance Initiative*, August 1999)

The overriding aim of risk allocation under private finance initiative (PFI) projects is to strike a sensible balance between risk transfer and value for money (VFM). This chapter analyses how that objective may be attained through the contractual structure of PFI/public-private partnership (PPP) projects, the competitive procurement process and the contract terms of the various project documents.

It is perhaps appropriate to begin with some general comments on the scope of this chapter: PPPs and risk management. The concept of PPPs covers a broad range of relationships and legal structures, including PFI projects (i.e. a PPP which involves private finance). The essential characteristics of PPPs are the involvement of both public and private sector parties and a relationship which includes at least some elements of partnering (i.e. more than simply an arms-length contractual relationship). Since May 1997 the Labour Government has re-badged PFI projects as PPPs, though the IPPR Commission's recent report on PPPs (*Building Better Partnerships*, June 2001) suggested that terms used in the public/private context were insufficiently precise. The scope of this chapter covers both PFI transactions and (since 1997) PPPs.

On the one hand, risk management involves a broad range of techniques required to identify, quantify, allocate, transfer and control project-related risks. However, the practice of risk management differs within the context of the principal objectives of public and private sector parties. For the public sector this is to identify, quantify and transfer appropriate risks based on the principles of VFM and allocation of risks to the party best able to manage them. On the other hand, for the private sector this is to assess and price transferred risks, to allocate and flow-down those risks to the appropriate consortium member, and to manage and control those risks.

This chapter will firstly consider the rationale for risk transfer in PFI, followed by some key stages in the historic development of risk allocation and risk transfer under PFI transactions.

7.2 Rationale for risk transfer

From an early stage, there were two principal reasons for the transfer of risk under PFI projects: VFM and accounting treatment.

VFM is required to offset the higher costs of private finance. Since private (or 'unconventional') finance for public sector capital projects will invariably be more expensive than public finance, there must be off-setting benefits to justify its use. Within the context of VFM, the following factors are relevant:

(1) Conventional finance: the government is able to borrow from the capital markets at lower rates of interest, because of the market's perception of the government as a good debtor. This derives principally from the government's unique power to impose taxes. The government also benefits from low transaction costs (e.g. lower broker's commission), the capacity to spread risks over a wide range of activities (public sector financing costs do not reflect the risks of individual projects) and a well-established and largely institutional market for government debt (e.g. gilts are neither subject to stamp duty nor liable to capital gains tax).

(2) Public expenditure: stringent obligations of prudence and propriety attach to the use of public money. In addition, public sector capital projects are subject to scrutiny both by the NAO and the (House of Commons) Public Accounts Committee. The NAO has powers to 'carry out examinations [VFM Examinations] into the economy, efficiency and effectiveness with which any Department, authority or other body ... has used its resources in discharging its functions' (Section 6(1) National Audit Act 1983).

(3) Off-setting benefits: the principal off-setting benefits under PFI projects are a combination of (1) transfer of risks to the private sector; and (2) efficiencies deriving from a more entrepreneurial approach and private sector management techniques.

Accounting treatment is required to keep assets off the public sector balance sheet. Previously, if a department had procured a capital asset (such as a new hospital, computer system or road) the full capital value of that asset would score against the department's allocated expenditure (i.e. its Public Expenditure Survey provision) for that year.

However, under PFI, if sufficient risks are transferred to the public sector, the asset could be treated (in effect) as off balance sheet for the public sector. The principal implications were, firstly, that payments made by the public sector under the PFI contract would be scored as 'revenue' (i.e. revenue expenditure) rather than capital expenditure, and, secondly, that payments for PFI services (including elements covering the capital cost of assets used to perform contracted services) could be spread over the life of the contract (re-timing of payments). Within the context of accounting treatment, the following factors were relevant:

(1) Balance sheet: reference to 'balance sheet' in this context is rather misleading since (at least at the time of launch of the PFI) there was no public sector balance sheet as such.
(2) PSBR: 'In cases where the great majority of the risk stays with the private sector, the capital value of the leased asset will not be counted against expenditure provision' (Treasury, *Guidance for Departments on Private Finance*, December 1992). The Treasury held that payments under PFI projects should not impact on the Public Sector Borrowing Requirement (PSBR) (i.e. the Government's cash deficit or difference between cash receipts and outgoings), the principal measure of the Government's financial position.
(3) Additional finance: originally it was thought that private finance for public sector capital projects would be additional to conventional funding. 'The measures which the Chancellor has announced will add to the total level of national investment . . . The private sector element will make additional resources available in the area concerned, and will allow the publicly financed part to be used to the full' (Treasury, *New Release*, November 1992). It is now accepted that private finance is not 'additional' as such, but must be funded out of departmental allocations over the project duration.

These separate reasons for risk transfer remain equally valid today, though more recently the Treasury have tended to emphasise more the VFM benefits of PFI transactions.

7.3 Early stages in the development of risk transfer

With hindsight it is easy to take for granted many of the familiar features of risk allocation and risk transfer that underpin PFI projects, forgetting that these evolved over a period of several years. We now trace some of the key early stages in the evolution of risk allocation and transfer under PFI projects.

The Treasury rules of 1981 established (for the first time) a link between the higher cost of private finance and the need for transfer of risks to make good the difference by making explicit the costs of bearing risks. The Treasury rules (also known as the Ryrie rules) arose from the report (issued on 28 September 1981) of a Working Party established (in early June that year) by the National Economic Development Council (NEDC). This was the first thorough-going review of the use of private finance for public sector capital projects. The Working Party, chaired by William Ryrie (a senior Treasury official), was made up of representatives from the Treasury, CBI, TUC, nationalised industries and NEDC. The report was NEDC's response to widespread concern that investments by the nationalised industries might be frustrated by the operation of 'external financial limits' (government imposed caps on the annual spend of each department) and a general belief that higher investment by the industries might benefit the UK economy.

The Working Party encouraged departments and the nationalised industries to explore new private sector financing options, and proposed a progress review by June 1982. The significance of the Treasury rules was:

- Risk transfer: clear recognition that the higher cost of market capital (compared with loans from the National Loans Fund) must be justified by off-setting benefits
- Best VFM: private finance must provide the most cost effective solution (i.e. best VFM for the taxpayer/citizen) based on a comparison with both publicly and privately financed options and taking account of any differences in financing costs. Thus, privately financed options must be compared against the best available publicly financed alternative.
- General application: though aimed at the nationalised industries, the rules came to have wider application, representing the Treasury's approach to private finance across the public sector.

Unfortunately, the rules were used in practice to frustrate private finance schemes. A revealing explanation of this process is set out in a paper (*The Opportunities for Private Funding in the NHS*, April 1993) by David Willetts MP (then a Treasury official). He confirmed that 'the Treasury's objective . . . was to stop such schemes . . . The conditions they set for privately financed projects were not intended to be met in practice'. Over time the Treasury rules became associated (for the private sector) with a negative approach to private finance.

In practice, the lack of progress in developing private finance (for public sector capital projects) during the 1980s was influenced by other factors as well, including:

- Recession: the Government's concern to control public expenditure during the recession of the early 1980s
- Privatisations: greater focus on the Government's privatisation programme. This began during 1983–1984 with BP and Britoil, followed by BT, Cable & Wireless, British Gas and British Airways over the next three to four years. In addition, privatisation (by removing certain nationalised industries from the public sector) reduced the pressure for direct access to private finance

- Political challenge: the Government's determination to combat avoidance instruments used by certain (particularly Labour-controlled) local authorities to evade public expenditure controls.

But how to make explicit the costs of bearing risks? Treasury guidance to departments (*Economic Appraisal in Central Government: a Technical Guide for Government Departments*. April 1991: better known as 'The Green Book', which superseded earlier guidance of 1984), set out guidelines to appraise and value risks. Annex F to the Green Book ('Private Finance for Potential Public Sector Projects') provided (for the first time) guidance on the appraisal of risks relating to private finance, as follows:

- Public sector comparison: the appraisal should include a comparison between the best publicly and privately financed alternatives to provide the relevant facility. 'The expected costs of risks transferred to the private sector should be added to the cost appraisal for the conventionally financed option'
- Benefits: the main benefits of risk transfer were stated to be 'the better incentives and rewards which it can provide for the private sector to supply better VFM'
- Large projects: for large or complex projects the distribution of risks between financiers, suppliers, users and taxpayers should be examined to achieve the most cost effective balance.

The Green Book (1991 edition) also established the principle that risks should be allocated to the party best able to manage them. It stated that 'sometimes it may be cost effective to transfer a risk to a third party by means of contractual arrangements with suppliers or by insurance. However, this should normally be confined to the transfer of risks to those who have a particular expertise in its management, or opportunity to reduce it. Transferring risk to another body less able to absorb it is unlikely to be cost effective'.

The 'private finance initiative' was eventually launched by Mr Norman Lamont (then Chancellor of the Exchequer) in his Autumn Statement (of 12 November 1992). The Statement introduced important changes to Government accounting for PFI projects. An early challenge for the PFI was to establish how much overall risk must be transferred to the private sector to justify the use of private finance.

Although the accounting changes also dealt with joint ventures and service-based procurements, at this early stage there was an expectation that leases would provide the most suitable vehicle for private finance. Thus, the accounting treatment of leases under the PFI rules was of particular interest: the new accounting rules established that (in relation to leases), to justify the use of private finance, 'the great majority of the risks' must remain with the private sector. The principal effects were, firstly, to take the relevant asset 'off-balance sheet' for the public sector (off-balance sheet); and, secondly, only the annual payments under leases would 'score' against a Department's PES provision (revenue payments). Thus, a public body could procure an asset without material impact on its PES provision.

The Chancellor's statement on leasing under PFI was succeeded by two Treasury guidance documents:

(1) Treasury *Guidance for Departments on Private Finance*, issued on 9 December 1992: this Guidance provided for best VFM and classification of leases. PFI leases must provide best VFM. Departments must show (through an appraisal of alternative leasing offers) that the advantages of leasing under PFI (such as efficiency gains and risk transfer) outweigh the higher cost of private finance.

(2) *Leasing Guidance for Departments* issued in May 1993: this Guidance applied to departments, non-departmental public bodies (NDPBs) and NHS Trusts, but not to local authorities. It set out methods for public bodies to assess the costs, risks and return necessary to produce a VFM comparison between leasing and direct purchase and to establish whether sufficient transfer of risk had occurred (to justify off balance sheet accounting treatment under the PFI). The principal features of the Guidance were (1) Basic test: to provide best VFM, the benefits of risk transfer and efficiency gains (from private sector management) must exceed the additional cost of private finance (NB. for this purpose public bodies were expected to make comparisons not only with direct purchase but also with alternative leasing offers). (2) Risk transfer: the main types of risk (relevant to the basic test) were identified as 'those which would be in place if an asset were owned', namely (under the Treasury analysis) residual value and obsolescence, loss of use (due to non-availability of the asset), repair costs and reduction/cessation of demand (by the public body to use the asset). Such risks would vary depending on the particular circumstances, nature of the asset concerned and identity of the public body lessee. (3) Service elements: but how to achieve sufficient transfer of risk to justify PFI accounting treatment? The Guidance provided that ... 'many leases bring with them elements of service, with a proportion of risk falling on and being managed by the [private sector] lessor'. This then appeared to be the way forward.

7.3.1 Suitable projects

How were public bodies to identify which public sector capital projects were suitable for the use of private finance? During the early years of the PFI, the Government struggled to persuade public bodies to test the use of private finance.

Accordingly (in his Autumn Statement of November 1994) Kenneth Clarke (then Chancellor of the Exchequer) introduced 'universal testing': first applied in the health sector, universal testing obliged all public bodies seeking Treasury approval for capital projects first to test the suitability of private finance (Treasury guidance, *Private Finance and Approval of Capital Projects*). In considering whether a capital project should be taken forward under private finance the public body must apply two tests: (1) Genuine transfer of risk: the initial test is whether, in principle, it is possible genuinely to transfer control and the associated risks to the private sector without disproportionate cost. Where this is thought to be so, a privately financed solution should be considered; and (2) VFM: 'the next question is whether it provides value for money ... Unless excluded by prescribed criteria, future invitations to tender (ITTs) should be on the basis of private finance. ITTs should indicate that adequate risk transfer will be one factor in awarding the contract, and encourage bidders to offer innovative solutions'.

The only permitted exceptions (to this general presumption) were circumstances where the transfer of control and risks was not feasible. Stated examples included:

- An overriding operational need for the public sector to provide the service (such as certain defence systems)
- No prospect (for policy or operational reasons) of transferring significant control and risk without disproportionate cost (such as repairs and maintenance works on government buildings)
- The project is covered by an existing contractual commitment.

Thus, projects suitable for PFI were those where genuine transfer of risk was both practicable and could be achieved to provide VFM.

7.3.2 Risk allocation

But how to apply the Treasury's general test (i.e. transfer of the great majority of the risks) to the broad range of individual risks and categories of risk that could affect a PFI transaction? Inevitably it took time to work out the ground rules for risk transfer and for any sort of market practice to develop. *Ad hoc* guidance relating to risk was issued in a piecemeal fashion over several years, principally in the form of isolated statements, as follows:

(1) Retained risks: it was perhaps more straightforward to focus on those risks which should properly be retained by the public sector (on the assumption that all other risks would be transferred to the private sector). 'We need to consider the best way to handle risks which the private sector may not be best placed to control. That includes planning and political risks, Parliamentary process and public enquiries' (Chancellor's speech to a CBI Private Finance Conference in May 1993).

(2) Timing: there was initial uncertainty about the optimum timing of transfer of risks. 'We must tackle the related issue of the points in a project's life at which competition and transfer of risk to the private sector should take place' (Chancellor's speech to a CBI Private Finance Conference in May 1993).

(3) Competitive process: how to persuade the private sector to accept unfamiliar risks? 'Departments should state which risks they are prepared to accept or negotiate on with bidders. The extent to which the private sector is willing to take on such risks may be a factor in a competition' (Private Finance Panel: *Guidance on Competition*, March 1994)

(4) Statement of risks: the allocation of risk and reward should be clearly defined and agreed in advance, with private sector returns genuinely subject to risk (Private Finance Panel: *The Private Finance Initiative: Breaking New Ground*, November 1993).

(5) Detailed guidance: the first concerted review of risk transfer under PFI appeared in the PFI Handbook (*Private Opportunity, Public Benefit – Progressing the Private Finance Initiative*), November 1995 (POPB). This was issued as part of Michael Jack's (Financial Secretary to the Treasury) 'PFI Action Agenda': an attempt to relaunch PFI which was faltering badly at the time. Some features of risk allocation and risk transfer in POPB were as follows.

(6) Genuine risks: Treasury insistence that the private sector must genuinely assume risk. 'The main requirement with projects of this type will be to ensure genuine transfer of risk to the private sector supplier. There should not normally be any guarantee or indemnities payable by public sector to private sector, explicit or implicit, to cushion the private sector supplier against unexpectedly low levels of demand or other forms of project failure'.

(7) Optimal risk transfer: transfer of risk is one method of achieving VFM, but based on an objective of optimal (rather than total) risk transfer. 'As a general rule, VFM can be expected to increase initially as risk is transferred to the private sector until the optimum point is reached at which all risks have been allocated to the partner best able to manage them. Any further risk transfer will lead to a decline in VFM'.

(8) Valuation: the amount and type of risk that needs to be transferred to achieve optimum VFM will of course vary from case to case and must be determined in negotiations with the bidders.

(9) 'Financially Free-Standing Projects': these are projects where the private sector seeks to recover costs entirely through charges for services to the (normally private sector) end user. In such circumstances the public sector will have minimal control and will find it difficult to manage the risks. It follows that few if any risks should be retained by the public sector under financially free-standing projects.

(10) Relevant risks: the normal categories of risk which 'are of sufficient general importance to be worth considering for the great majority of PFI projects' include design and construction risk; commissioning and operating risks; demand risks; residual value risks; technology/obsolescence risks; regulation and similar risks; and project financing risk.

(11) Retained risks: risks normally retained by the public sector include the risk of a wrongly-specified requirement and the risk of criticism (arising from failure of a public service).

POPB was supplemented by further guidance, the *PFI Guidelines for Smoothing the Procurement Process* of April 1996 and *Risk and Reward in PFI Contracts: Practical Guidance on the Sharing of Risk and Structuring of PFI Contracts* of May 1996. The latter, in particular, focused on those risks which might be shared between public and private sector parties.

7.4 Recent developments

Early concerns that a change of government might impact adversely on PFI proved unfounded. The new Labour Government (elected in May 1997) issued a *12-Point Plan for Partnership* and Geoffrey Robinson (then Paymaster General) announced a 'Review of Private Finance Machinery and the End of Universal Testing'. The 12-point plan included commitments: to ensure that the guidance on risk transfer and VFM (including templates) is kept up to date and to develop a clear and consistent policy on generic risks (e.g. the approach to changes in Government health and safety policy or the treatment of contaminated land).

On 8 May 1997 Malcolm Bates was invited to conduct a 'speedy review of the PFI process'. He submitted his report on 13 June: as quickly as 23 June 1997 all 29 recommendations in his PFI Review were accepted by Ministers. Two recommendations in particular were relevant in relation to risk transfer: those regarding model contract terms and those regarding accounting treatment.

7.4.1 Model contract terms

Though the Private Finance Panel had published earlier guidance on PFI contract terms (*Basic Contractual Terms* of October 1996 and *Further Contractual Issues* of January 1997), there was a need for more detailed guidance and ideally standardisation of contract terms. The Bates review noted that 'sets of model conditions/clauses should be produced for each Government Department in consultation with the Taskforce and the construction industry, operators and funders'. Treasury Taskforce guidance on project agreements (*Standardisation of PFI Contracts*) was issued (as a Conference draft for consultation) in September 1998. There were three principal objectives:

- Risk allocation: to promote a common understanding as to what risks are included in a standard PFI project
- Uniform approach: to allow consistency of approach and pricing across a range of similar projects
- Cost/time reduction: to reduce the time and costs of negotiation and confine negotiations to deal-specific issues.

The Guidance was published in final form in July 1999 and has proved to be the prime reference work for all PFI practitioners. However, the two principal procuring Departments (MoD and DoH) have since published their own separate guidance and standardisation agreements: (1) NHSE: *New Guidance and Standard Agreement* of December 1999; and (2) MoD: *Standardisation of (MoD) PFI Contracts* (Volumes 1 and 2) of September 2001 (Version 2).

7.4.2 Accounting treatment

Insufficient risk transfer under a PFI project could potentially both undermine VFM and leave the asset on the public sector balance sheet. But what test should apply to determine whether there has been sufficient transfer of risk? 'The accounting treatment of PFI transactions is a serious concern to the private sector, but there is no immediate prospect of agreement in the accounting professions. The Treasury therefore should issue guidance pending the deliberations of the Accounting Standards Board' (1st Bates Review). Several factors in accounting treatment were:

- Accounting standards: whilst the Government and Treasury are principally responsible for public sector accounting practice, the Accounting Standards Board (ASB) has responsibility for issuing accounting standards to the UK private or commercial sector
- Need for uniformity: by definition PFI transactions involve not only a public sector purchaser, but also a private sector provider of finance and services. Any difference in approach to the separate (public and private sector) sides of a PFI transaction might result, e.g., in failure to 'recognise' (i.e. include) assets on the balance sheet of either party. Thus, both the Treasury and ASB were keen to agree uniform accounting treatment for PFI transactions.

As concerns commercial accounting practice; the most relevant commercial accounting practice was:

- SSAP 21: Standard Statement of Accounting Practice 21, Accounting for Leases and Hire Purchase Contracts which was issued in August 1984 and implemented in 1987. SSAP 21 distinguishes between finance leases and operating leases by focusing on the substance of the transaction (as opposed to the form of the lease contract)
- FRS 5: Financial Reporting Standard 5; Reporting the Substance of Transactions which was issued in April 1994. The ASB's overriding objective was to ensure that the commercial substance of transactions is properly reported in an entity's financial statements. Though FRS 5 could potentially apply to PFI, it made no express reference to PFI transactions.

The Treasury issued interim guidance (*How to Account for PFI Transactions*) in September 1997 and the ASB responded with an Exposure Draft (a consultation paper) in early December 1997. There was much common ground between their separate approaches: the test to determine which party (public or private) should recognise the asset on its balance sheet must be based on an overall analysis of the risks. But should such risks be all risks inherent in the transaction or only those which relate to the property or asset? Since (under PFI) most service-related risks are transferred to the private sector operator, excluding some of those risks could potentially prevent certain projects achieving sufficient risk transfer (to satisfy the off balance sheet test). The ASB adopted a stricter test to determine which service elements should be assessed for that purpose; essentially these were confined to property-related risks.

Finally on 10 September 1998 the ASB issued its Application Note to FRS 5. In practice the ASB made few changes to the Exposure Draft. The Application Note further clarified issues of separability (i.e. exclusion of those risks relating only to services) and those variations in profits and losses that are relevant to the FRS 5 test.

In January 1999 the Treasury responded with a Technical Note (in the form of a consultation draft) to provide additional practical guidance 'to ensure the over-arching principles of the [ASB's Application Note] are consistently applied'. Following lengthy consultations between the Treasury and ASB, the Treasury issued its (revised) Technical Note No. 1, *How to Account for PFI Transactions* in July 1999. The principal outcomes of this process were as follows:

- Uniformity: more consistent accounting treatment (by public and private sectors) for future PFI transactions
- Tougher test: a more rigorous test (based on FRS 5) designed to exclude elements (of the PFI contract) relating solely to services
- Key risks: demand risk and residual value risk will have particular significance in future PFI projects.

The accounting debate between the Treasury and ASB highlights the distinction between VFM and the transfer of risk necessary to satisfy the FRS 5 test (VFM evaluation takes account of construction phase risks, pure service elements and operational efficiencies, each of which is ignored by the FRS 5 test).

7.5 Legal aspects of the management of risk under PFI projects

Some legal aspects of the management of risk under PFI projects could be described under the following headings:

- The contractual structure of PFI projects and how that may assist risk allocation
- The need to identify and allocate risks during the procurement process
- The techniques used to allocate and control risk under the project agreement
- Flow-down of risk under the consortium structure.

7.5.1 PFI contract structure and risk allocation

Successive governments had expressed concerns about cost and time overruns on publicly funded and managed capital projects. At an Adam Smith Institute Conference in March 1995 Sir George Young (then Financial

Secretary to the Treasury) analysed the feast-and-famine syndrome of public sector infrastructure projects. Common features were:

- Budget cuts: the tendency for departments to siphon funds from capital budget to meet any crisis of current expenditure, usually justified on the basis that capital spending is a commitment to the future
- Random nature: the random and unplanned nature of public sector capital projects, which lacked the necessary price discipline to enable government to prioritise spending
- Cost overruns: the tendency (once a public capital project had been approved) for budgets and timescales to overrun almost as a matter of routine. Examples frequently quoted at that time were the British nuclear programme, which had proved to be uneconomic only when the Government attempted to privatise the industry; and the Humber Bridge or 'Bridge to Nowhere' project.

In this context, the Private Finance Panel concluded that 'the public sector's record in, for example, the design and construction of capital projects is poor. Time and cost overruns are common. Part of the reason lies in the attitudes and culture of the public sector' (Private Finance Panel, *The PFI: Breaking New Ground*, November 1993). Typically the principal causes of cost and time overruns on public sector-managed projects are:

(1) 'Gold-plating': project design is usually handled by officials whose interest is to design to high technical standards rather than to strike a sensible balance between cost, return and risk. There is little incentive for public officials to take commercial risks.

(2) Variations: designs are often over-engineered, and post-contract changes to design and other specifications create a route for costs to rise. Contractors tend to bid low to win public works contracts, but then charge high prices for any contract variations. 'Capital projects are too often hijacked by technical specialists when they should be run by professional project managers. The scope this creates for expensive last-minute changes to specifications is enormous' (speech by Norman Lamont, then Chancellor of the Exchequer to a CBI Private Finance Conference in May 1993).

A comparison of the role of the public sector at different stages of the procurement process for conventional capital asset procurement and PFI procurement respectively (see Private Finance Panel, *Risk and Reward in PFI Contracts: Practical Guidance on the Sharing of Risk and Structuring of PFI Contracts*, May 1996) illustrates some key differences in risk profile:

Project stage	Conventional procurement	PFI procurement
Design	Procures detailed design	Specifies required services (no responsibility for design)
Specification	Seeks priced tenders to build	No responsibility for construction (private sector to manage/pay for construction)
Construction	Party to/manages construction contract Staged payments during construction	Pays for services (but only after completion of construction)
Operation of asset	Operates asset	Monitors service performance

'The optimum risk transfer package – that which maximises VFM – will vary widely from contract to contract and between different types of service. As a general rule, PFI schemes should always transfer to the supplier design, construction and operating risk (both cost and performance). Demand and other risks should be a matter of negotiation...' (POPB). We now consider how the structure of PFI contracts may impact on some of the principal project-related risks.

Design and construction risks

Design and construction risks should always be transferred under PFI projects. Certain features of the PFI structure should assist the transfer of design and construction risks to the PFI contractor (NB: sometimes a wholly-owned subsidiary company, but more often a special purpose vehicle owned by a consortium comprising several unrelated companies), as follows:

(1) Fixed price: the Unitary Charge will be agreed up-front, preventing the contractor from passing-on cost overruns. However, in practice the Government has shared cost overruns on some major IT PFI projects which ran into difficulties: recent examples include ICL Pathway (the failed Post Office/DSS benefits swipecard project, which led to public sector losses of nearly £1 billion and write-off of £150 million development costs by ICL) and NIRS2 (the National Insurance social security payments IT system where the Government declined to seek further compensation from Andersen Consulting).

(2) Design risk: the contract will specify a service requirement rather than describe the asset or property required to carry out the services. Thus, the contractor must take responsibility to develop a design that will satisfy the contracted service requirements.

(3) Deferred payment: services will not normally commence until the PFI asset is completed and commissioned and the Authority will make no payment until such time. Thus, the contractor will bear the risk of time

overruns, which may also involve the payment of liquidated damages (LDs).

(4) Operations: the contractor not only provides the asset, but also operates it during the contract period. The contractor will be responsible for the balance between design quality and service provision necessary to meet contract requirements throughout the contract period. Relevant features include life cycle planning for buildings and equipment to meet contracted requirements throughout the project duration.

In terms of design changes, a fundamental principle of PFI is that the contractor should assess how best to satisfy the contract specification, without interference from the authority. This involved a cultural change. Difficulties arose in several early-stage projects, where the authority sought to influence building design, which could have compromised the risk profile and led contractors to seek sharing of design risk. However, the opportunity for an authority to influence (but not to bear the risk of) design is now dealt with through a combination of:

- Consultation: the authority will normally seek an opportunity (without taking design risk) to comment upon outline design at an early stage
- Detailed specification: certain key design features (which are essential to the authority) can often be built-in to the detailed specification
- Change control: exceptionally, an authority may request changes during the design and construction phase (e.g. where necessary to reflect a change in legislation).

Commissioning and operational risks

The construction phase of a PFI project will normally be relatively short by comparison with the services or operational phase (typically of 20–30 years duration). Transfer to the contractor of most commissioning and operational risks is a key element of PFI projects. Relevant features of these risks are:

(1) Commissioning risks: before the underlying asset is accepted and the services can commence the asset must satisfy a variety of technical tests. This commissioning process is of great importance to authority, contractor and the project funder and will normally be dealt with by experts (technical advisers) engaged either by the individual parties or appointed on a common basis (typically for lower value projects).

(2) Design/operational balance: the balance of overall costs between construction/design and cost of hard/soft services will be a key strategic decision for the contractor. In theory, higher spend on design and quality of buildings or assets should reflect in a lower cost of services and vice versa. This impacts in particular on the interface between building sub-contractor and services sub-contractor: often an interface agreement will be required to allocate responsibility between those parties and to regulate hand-over issues in the lead-up to commissioning.

(3) Control of services: payments under the PFI project agreement (the Unitary Charge) will be adjusted (by use of a formula, known as the

Payment Mechanism) to reflect the quality of services provided by the contractor. It follows that to ensure the project payment stream remains intact, the contractor will wish to manage and control service performance. Two issues to consider under control of services are retained services and non-core services. In relation to retained services: 'Any approach to a PFI services contract should be one of the public sector retaining as little operational control and responsibility as possible' (Private Finance Panel, *Risk and Reward in PFI Contracts*, May 1996). In some areas the Government has decided, as a matter of policy, that provision of core services must be retained in the public sector. The most politically contentious examples are clinical services in the NHS and teaching in the state education sector. Difficulties may arise if the contractor is unable to exercise sufficient management control over staff carrying out the non-core services. In satisfaction of an election manifesto commitment, the Government is currently testing the practicality of retaining support service staff within the NHS. Pilot projects are in place on PFI hospital schemes at Wallsgrave, Havering and Stoke Mandeville. The outcome is likely to be a complex and somewhat artificial distinction between NHS managers (whose employment will transfer to the contractor under TUPE) and other NHS staff (who will opt-out of the TUPE transfer and remain employed by the NHS). It remains to be seen whether and (if so) how this will work in practice.

Demand risk

Private Finance Panel guidance (*Risk and Reward in PFI Contracts*, May 1996) concluded that 'when preparing an ITN the public sector should almost always at least offer the private sector bidders the opportunity to take volume/demand or usage risk within the unitary payment structure... The issue is very often not one of whether the private sector is willing to take it, but the extent to which volume or demand risk can be shared between the private and public sectors as part of the VFM equation'. Demand risk can arise in different circumstances, including:

(1) Financially free-standing projects: these are projects where costs must be recovered entirely through charges to the end (usually private sector) user. However, demand or volume risk can be difficult to predict, as the following examples illustrate:
  - Queen Elizabeth II Bridge: this toll bridge project was signed in 1993. The concession period is 20 years or such earlier date as the Contractor accumulates sufficient revenue to exceed the total debt outstanding. Project revenues have already reached that threshold
  - Royal Armouries Museum: failed to generate sufficient income from visitor numbers to meet the contractor's operating costs and service project debt. The contractor was obliged to re-negotiate the original deal: the authority took back responsibility for operating the Museum and left the contractor with only a limited range of support services
  - DCMF prisons: the Home Office has a duty to select the prisons to

which inmates are sent. At an early stage HMPS tested whether bidders for DCFM prisons were willing to accept any 'occupancy risk', but concluded that, because the private sector is unable to control the risk and would therefore charge a high price for it, risk transfer could not provide VFM.

(2) Third party revenues: often there is scope for a contractor to use PFI assets outside contracted PFI services, and thereby to generate third party income. Examples include community use of school buildings outside school hours, use of hospital beds/wards for private patients, use of equipment outside contracted requirements and so on. A variety of techniques can be used to transfer or share that risk: these include sharing income produced (perhaps subject to a guaranteed minimum payment by the contractor), sale of surplus land or assets at a designated value and so on.

Legal and regulatory risks

Legal and regulatory risks typically arise in the following context:

(1) Conditions precedent: the PFI project agreement and funding documents will often contain conditions precedent (CPs), which must be satisfied before the agreement and related funding become effective. CPs often include outstanding consents or approvals and frequently will include a requirement to obtain detailed planning permission for new PFI buildings. Normally an authority will first obtain outline planning permission, but on-going risks for the contractor include:

(2) Judicial review: the possibility of on-going challenge to the planning authority's decision or indeed the authority's initial decision to award the PFI contract. The most highly publicised recent example was Transport for London's unsuccessful challenge (in June 2001) to London Underground's selection of preferred bidders for its Tube infrastructure PPP contracts.

(3) Grouped schemes: the risk that detailed planning permission may be refused for some buildings or aspects only of a larger grouped PFI scheme. The contractor and project funder will wish to avoid or minimise exposure to risks which they are unable to control: typically such conditions must be satisfied as a pre-condition to project funding. If a key condition cannot be satisfied and the project is abandoned, the costs of detailed design may be at risk: these can be substantial and could either be borne wholly by one of the parties or shared.

(4) Future changes: the formula to allocate risk for future changes of law is now fairly well established. Both TTF and NHSE Guidance provide (in broad terms) that the contractor must bear the risk of any legislative change foreseeable at the date of contract signature. But in relation to future (and unforeseeable) legislative changes, the authority will bear sector-specific and PFI discriminatory changes, whilst the contractor bears the risks of general changes of law.

Technology or obsolescence risk

The nature of this risk will vary according to the time during the project period that the risk arises:

- Pre-commissioning: the risk of failure to satisfy technical specifications is essentially a design and construction risk. This is the period of maximum risk exposure on most large IT projects
- Post-commissioning: the risk that a capital asset becomes obsolescent or that new technology will provide a better solutions and/or service.

Relevant factors in relation to technology and obsolescence risks are:

- Specification: provided the service continues to meet the contracted specification, the contractor may be unconcerned that the underlying technology has become outdated
- Contract period: for IT projects or those with a high technology content, the project duration will often tend to be shorter than for accommodation-based projects. This reflects the authority's expectation that it will need to provide for renewal or upgrade of technology at appropriate intervals
- VFM: the authority may wish to incentivise the contractor to provide upgraded or refreshed technology at regular intervals, based on an assessment of VFM and/or the possibility of shared savings from use of new technology. This could be achieved through use of an external benchmark to reflect developments in IS/IT software or linkage to a residual value for the asset on expiry of the project term.

Various other categories of risk will also need to be allocated under most PFI projects: these include residual value risk, project financing risk and so on.

7.6 Preparation for the project and risk analysis

There is a special need for careful and robust planning during the early formulative stages of any PFI project. The reasons are self-evident:

- Duration: with a typical project life of between 20 and 30 years, it is essential to plan ahead (so far as possible) for unforeseen contingencies. This involves planning for and appraisal of risk
- Project cost: the procurement is of services (rather than assets), defined in output terms (i.e. specifying requirements rather than the means of achieving them) and over an extended period. In broad terms around two thirds of overall PFI project costs will normally relate to services rather than up-front capital expenditure
- Approvals: the project not only satisfies tests of affordability and VFM, but also receives official sign-off at various stages.

Treasury guidance to public authorities is designed to provide a consistent and methodical (stage-by-stage) approach towards preparation for and management of PPP projects. Over time, well-defined procedures have developed across most PPP sectors. Some common features are discussed as follows.

7.6.1 Options review

This answers the question 'Is the project both viable and suitable for private finance?' The following factors are relevant:

(1) Prior options: this analytical approach developed in relation to Next Steps (or Executive) Agencies, but is equally applicable to services or activities provided under PFI projects. 'Does this activity need to be done at all? If it does, could it be left entirely to the private sector? If not, what role do the public sector most usefully play: an enabler; a purchaser of private sector services; or, if those are not feasible and do not provide VFM, an asset owner and service provider in its own right?' (POPB).

(2) Universal testing: during the period from November 1994 to May 1997 the policy of universal testing applied (see above) to all PPP projects. The concept was designed to promote wider use of private finance, but gradually became discredited as unsuitable projects were tested and then abandoned. This led to lengthy time delays and bidder frustration, with an over-congested list of projects competing for funding approval. 'I am … announcing an end to universal testing for private finance potential. That has been a recipe for frustration and delay and works against the concept of prioritisation which we want to build into the process. Departments should not spend time and money trying to develop models for private finance where these will not work' (Geoffrey Robinson: announcing the 1st Bates Review on 8 May 1997).

(3) Suitable projects: capital projects suitable for private finance are those which provide VFM and where genuine transfer of risk is available without disproportionate cost. 'Privately financed projects offering the greatest potential for VFM gains are likely to [include] those with the most scope for transferring manageable risks to the private sector' (Green Book, 2nd edition, 1997). A recent example of projects tested for suitability of PFI and held not to provide VFM was the MoD Heavy Trucks programme: in March 2001 Baroness Symons (Defence Procurement Minister) announced that three major defence vehicles procurement projects (Future Fuel Vehicles, Future Cargo Vehicles and Future Wheeled Recovery Vehicles respectively), worth more than £1 billion in aggregate, would proceed under conventional funding rather than PFI.

7.6.2 Identification of risks

The Treasury approach to risk appraisal and management is set out in the Green Book. A first step is to identify those risks likely to affect the chosen project, as follows:

- Risk Analysis Matrix: the authority should develop a list of the 'various risks and uncertainties to which particular project options are exposed, together with an assessment of the likelihood of their occurring, and the impact on the outcome of the project' (Green Book)
- Major risks: next the authority should seek to identify the principal risks, based on information from earlier projects, official reports and other sources.

7.6.3 Risk quantification

Once the principal risks have been identified, the authority should seek to clarify and quantify those risks. 'The identification and costing of risks is likely to be one of the more problematic components of the public sector comparator (PSC). However, such costing is essential if an unbiased PSC is to be calculated' (Green Book, Annex D). Various techniques are used to quantify and value risks, as follows:

- Sensitivity analysis: is the most widely used technique and involves calculation of how changes to particular assumptions might affect net present values, total costs or other project outcomes
- Scenario planning: is used in relation to strategic decisions (such as planning for an investment programme or a large and complex project) and typically involves a comparison between several different scenarios
- Monte Carlo analysis: is used where the basic assumptions (on which an analytical model is normally constructed) remain uncertain. Monte Carlo links such assumptions to probabilities, which allows key outputs to be tested (NB: an example of this technique is used in Annex A to the Treasury's *Technical Note No. 1: How to Account for PFI Transactions* of June 1999)
- Decision rules/trees: decision rules involve setting artificial measures to assess the viability of competing options and then testing the options against those measures or rules. Decision trees are graphical representations used to analyse certain projects, typically those which involve a stage-by-stage process
- Discount rate: the Treasury recommend that adjustments to discount rates should not be used as an appraisal technique to reflect risk in central government.

7.6.4 Risk control

Various procedures and techniques have been used to control and minimise the impact of risk under PFI projects, as follows:

(1) Standardisation guidance: Taskforce guidance on *The Standardisation of PFI Contracts* was published in July 1999 (TTF Guidance). In our view TTF Guidance has been instrumental (and indeed is the single most important factor) in creating consistency of market practice and allocation of risks under PFI projects.

(2) Sector guidance: where PFI projects are launched in a new market sector, the central authorities will often commission guidance to assess whether such projects have special characteristics and, if necessary, to adapt TTF Guidance as appropriate. Standardisation guidance has been produced across a wide variety of market sectors including:

- The big procurers: *NHS Standardisation Guidance and Project Agreement* (NHSE, 1999); *Standardisation of Ministry of Defence (MoD) PFI Contracts* (Volumes 1 and 2) Issue 2 (MoD, 2001); and draft *Standardisation of PFI Contracts for Local Authorities* (DTLR and 4Ps, 2001)
- Sector-specific: includes *Student Accommodation in Higher Education* (DTLR, 4Ps and OGC, October 2000), *Social Housing PFI* (HEFC and DFES, 2001), and *Simplifying PFI in Schools* (DFES, PUK and OGC, September 2001)
- Pending guidance: further guidance is currently in preparation in relation to street lighting and leisure PPP projects.

(3) Pilot projects: similarly a pilot project will often be used to test whether a new sector is suitable for private finance and to assess any special features. Recent examples include the first wave social housing PFI schemes, leisure PFI (tested at Sefton MBC) and street lighting PFI schemes (at Walsall, Islington and Sunderland).

(4) Market sounding: before a project is formally advertised, the authority will often conduct an informal market sounding exercise to gauge the private sector's appetite and perhaps to test early reaction to any special features or planned transfer of risks. Market sounding has been conducted recently in relation to the following PFI projects:

- Investures: plans for NHS Estates' surplus properties to be developed through a PPP were adopted only after the original proposals for an auction failed to secure best VFM
- MoD barracks: plans to re-develop MoD's UK barracks either as a single £1 billion serviced-property PPP or as two separate PPPs (Project Connaught and Project Allenby respectively) were drawn-up in Spring 2001 following informal soundings with PFI institutions
- National Roads Telecomms project: was launched by the Highways Agency in Autumn 2000 only after extensive industry consultation
- DERA: MoD's original plan to privatise DERA was modified after objections from the US Defense Department.

(5) Due diligence: detailed early-stage planning and preparation (due diligence) should assist an authority to identify any project-specific features, obstacles and risks. The due diligence process should assist an authority to develop its output specification, risk matrix and procurement documents (typically both the Information Memorandum, which provides outline project details at pre-qualification stage, and Invitation to Negotiate (ITN)).

(6) Risk transfer: the authority may seek to transfer risk to third parties through a combination of insurance and contract terms. We consider these elements in greater detail below.

(7) Other techniques: other techniques recommended by the Treasury to control and minimise exposure to project risks include the use of flexible design (e.g. hospital buildings adaptable to the changing requirements of acute healthcare and advances in medical treatment), the use of standard designs (i.e. design tested and approved on earlier projects) and more flexible project duration (e.g. use of a shorter contract term for IT projects, where technology may become outdated) or break options (allowing an authority to take back a service at one or more future dates: as with the latest DCMF 5 prison projects at Ashford and Peterborough respectively).

7.6.5 Approval process

An early criticism of PFI was lack of central control over PFI spending. The decentralised nature of PFI and absence of any system to monitor forward PFI commitments had implications for Treasury control of public expenditure (Treasury Select Committee: 6th Report on PFI of April 1996). Whilst concerns remain about the capacity of PFI/PPP projects to silt-up future spending allocations, the process is now under better control, in terms of:

(1) Spending priorities: at a high level government spending priorities will be set out in the Comprehensive Spending Review (covering the three-year economic cycle) and longer-term investment plans, such as *Transport 2010: the Ten Year Plan* (July 2000), the NHS Plan of July 2000, the *New Deal for Schools* (a £7.8 billion three-year capital spending programme announced in September 2000) and the Strategic Defence Review. These long-term plans flow down to departmental PES provisions through the annual Budget statement.

(2) Project approvals: departmental spending (both conventional and PFI) is based on annual PES provisions. Local government spending on PFI is controlled by the sponsor department and cross-departmental Projects Review Group (NB: established following the 1st Bates Review of June 1997). PRG approves local government PPP projects in tranches every three to four months or so.

(3) Project-specific approvals: the hierarchy of approvals necessary for most PFI projects is designed to test the robustness of early-stage project planning and in particular VFM, affordability and related transfer of risks. The process is now supplemented by OGC's Gateway Review process and (for local government) by similar reviews to be conducted by the new Strategic Partnering Taskforce.

7.7 Risk transfer and the procurement process

'Competition is central to public procurement, both as a means of securing VFM and to help guard against corruption or the appearance of it' (POPB). The Treasury and other executive bodies involved in the development of PFI have used the competitive public procurement process to refine risk allocation and test market reaction to the transfer of new risks under PFI projects. There are a number of background features to this process.

7.7.1 EU procurement rules

The procurement by public bodies of goods, works and services is governed both by general EU Treaty provisions (relating to freedom of movement) and directives which regulate procedures for the award of major government contracts. Those directives have been re-enacted in English law by a series of regulations, which include the Public Services Contracts Regulations 1993. The main features of the regulations are:

(1) Negotiated procedure: 'the use of the negotiated procedure for all PFI projects is strongly recommended ... Although there are restrictions on the use of this procedure, they are unlikely to prevent its use for most PFI projects' (POPB). The rationale is set out in Regulation 10 (2) (b) (of the Services Regulations) which provides for use of the negotiated procedure 'exceptionally, when the nature of the services being provided, or the risks attaching thereto, are such as not to permit prior overall pricing'. Increased standardisation of contract terms and risk profile (by commoditising risk transfer) however, threatens to undermine the principal justification for use of the negotiated procedure for UK PFI projects. In addition, the EU Commission has long objected to the systematic use of the negotiated procedure for PFI projects in the UK. Formal challenges to use of the procedure have been submitted in relation to the Pimlico school (since abandoned) and Ipswich PFI projects. Following from this, a new consolidated EU public procurement directive threatens to undermine use of the negotiated procedure for PFI projects.
(2) Single tenders: early discussion about the use of single tenders for PFI projects has not in practice resulted in their widespread use. 'As with any other public procurement, the normal presumption should be that PFI projects will be competitively tendered' (POPB).
(3) Amended specifications: sometimes an authority may need to amend the project specification or risk profile during procurement. This could undermine the integrity of the procurement and require that the PFI process is unscrambled and started afresh. Difficult issues arose in relation to PFI projects that were subject to lengthy delays (e.g. several first wave hospital PFI schemes) and to those projects where a material change had occurred (e.g. the Home Office HQ PFI project).

7.7.2 Procurement guidance

Various procurement procedures have been developed to assist public sector managers to handle sector-specific projects. These include POISE (i.e. 'Procurement of Information Solutions Effectively'), developed by NHS Supplies for NHS IM&T projects, and OGC (Office of Government Commerce) guidance across a range of topics, including construction procurement and e-Government.

7.7.3 Risk allocation

How best to deal with risk allocation and transfer during the procurement process? Various techniques are used, including:

- Market sounding: the market sounding process can be used to test the private sector's reaction to new or unusual risks (see above)
- Refinement of risks: where the project scope or risk profile remains to be finalised, bidders may be asked to bid against an indicative risk profile and invited to submit alternative bids on other bases. A pre-ITN or ISOP (i.e. Invitation to Submit Outline Particulars) stage may be introduced into the procurement process: this is often used to test bidder ideas on large, one-off projects or to cut-down a long list of pre-qualified bidders (prior to ITN stage)
- Risk Register: the ITN should 'tell bidders clearly what risks they would be expected to take on and what risks are negotiable'. The ITN will often incorporate a risk register listing individual project risks and either making provisional allocations of risks between the parties or seeking bidder proposals on risk allocation.

7.7.4 Risk transfer

The objective of risk transfer is to achieve a sensible balance between transfer of risks and VFM, based on the principle that risks should be transferred to the party best able to manage them. It is expected that the risk transfer strategy will deal with the following factors:

- Market practice: bidders will normally expect risk allocation to reflect positions achieved on earlier projects in the same PPP market sector. Risk allocation which is out of line with earlier practice will be regarded as 'off-market' and is likely to be resisted on that basis
- New risks: an authority may wish to test market appetite to manage new risks. Typically the authority will seek bidder comments on the proposal combined with a risk-specific cost. That cost must be assessed for VFM. Recent examples include volume risk under leisure PFI projects and risk of latent defects in existing buildings under PPP refurbishment schemes
- Funder involvement: funders will often take the lead, through the funder due diligence process, in commenting on any proposal to transfer a new or unusual risk.

7.8 Risk transfer and the project agreement

The Green Book states that 'risks to one party to a transaction can be reduced by transferring them, at a cost, to another party, either by the terms of contract between suppliers and customers, or by insurance' (Green Book, 2nd edition, 1997). Risk transfer under PPP projects is normally achieved through a combination of the contract terms and insurance.

7.8.1 Project documents

Risk allocation agreed during the procurement process will need to be set out clearly and expressly in the terms of the various project documents. The following factors should be considered:

(1) TTF Guidance: whilst the allocation of principal risks between authority and contractor is largely set out in TTF Guidance and related sector-specific standardisation guidance, the flow-down of risks between the various consortium members is based largely on PFI market practice.

(2) Performance Mechanism: 'The Payment Mechanism is at the heart of the contract, as it puts into financial effect the allocation of risk and responsibility between the authority and the contractor. It determines the payments the authority makes to the contractor and establishes the incentives for the contractor to deliver exactly the service required in a manner that gives VFM' (TTF Guidance). Various Payment Mechanism structures have been used across different market sectors, depending whether the project is based on availability, services and/or usage or a combination of the same. Payments under a PFI project are made in the form of a Unitary Charge for the services, which is not made up of single independent elements relating to availability of performance. The Unitary Charge is paid only to the extent that services are available and the Payment Mechanism enables deductions to be made for sub-standard performance.

(3) Flow-down of risks: the flow-down of project risks between the consortium members is now well established in most PFI market sectors. The overriding requirement of authority and project funder will be to ensure that only a very limited range of risks are left with the contractor (consortium SPV). Thus, construction and design risks will almost exclusively be borne by the building sub-contractor and conversely the operational risks will be borne by the services sub-contractor. However, there are grey areas of uncertainty or shared responsibility and these are often dealt with by a separate interface agreement between the sub-contractors. Similarly, the services sub-contractor may act simply as manager of service provision where the day-to-day services are provided by other sub-contractors: here the allocation of risks between the two parties will need to be dealt with under the relevant sub-sub-contract.

(4) Role of the project funder: over time the project funder has been crucial in developing an agreed allocation of risks between authority and contractor and between the consortium parties. Risks which funders are not willing to underwrite are often termed 'unbankable' and funders will normally resist any attempt by an authority to alter an

established risk allocation (which will be treated as off market). Funders, however, have gradually become more comfortable with the contractual structure and risk profile of PPP projects, but continue to carry out stringent due diligence, particularly in relation to complex, one-off projects and pathfinders in new PPP sectors. Although PFI/PPP projects have traditionally been funded through bank loans, bond finance is becoming increasingly popular (particularly for larger value PPP projects).

7.8.2 Insurance

According to the Green Book 'the Government, in general, is a self-insurer. This is mainly because the Government can spread risk widely, and it does not need to insure simply to protect corporate financial viability. On the other hand, insurance can be cost-effective in some circumstances, where, for example, it leads to more careful examination of risks and requirements, by the insurer, of measures to reduce risks'. A need for insurance is emphasised in the TTF Guidance: 'Under PFI a greater range of risks is transferred to the private sector. The financing arrangements behind most contracts and the need to ensure continuity of service means that self-insurance for the full range of risks is not a practical option for the contractor. Indeed insurance requirements of the senior lenders will usually be extensive'.

As the PFI market matures, the insurance industry has developed insurance packages covering the principal risks under a PFI project. These will normally include construction risks, third party liability claims, material damage claims and employers' liability insurance. In addition, the contractor may seek to insure wider risks designed to protect the project cash flow: these include insurance against latent defects, business interruption, strikes and industrial action and advance loss of profit. PFI insurance packages typically involve waiver (by the insurer) of subrogation rights against co-insured members of the consortium and will cover the client authority as co-insured.

If a risk covered by a contractually-required insurance was previously insurable but becomes uninsurable, the consequences will depend upon the type of risk involved and whether either party was responsible for the uninsurability. Insured risks potentially affected include terrorism, vandalism and employers' liability. Terrorism cover has become widely 'unavailable' following the September 2001 attacks on New York. However, there are precedents for the government acting as an insurer of last resort (e.g. following the IRA Docklands bomb and more recently support for the UK airline industry following the September 2001 terrorist attacks). Costs of arson damage at schools through vandalism have increased dramatically over the last two to three years while employers' liability has become more difficult to obtain and substantially more expensive for building contractors engaged on dangerous activities (e.g. scaffolding and demolition).

7.9 Conclusions

Economic and political developments as well as trends in the PPP market are likely to have an impact on risk transfer and management in PFI projects in the future.

As concerns risk assessment, the perception and evaluation of risk in PFI projects will change in response to market developments. The recent debate on re-financing gains (available under many early-stage PFI projects) highlights how funder perception of risks under PFI projects has changed in recent years. For example, the Government's involvement in the railway administration of Railtrack may lead the market to reassess the risk that government may abandon failed projects. However, Ministers have sought to draw a distinction between Railtrack privatisation (which involved no direct contractual commitments by the Government) and PFI projects (based on the provision of services under a long-term contract). In relation to risk valuation Treasury officials are working on important changes to the Green Book (*Appraisal and Evaluation in Central Government*) designed to encourage wider use of private finance. The Treasury is concerned that current public accounting standards (reflected in the public sector comparator test) discriminate against PPPs: projected changes include reassessing the impact of delay in completion and cost overruns, as well as quality of service and design, under public sector managed projects.

As concerns insurance cover, the ability of contractors to control project risks through insurance may be affected both by external events and price pressures in the insurance market. For example, following the September terrorist attacks in New York, cover against risks of terrorism is likely to be unavailable (uninsurable) for the foreseeable future.

In late October 2001 the Home Office published estimated figures for arson attacks on public buildings. These demonstrate that the cost of arson damage at schools has increased from £31 million in 1995 to £65 million last year and an estimated £87 million this year. Obviously, this will create price pressure. Whilst it is difficult to develop accurate information about arsonists and their motives, Zurich Municipal (the principal insurer of schools) has sent guidance (*The Design and Protection of New School Buildings and Sites*) to all LEAs and PFI contractors highlighting the opportunity presented by the PFI school building programme to design-out vulnerability to arson.

In addition, premiums in the construction sector are expected to increase in response to poor safety records and higher compensation awards for building accidents. Rates for employers liability insurance will increase, particularly in high-risk construction trades such as demolition and scaffolding.

New risks will invariably emerge as PFI projects are extended into new sectors. However, as risk management becomes more sophisticated, the PFI market may be willing (at a price) to contemplate the transfer of previously unwelcome risks. New risks will relate to new sectors, new performance measures and new services.

An example of new risk under new sectors is volume risk in leisure PFI projects (e.g. Barclays PFI's willingness to fund Sefton MBC's recent pathfinder project may lead other funders to 'take the plunge!'). Similarly, on the first wave pathfinder schemes (currently 'on hold' as DTLR considers Tender submissions in the context of available PFI Credits) a key practical difficulty has been to produce intrusive surveys of the state and condition of properties that are sufficiently detailed to underpin the transfer of latent defect risk to the contractor.

In relation to new performance measures, new performance targets are likely to focus increasingly on outcomes. In the past contractors and funders have tended to resist outcome-based contracts, since a contractor's ability to achieve certain outcomes will often depend on factors outside the contractor's control or ability to influence. Prospective candidates include street lighting for reduction in crime; DCMF prisons for reduction in rates of re-offence; schools PFI for achievement of improved educational attainment standards; and hospitals PFI to achieve Government targets to reduce waiting times, improve recovery rates and speed throughput of patients.

As the barriers between public and private services become blurred, so the private sector is likely to move into new areas of service provision (new services), some of which were previously regarded as 'no-go' areas. Recent research and strategic studies (e.g. the IPPR Commission report on PPPs) suggest that such artificial distinctions should be eroded or removed. Recent examples of new services include:

- Civilian jailers: support services staff handling custody and suspect processing are to be employed by Equion (formerly Laing-Hyder) under the recently-awarded £120 million 25-year DBFO contract to provide operational buildings for the Metropolitan Police. This offers the potential to release police officers (currently covering civilian shortages) to return to front-line operational duties
- Sponsored reserves: support staff under certain defence PFI contracts who are trained as Territorial Army reservists and will switch to that status when involved in front-line activities
- Warship support: MoD plans to involve private companies in waterfront work, involving the routine support and maintenance of warships in port (NB: previously private sector work was confined to longer-term refit and repairs).

Changes in the public procurement process may have an impact on risk allocation and management. Some important developments relate to the new EU directives and Gateway process. The EU Commission's proposals to consolidate the Public Procurement Directive would prevent public authorities from negotiating with a preferred bidder on an exclusive basis, and instead seek a position where all bidders negotiate until contract award. Clearly this could have a material impact on the PFI market in the UK. Also, launched in February 2001, the Gateway Review process is regarded as the most significant innovation in OGC's reform programme for public procurement. It is based on private sector management disciplines. This should allow Government officials to become more expert in managing major projects and will provide real-time information about progress on current projects.

The IPPR Commission on PPPs recommended that authorities should experiment with a wider range of service providers under PFI projects, including other public authorities, not-for-profit organisations and joint ventures. The use of such new structures may impact on risk allocation and risk profile between authority and contractor. Such new structures include the increased use of companies limited by guarantee (CLGs) and other forms of not-for-profit vehicle for PPP projects (such as NATS PPP and the proposed CLG for Railtrack) where profits from the venture will be re-invested in the project rather than distributed to consortium shareholders. Another new structure is joint ventures. Local authorities are currently experimenting with revised structures for PPPs, involving an authority itself taking a minority shareholding interest in the consortium SPV (e.g. the Kirklees PFI schools and Walsall LEA intervention).

A recent report by Rothschild Bank, which concluded that DBO projects using conventional finance could provide a better VFM solution than PFI, illustrates how the PFI market has come to accept risk allocation and transfer as normal practice in public sector capital projects.

References

Accounting Standards Board (1994) *FRS 5: Reporting the Substance of Transactions.* April. ASB, London.

Accounting Standards Board (1997) *Application Note to FRS 5 Exposure Draft* (for consultation). December. ASB, London.

Accounting Standards Board (1998) *Application Note to FRS 5.* September. ASB, London.

Bates M. (1997) *1st Bates Review.* June. DETR, London.

DfES (2000) *New Deal for Schools.* September. DfES, London.

DfES, PUK and OGC (2001) *Simplifying PFI in Schools* (draft guidance). December. DfES, London.

DTLR (2000) *Transport 2010: the Ten Year Plan.* July. DTLR, London.

DTLR 4Ps and OGC (2000) *Student Accommodation in Higher Education.* October. DTLR, London.

DTLR and 4Ps (2001) *Standardisation of PFI Contracts for Local Authorities* (draft for consultation). August. DTLR, London.

HEFC and DfES (2001) *Social Housing PFI.* September. HEFC, London.

IPPR Commission (2001) *Building Better Partnerships.* June. IPPR, London.

MoD (2001) *Standardisation of Ministry of Defence (MoD) PFI Contracts* (Volumes 1 and 2). September. MoD, London.

National Audit Office (1999) *Examining the Value for Money of Deals under the Private Finance Initiative.* August. HMSO, London.

NHS Executive (1999) *NHS Standardisation Guidance and Project Agreement.* December. HMSO, London.

NHS Executive (2000) *NHS Plan.* July. HMSO, London.

Private Finance Panel (1993a) *The PFI: Breaking New Ground* (revised edition). November. HMSO, London.

Private Finance Panel (1993b) *The Private Finance Initiative: Breaking New Ground.* November. HMSO, London.

Private Finance Panel (1994) *Guidance on Competition.* March. HMSO, London.

Private Finance Panel (1995) *Private Opportunity, Public Benefit – Progressing the Private Finance Initiative* (the 'PFI Handbook' or 'POPB'). November. HMSO, London.

Private Finance Panel (1996a) *Basic Contractual Terms.* October. HMSO, London.

Private Finance Panel (1996b) *PFI Guidelines for Smoothing the Procurement Process.* April. HMSO, London.

Private Finance Panel (1996c) *Risk and Reward in PFI Contracts: Practical Guidance.* HMSO, London.

Private Finance Panel (1997) *Further Contractual Issues.* January. HMSO, London.

Private Finance Panel (1997) *Sharing of Risk and Structuring of PFI Contracts.* May. HMSO, London.

Treasury (1991a) *Economic Appraisal in Central Government: a Technical Guide for Government Departments* (the 'Green Book'). April. HMSO, London.

Treasury (1991b) *Private Finance and Approval of Capital Projects.* HMSO, London.

Treasury (1992a) *Guidance for Departments on Private Finance.* December. HMSO, London.

Treasury (1992b) *New Release.* November. HMSO, London.

Treasury (1997a) *Economic Appraisal in Central Government: a Technical Guide for Government Departments* (the 'Green Book': 1997 update). HMSO, London.

Treasury (1997b) *How to Account for PFI Transactions* (draft guidance). September. HMSO, London.

Treasury (1999) (revised) *Technical Note No. 1: How to Account for PFI Transactions.* HMSO, London.

Treasury Taskforce (1998) *Standardisation of PFI Contracts* (conference draft for consultation). September. HMSO, London.

Treasury Taskforce (1999) *Standardisation of PFI Contract Terms* (final version). July. HMSO, London.

Zurich Municipal (2001) *The Design and Protection of New School Buildings and Sites.* November. Zurich Municipal Switzerland.

8 Applications of risk management strategies in public-private partnership procurement

Evelyn McDowall

8.1 Introduction

This chapter offers a strategic overview of the progress of public-private partnership (PPP) projects which incorporate facilities management (FM) services as a major component of the scope from the technical adviser's perspective. It assesses how risk is managed throughout the life of the project. In addition, the chapter highlights the differences in approaches to risk management between hard and soft FM services.

While the long-term success of PPP contracts cannot be effectively established until some projects have run their course, the chapter reviews whether there is evidence to suggest that the introduction of policies and procedures to projects currently in development could reduce risk and ultimately reduce the cost of projects.

8.2 Developments in risk management in PPP

To establish how to manage risk effectively, it is important to identify the risks in a PPP project, their sources, who controls them and at what stages and how management actions can be enforced to control or minimise risk events and consequences throughout the life of a project. The starting point is the development of a risk allocation matrix which identifies who is likely to be responsible for the main risk categories throughout the life of the PPP project, usually some 25 to 30 years. An example is shown in Table 8.1.

Although the risk allocation between the parties may be different for every PPP project, there requires to be substantial risk transfer, particularly in the areas of design, construction and operations for a PPP procurement route to be considered. The risk profile will be further developed for each risk category to establish a comprehensive risk allocation matrix for all aspects of the project. The further development of the risk allocation matrix is undertaken when the scope and constraints of the project are known.

The following sections map the treatment of risk at key stages of a typical

Table 8.1 *Risk allocation matrix.*

Risk category	Client responsibility	Contractor responsibility	Shared responsibility
Demand	✓		
Design		✓	
Construction		✓	
Operational		✓	
Residual value		✓	
Third party income			✓

PPP project, referring to published guidance where relevant, to establish the framework in which the key players make decisions about possible risks.

8.3 Option appraisal

Risks are inherent in the project, regardless of the preferred procurement route from the outset as the parameters of the scope of the project are established. There is a wide range of risks that must be considered, but physical risks tend to predominate, e.g. type of building, number of sites, building complexity. The proposed location may carry inherent risks, e.g. land contamination, environmental sensitivity, land values, existing buildings. The treatment of risk at this stage of the process is governed by 'the Green Book' (Treasury, 1998) and is supplemented by guidance from sector-specific funding bodies on the scope and depth of option appraisal required to support an Outline Business Case (e.g. Treasury, 1995; Treasury Taskforce, 1999a; 4Ps, 2000c).

Consideration of the PPP procurement route introduces the requirement for the client to review its service requirements as well as the physical environment. The impact of operational risks as well as traditional construction risks have to be considered, bringing a much longer term perspective to the project than the traditional approach. FM and life cycle costs and risks are principal elements of the investment appraisal.

In many PPP projects there are incentives to have a multi-site contract, especially where single sites are relatively small and would not otherwise lend themselves to the PPP procurement process. There are also financial incentives where procurement costs are reduced as the number of sites increases. Multi-site contracts can afford the operator economies of scale and opportunities to spread risks within the portfolio, especially where there are similarities within the portfolio, e.g. building age, size, user population. These risks have to be offset against greater difficulties in managing multi-site environments and delivering services to a diverse range of users and sites.

Many of these parameters, once chosen, are difficult to change, although PPP by its nature affords bidders the opportunity to submit variant bids that challenge the status quo where they believe alternative parameters would offer better value for money and an alternative risk profile.

Although in principle the operational risks will require to be transferred to the PPP operator, the detailed risk allocation matrix developed for the project may identify areas where it is not in the interests of the client to transfer the risk. An example of an operational FM risk allocation matrix is shown in Table 8.2.

Although the risk allocation matrix shows the relative responsibilities of each party, this position is not fixed and can change during the procurement process as the negotiation between the parties develops. The most significant areas of flexibility between the parties have been seen in the following areas.

8.3.1 Demand risk

In PPP projects the client remains responsible for some risks throughout the contract period. Examples include demographic change and demand risk. It is now well-established practice that the client is in the best position to predict the demographic profile of the users of PPP services, based on its knowledge of the core functions of the services it runs. The private sector only takes risk where best value can be demonstrated.

Demand risk is becoming shared or transferred to the private sector in some services where there are incentives for the service provider to achieve greater risk transfer. An example is in catering services where the private sector is being encouraged to take demand risk on meal uptake, especially in areas such as free school meals. The private sector has been slow to respond to this aspect of PPP because demand for customer-driven services is more volatile than building-related services such as maintenance. The client is not always able to provide sufficient documented evidence of the historical pattern of service delivery to provide a benchmark for the service improvement required of the service provider. However, PPP can offer the right environment for effective risk transfer because the service provider controls the design of the restaurant, kitchen and service delivery methodology.

Core business demand risk has had even slower uptake than FM service

Table 8.2 *Examples of operational FM risk allocation.*

Operational risk	Client	Contractor	Shared
Changes to core business	✓		
Service demand	✓		
Non-availability of service		✓	
Underperformance		✓	
Cost changes		✓	
Health and safety		✓	
Default of service operators		✓	
Theft and vandalism			✓
Energy			✓
Legislative change			✓

demand risk. Clients are often reticent to offer aspects of what they perceive to be core business to the private sector. Such decisions are seen to be highly politically sensitive. In some areas, however, there is progress, with Information and Communication Technology (ICT) service providers offering to contribute to learning gain in schools where their services are implemented. If demand risk is transferred, the contract has to include robust and objective methods of measurement to ensure that changes in demand and service uptake can be measured and rewarded.

8.3.2 Energy

In early PPP projects, the emphasis has been on the operator taking responsibility for energy consumption and tariff levels. More recent projects have shown a move towards a split in responsibility for energy with the operator taking consumption risk and the client taking tariff risk. Possible options are shown in Table 8.3.

8.3.3 Theft and vandalism

Although clients may prefer to transfer all possible risks to the operator, it may not be cost effective to do so. One example relates to theft and vandalism, where it has been found that the client may have far greater control and incentive to manage the risk of theft and vandalism during the day when they are occupying the buildings. Meanwhile the operator is in a better position to

Table 8.3 *Options for energy risk allocation.*

Option	Consumption risk	Tariff risk	Application
1	Operator	Operator	New build scenarios where operator has control over design and energy efficiency decisions Operator has greater or equivalent purchasing power than the client – particularly for single buildings, e.g. headquarters
2	Operator	Client	New-build multi-site scenarios where operator has control over design and energy efficiency decisions Client has greater purchasing power than the private sector – particularly relevant in local authority schemes
3	Shared for agreed period and benchmarked; thereafter operator	Client	Refurbishment in multi-site scenarios where operator's energy efficiencies are controlled by the design and layout of existing buildings. Client has limited information about the energy peformance of the existing portfolio to enable the operator to quantify risk

Table 8.4 *Example of risk allocation for theft and vandalism.*

Theft and vandalism in schools	Responsibility	Management systems	Physical systems
During school hours	Client	School policy Disciplinary procedures Patrols by lunchtime monitors Teacher supervision during class-time	
Outside school hours	Operator	Patrols of buildings Call out procedures	Alarms CCTV Locks and access systems

manage the risk out-of-hours. Table 8.4 demonstrates the management systems available to each party in support of the decision to share responsibility for theft and vandalism over a typical day in a school scenario.

8.4 Outline business case

The Outline Business Case (OBC) builds on information developed for the option appraisal. It uses the options identified in the option appraisal to establish the most cost effective procurement route – PPP or traditional procurement. Guidance on development of the OBC stage of a PFI project has become generally accepted with industry recognised risk allocation matrices and methodologies for quantifying risk (Treasury Taskforce, 1998a; 4Ps, 2000c; McDowall, 2000b). Types of risk include design risks, construction risks, availability and performance risks, operating and life cycle replacement risks, demand/funding risks and residual value risks.

Financial modelling of the parameters of the project will determine a projected 'unitary charge' that the client will pay over the contract period for the services and the Net Present Value (NPV) of the services to compare the PPP and PSC options. The unitary charge and NPV are used by the client to determine the affordability limits of the project. The quality of the base data has a significant impact on the robustness of the projected figures. Factors which impact the risk of cost increases are the sources of data and whether they reflect market-tested rates, accuracy of demographic projections, and accuracy of space requirement projections. A reference project that proves the technical feasibility of a proposed option reduces private sector risk.

In addition, the public sector client has to demonstrate value for money. Arthur Andersen *et al.* (2000) identified a number of value for money drivers of PPP projects. Risk transfer scored the highest amongst the issues identified. The contractor's ability to develop procedures and methods to manage the risk must accompany the transfer of risk by the client to the private sector. Otherwise the trade off between value for money and risk transfer cannot be optimised.

At the OBC stage the risk allocation matrix is expanded, building on the allocation of risks identified during option appraisal to examine the impact should risks become reality, the probability of occurrence and the value of risks should they occur. It is current practice for a risk matrix to be developed for the Public Sector Comparator (PSC), identifying those risks which would normally be retained by the client had a traditional procurement route been chosen but which would transfer under PPP. The development of a project-specific risk matrix should be undertaken with regard to the government guidance applicable at the time and to the specific requirements of funding bodies. An example of the development of a risk matrix is shown in Table 8.5.

The risk matrix represents a strategic view of risk allocation and cost. The bidders for the PPP project will develop more detailed risk matrices, reviewing risks on a service-by-service basis, identifying their most significant areas of exposure to risk e.g. recruitment and retention of staff, industrial action, staff cost increases, obsolescence of equipment, potential cost increases above inflation. Their analysis of the same risks may differ from the client's, as they take a commercial view based on the information available to them from the project under review and the market as a whole.

8.5 Procurement

The bidding process in PPP projects is used to establish an affordable and value for money solution to the client's brief for long-term service delivery. There are many factors that can influence the outcome of the bidding process including the management of the process itself, the affordability limits set by the client, synergies between design and FM, and the strength of contract documentation. The following sections examine the contribution that these issues may make to the project risk profile and how they may be managed.

8.5.1 Mitigating risk through the bidding process

The bidding process for PPP can have a significant impact on the risk profile and affordability of a project. A highly competitive bidding environment can ensure that the client achieves value for money but there are dangers. Bidders may take commercial decisions to reduce their costs artificially to make them more attractive to the client in the hope that they will be able to reinstate some costs if they are chosen as the preferred bidder. They may also use the project to buy into a new market sector. The OBC can be used as the benchmark to test bidders' assumptions and highlight where bidders may be artificially reducing costs. Although the client will be seeking the best value for money bid, they have to be confident that they are not introducing higher risks of operator failure in the future.

Table 8.5 *Example of risk calculation matrix.*

Operational risk/cost changes	Definition	Responsibility	Probability ranking (1 = remote chance; 6 = virtually certain)	Probability (%)	Possible value	Basis of possible value calculation	Most likely value (probability × possible value)
Replacement costs	Risk that life cycle costs change other than in accordance with inflation	Operator	3	50	£140 000	20% per annum on average annual cost of £700 000	£70 000
Maintenance costs (hard services)	Risk that preventative and reactive costs change other than in accordance with inflation	Operator	3	50	£170 000	20% per annum on average annual cost of £850 000	£85 000
Soft service costs	Risk that operating costs change other than in accordance with inflation	Operator	3	50	£150 000	10% increase per annum in operating costs of £1.5m	£75 000

As identified in the option appraisal, quality of information will contribute to the risk profile. During the procurement phase bidders will be provided with access to the client's database and assumptions about the scope of the project. Disclosure is important because it allows bidders to make their own judgment about the quality of information provided and thereafter commission additional information e.g. physical surveys. It is also essential that the client divulges accurate information on transferring staff so that the operator can meet its statutory obligations under the Transfer of Undertakings (Protection of Employment) Regulations (TUPE) where it is deemed to apply (Treasury Taskforce, 1998b). Failure by the client to disclose information can lead to bidders commencing litigation to recover costs.

Many PPP projects, by their nature, require consultation processes to be carried out as part of statutory requirements, e.g. planning applications, school closures. The outcome of such processes can have an impact on the procurement process by lengthening the procurement timetable. Clients have to establish the risk of not achieving the required results from such statutory requirements and the impact on the procurement process if they fail. To reduce the risk to bidders and maximise the incentive for the private sector to participate in bidding, some clients have designed the procurement timetable to ensure that statutory consultation requirements are completed before the bidding process begins.

In some circumstances the timetable for statutory requirements has to run in parallel with bidding. Bidders have to make their own judgments about the likelihood of delays or rejection of proposals. They may, in addition, carry out their own consultation procedures where variant bids are being considered, increasing the risk of delay to contract signature.

Delays in the timetable can have a considerable impact on procurement costs and risks. In addition to the consultation requirements discussed above, typical events that prolong the procurement process are land title issues and planning. Early identification of these issues and action by the client can minimise the risk of delay and increased costs. The proposed construction timetable to meet client delivery dates also has to be borne in mind, e.g. start of school year, so that the operator is not under undue pressure to deliver to unreasonable timescales.

The introduction of a 'Best and Final Offer' (BAFO) can significantly contribute to the short, and long-term risk profile. During this stage of the negotiation a shortlist of two to three bidders submit revised proposals based on requirements set out by the client to clarify all commercial aspects of the bids and confirm affordability. It can minimise potential delays at the preferred bidder negotiation stage by identifying all commercial points and testing them in market conditions.

8.5.2 Affordability

The bidders may succumb to pressure to meet clients' affordability limits that may not be achievable. The client must have confidence in its affordability threshold to ensure that bidders are not artificially lowering costs at the expense of risk in the future. The procurement process must be designed so that problems with affordability are adequately explored and that the competitive bidding environment exposes those factors that are contributing to the affordability issues. Bidders may offer variant bids which are perceived to offer better value (and affordable) solutions. In such cases the scope of the original project should be used as a benchmark to ensure that there is a clear understanding of what is being offered as an alternative to the original solution, e.g. lower space standards, a more lenient payment mechanism, alternative sites, new-build v. refurbishment.

8.5.3 Developing the contract documentation

While early PFI deals took a long time to reach contract close because of the extent of legal precedence, the publication of guidance for standardising PFI contract terms has brought substantial clarity to the transfer of risk through typical PFI contracts. Guidance has been provided on general issues such as contract termination, employment issues and step-in rights (Treasury Taskforce, 1999b). The contract documentation still requires significant technical, commercial and legal input because of the scope and complexity of deals, incorporating a diverse range of subject areas from employment to property, finance and energy management to establish a commercially acceptable but comprehensive deal.

The development of the specification is a tool whereby the client sets out the scope of service and for every aspect of the service to be delivered. The risk transfer requirements are inherent in the scope, and performance standards set out in the specification. The style of the output specification document can be used as a way of ensuring that the service provider assumes the risk to be transferred. Types of risk include size of accommodation, methods of service delivery, quality and longevity of materials. Information relating to the client's accommodation requirements (including space and functionality), design standards, FM service delivery needs are essential elements in the output specification.

A flexible and unprescriptive style provides the channel for risk transfer to the private sector but also allows the private sector to take a non-traditional approach to the service delivery methodology. It also allows the operator to consider alternative methods of service delivery over time.

The client plays an important role in ensuring that risk is adequately transferred to the private sector through the output specification. The onus is on the client to identify critical success/failure factors to enable bidders to qualify their exposure to service shortfalls. To establish the severity of potential payment deductions bidders will assess each performance standard for the likelihood of failure, mitigation methods and costs.

On receipt of the output specification the bidders will begin to design a service delivery and management regime to meet the specification. Part of the design process will include a review of the risks in delivering the service.

Some risks will relate to the physical delivery of the services to meet the particular client's needs. Others will relate to the payment mechanism which is unique to PPP projects. A typical extract from an output specification is shown in Tables 8.6 and 8.7.

Table 8.6 *Example of client's FM requirements in the output specification.*

Specific requirement: grafitti	Performance standard	Priority (Priority 1 = health & safety/legislative compliance; Priorty 5 = minor issue, seldom affecting core business service delivery)	Rectification period
The Contractor must remove graffiti inside and outside of the buildings upon notification of occurrence/opening of building/when evident	No failure to remove graffiti	3	5 days

Table 8.7 *Operator's risk matrix.*

Specific requirement	Service methodology	Risks	Mitigation
The Contractor must remove graffiti inside and outside of the buildings upon notification of occurrence/opening of building/when evident	*Pro-active:* • Regular patrol of inside and outside of buildings, noting evidence *Reactive:* • Respond to notification via helpdesk • Inspect graffiti and assess appropriate removal methodology • Obscure from public display as temporary measure • Undertake remedial action (e.g. paint over, chemical treatment etc.) • Report completion of work via helpdesk	Failure to remove graffiti within agreed timescale: • Underestimate frequency and scale of problem • Graffiti cannot be removed using known methodologies • Specialist removal contractor not working within Operator's performance/ payment system	• Increase staffing • Provide additional on-call support for peak periods • Agree longer rectification periods above predetermined frequencies • Agree longer rectification periods where specialist treatment is required • Pass risk to specialist sub-contractor on back-to-back contractual terms • Increase checking and monitoring to ensure compliance or alternative solution in the event of poor performance

8.5.4 Developing the payment mechanism

The principle governing risk in payment mechanisms in PPP projects is that the client does not pay for the service unless it has been delivered and to the required quality standard. The client pays a unitary charge throughout the contract term that cannot be altered unless a 'change' is agreed between the parties.

While payment mechanisms may be developed in conjunction with guidance (Treasury Taskforce, 1999b; 4Ps, 2000b), the monitoring of change to maintain the negotiated allocation of risk is relatively unstructured. While individual changes will be documented through recognised change procedures, the trends and step changes in risk allocation may go unnoticed over long periods.

A typical payment mechanism will have the following components:

(1) Differentiation in penalty between availability and performance failures.
(2) Availability failures are usually limited to those that have significant impact on the ability of the client to use the facility, e.g. health and safety or other statutory compliance issue. Unavailability may be deemed to affect only a certain area of the buildings, or should unavailability become material regardless of the area affected, the whole building may be deemed unavailable, e.g. all toilets not working. A financial deduction is usually made in the event of any availability failure. The whole of the unitary charge is at risk for availability failures.
(3) Performance failures are usually those that do not have a material impact on the buildings remaining operational but which would, if not rectified, have an impact over time. Performance failures are prioritised so that the severity of deduction is related to the impact of failure on the client. A number of performance failures may be allowed before financial deductions are made and deductions are applied on a sliding scale depending on the priority of failure. Performance standards are likely to have a rectification period to allow the contractor to undertake necessary works without a financial penalty being enforced. Approximately 30% of unitary charge is usually at risk for performance failures.
(4) Performance and availability deductions are treated as mutually exclusive so that the operator cannot accrue performance failures in an area which has been deemed unavailable.

8.5.5 Escalation in the event of repeat failures

Provisions are usually made to increase the financial penalty by a factor (usually 50%) where it can be demonstrated that the contractor is continually failing to rectify problems. While this may be seen by the operator to be a double penalty when he is having a period of poor performance, the client uses the ratchet effect to quickly highlight fundamental failings by the operator.

In the event of performance falling below predetermined levels there are usually provisions for a Warning Notice to be issued. If this event occurs it signals to the operator, and its funders, that significant improvement in the service is required. A rectification plan to resolve service delivery will be a prerequisite for the continued operation of the contract. In the event of an agreed number of warning notices, the contract can be terminated.

The operator's risk of availability and performance deductions depends on the clarity of definition set out in the output specification. This includes the priority given to a service, possible rectification period to correct service shortfalls without a penalty being applied, the number of failures allowed within a time period without service failures and the severity of payment deductions for repeated failures within an agreed time period.

The payment mechanism must be seen to be equitable to prevent unnecessary risk premiums by the bidders. The severity of deductions should relate to the service failure. Buffers should be built in to prevent payment deductions from becoming too penal so that banks will accept the risk. Only in exceptional circumstances, where a risk is critical to the core business, should no buffer be provided. An example is the requirement for the FM operator to prevent the presence of unauthorised persons in a building. In such circumstances, the operator must be able to develop a service methodology which minimises the risk of occurrence, e.g. physical measures (locked doors, well-lit entrances), monitoring (remote CCTV) and management methods (admission arrangements for authorised personnel, training). Otherwise the contractual requirement may be seen to be unenforceable.

To test the bankability of the risks within the deal the bidders will make a risk assessment of each service line in the output specification, calculate frequency, cost of failure and mitigation. During the procurement stages part of the mitigation by bidders will include reducing the onerous nature of the link between the output specification and the payment mechanism. This may be done by lowering priorities, changing availability to performance failures (attracting lower penalties) and rewording the service scope and standards definitions to narrow the potential for failures to occur.

The bidder will review each service requirement in the output specification and assess the financial impact of failures on the service, taking account of mitigation factors. An example of a bidder's risk assessment is shown in Table 8.8.

Where the monetary values of predicted failure rates are unacceptably high the bidder may require to review its service delivery and management methodology. It may be cost effective to increase staffing levels or apply alternative delivery methods to reduce the predicted frequency of failures and the related cost.

While the onus may be on the operator to identify the cost impact of service failures, the client has to carry out a similar exercise to ensure that the output specification and the payment mechanism work in tandem. Weighting factors may have to be applied to areas of buildings and types of service to ensure that the operator delivering the full scope of service is not in a position to prefer to accept the payment deduction rather than rectifying service failures.

Table 8.8 *Example of bidder's risk assessment of output specification requirement.*

Specific requirement: slip resistant floors	Deduction per failure (based on formula in the payment mechanism)	Predicted frequency per month	Annual value	Mitigation	Cost of mitigation
Floor surfaces shall be clean such that they remain slip resistant when there is dampness or water spillage	Say £50	5	£3000	Industry recognised cleaning method	Included
				Use of cleaning agents compatible with flooring material	Included
				Training in cleaning methods and use of chemicals	1 day × 10 cleaners per annum @ £200 = £2000
				Monitoring and checking regime	Included
				Life cycle replacement policy	Check frequency and specification within LCC fund

Wherever possible the client should be allowed to use assets and services in the event of a service failure, even if they are operating at unacceptable performance levels for a period of time, e.g. broken window leads to temperature in a room falling below acceptable levels. The 'unavailable but used' concept has become an accepted principle in PPP projects to make the payment mechanism more commercially acceptable and affordable to bidders and provides a proportion of the service where alternatives at short notice are difficult for the client to find.

Some predetermined events should be agreed to be outside the scope of the payment mechanism, e.g. planned maintenance, to reduce risk premiums and deliver an affordable service. These events should be strictly limited but may relate to the type of operating environment.

8.5.6 Linking design and FM

One of the drivers for the development of PPP as a viable procurement route has been assumed to be the opportunity to bridge the gap between traditional construction and FM roles so that there is continuity between design, construction and FM services. It is assumed that bidders will take the opportunity to reduce the whole life cost of buildings by optimising the relationship between capital, revenue and life cycle replacement costs for building fabric and services.

There may be several ways for bidders to approach the whole life cost of operating the hard FM services. Table 8.9 shows two alternative models and the possible consequences for risk. Model 1 assumes that the bidder is able to specify the most durable materials and components (which may be more expensive) and offset the expense against longer life spans. In practice the operator may not be able to adopt Model 1 as the client may not be able to afford the resultant unitary charge from a front loaded cost profile. The extra capital cost may not offer lower life cycle values. At the other extreme, Model 2 is highly incompatible with the ethos of PPP projects and typical payment mechanisms as the client will want to reduce the risk of disruption through premature failures of building components.

Bidders will usually offer a compromise between the two models, with choices made for each element of the building balancing capital expenditure, life cycle replacement and maintenance. This is particularly relevant on refurbishment projects. The bidder may not gain from always specifying superior components. Average boilers may have an average life span of 10–12 years, while the best boilers may have a maximum 15-year life. Regardless of the choice of boiler, the bidder will have to undertake two replacements over a 30-year period as the bidder may also have to assure some 5 years residual life in the building at the end of the concession. The additional initial cost may not give a long-term saving. Consideration must also be given to the level of maintenance required as a good practice methodology to achieve or extend the predicted life expectancy.

In addition to consideration of life cycles of the main components of the building the bidder will have to consider how maintenance activities and replacements will be achieved. Lower overall life cycle costs are likely to be

Table 8.9 *Alternative models for costing 'hard' FM services.*

Model	Construction value	Life cycle replacement value	Reactive maintenance value	Operational risks
1	High	Low	Low	• Disruption to client – low • Risk of performance and availability failures – low • Predictability of life cycle replacement – good
2	Low	High	High	• Disruption to client – high • Risk of performance and availability failures – high • Predictability of life cycle replacement – low

achieved where replacements are easy to undertake when they are required. Clients and the funders of the bidders are also interested in the life cycle requirements because of the potential impact on performance and availability deductions through the payment mechanism. The client wants to be assured of continuity of service delivery and minimal disruption from maintenance activities. The banks want to be assured that the life cycle replacement fund is sufficient to ensure that performance and availability failures are kept to a minimal level and that the step-in rights of the client are not invoked because of lack of maintenance or replacement of parts of the building.

While the optimum balance between construction, life cycle and maintenance decisions remains a strategic goal, the bridging of the gap between design and FM has been limited. Internal relationships within the Special Purpose Vehicle (SPV) can replicate traditional problems experienced in traditional construction with gaps between the designers and the constructors and little involvement of FM until late in the design and construction processes. The management of the SPV have to take a pro-active role in reconciling capital, life cycle and revenue costs relating to the building fabric and services to mitigate this risk.

8.5.7 Changing assets into services

The design of PFI projects introduced the concept of treating assets as services – the client should only use and pay for those aspects of buildings they required for their core business. This has opened opportunities for PFI operators to sell unused time in buildings to third parties. To counter this opportunity there has been opposition to, and restriction of, many income generation schemes because of perceived risks. In schools, for example, the type of organisation able to use school buildings is subject to vetting and local authority approval. Perceived security risks limit the use of government buildings by others, even if they belong to other government departments.

Typical PPP projects include a number of 'soft' FM services, such as cleaning and catering. These services can be subject to greatest scrutiny regarding their inclusion in the scope of the project. This has been prompted from a requirement for the client to demonstrate long-term value for money. Strong arguments on behalf of the PPP operator and the existing operator of catering services in value for money terms has resulted in this service being excluded from the PPP on many projects. A summary of the arguments is shown in Table 8.10. The case for excluding cleaning from PPP projects is less clear cut because of the intrinsic link between cleaning (frequency, methodology and use of cleaning agents) and the life expectancy of elements of building fabric. The scope of soft services in any project will depend on project-specific factors and should be subject to rigorous review.

The most significant issue for the PPP operator is the impact on performance and availability criteria where there are other FM-related operators within the same building. The responsibilities between the parties must be

Table 8.10 *Assessment for including catering in PPP projects.*

Case for including catering in PPP	Case for retaining catering with existing operator
Long-term investment required by operator in design of kitchen, restaurant, servery etc. to gain customers and generate income	Commercial view of catering may not be appropriate for the client base, e.g. hospital patients
Revenue generation contributes to reducing the unitary charge Predictions of levels of revenue generation is at the PPP operator's risk	Catering service may be delivered to other users outside the scope of the PPP project. Alternative source of delivery may be required. Split of resources may be difficult to achieve
Integration of all services with single point of responsibility within a building	Recent investment by existing operator in kitchens, restaurant, marketing, management systems etc.
New service with no previous or incumbent operator	Impact on failure of other services within PPP project is low
Economies of scale achievable through single management interface	Cost advantage of PPP operator of the same service is low
Commercial perspective on service delivery to increase number of customers and spend per visit	

carefully designed to ensure that an overall effective service is delivered to the client who will have to play a role in managing the interface.

Although the costs of the contract as a whole are indexed annually according to an agreed inflation factor, soft service costs may increase at a different rate. This is because soft FM services by their nature have a significant resource element which is highly volatile to cost changes over time. Soft FM costs are highly dependent on wage rates, staff turnover rates and ability to recruit staff. Because of the long-term nature of the contract, bidders would have to build in significant risk premiums to take account of fluctuations in staff costs, creating affordability issues for the client. In practice, market testing and benchmarking regimes are built into the contract so that the FM operator is only carrying this risk for five to seven years. Typically the FM operator has to demonstrate whether the cost of providing the service has risen by undertaking an open book tendering exercise for the services.

8.6 Operation

While some government guidance has been developed to assist clients to manage the risks of operating PPP contracts (e.g. Treasury Taskforce, 1998c) there is little clear evidence of the key risks to be managed long term, other than those identified through the terms and conditions of contract. 4Ps (2001) has reviewed the progress of a number of school PFI deals to develop industry guidance and improve the PFI procurement process. It identified a number of common risk issues that can arise. These issues and other anecdotal evidence from practical experience show that effective action at during procurement process and during mobilisation can address potential risks. Long-term effective management remains an important aspect of minimising risk (McDowall, 2001).

8.6.1 Service delivery

The delivery of services can commence either when the building works required to support the delivery of services is complete or during a 'transition' phase while buildings are being constructed or refurbished. Delivering services during the transition phase can create problems as the FM operator is responsible for the liabilities inherent in the physical environment without necessarily having the right level of control. The FM operator is usually delivering services in buildings that require major investment to achieve the output specification standards (McDowall, 2000a). A highly reactive mode of operation can develop – in many circumstances, this is not in the best interests of client or the FM operator. The relationship with the client can be put under strain where the FM operator is constantly seen to be under-performing. Where services are delivered during the transition phase a lower specification may be used to allow the FM operator to perform. The payment mechanism may not be onerously applied during this stage (see Monitoring the contract).

Successful on-site service delivery in any project is highly dependent on the calibre of FM operator staff and the continuity of staff by both the client and the operator. In traditional FM contracts the client can have a major role in determining the key personnel to work on a project. PPP projects reduce this control mechanism because the client can only monitor outputs and the FM operator cannot guarantee continuity of staff over such long-term contracts. The SPV and its funders have similar interests to the client in how the contract is managed. It may be able to impose conditions through the FM sub-contract to determine the quality of the management of the contract and initiate management changes where necessary to address service performance issues before the client has the need to impose warning notices.

The continuity of staff between procurement and contract monitoring is critical but can be reduced by the time lag between contract signature and on-site delivery and monitoring of FM services. Unless transitional FM service delivery is included, a two-year time lag can be introduced. Careful attention needs to be given to the documentation developed during the mobilisation of the project to minimise the loss of continuity. Ideally this should be agreed in principle prior to contract signature to ensure that the client has sufficient confidence in the FM operator's management procedures.

Staff transfers were a major issue for clients and FM operators during early PPP contracts. As contractors have become more experienced in using the PPP procurement process they have developed well-defined staff transfer processes to ensure compliance with the TUPE Regulations and to provide the information necessary to ensure that staff are offered appropriate incentives to transfer, whether or not TUPE applies (Treasury Taskforce, 1998b).

In many local authority deals staff do not automatically transfer but the client offers them redeployment within the organisation. This is manageable where the deal is only a small proportion of the total support service delivery of the organisation but is more difficult to manage in the larger schemes. In the health service, pilot projects are being developed to allow support service staff to remain employees of the public sector while working under the management of the private sector. The development of models is currently in its early stages. In other projects, some aspects of the services are excluded from the scope of the project from the outset. Members of the support services staff continue to work for the client body. In such cases, effective interfaces have to be designed to ensure that risks are not inadvertently transferred back to the client organisation. Examples include catering and cleaning.

While staff transfers may not be compulsory in many deals there are advantages and disadvantages to the FM operator. Where staff transfer, the FM operator has an established workforce that has knowledge of the client operation and may be able to deliver effectively from the commencement of the contract. The FM operator may, however, wish to bring new skills and methods to an existing workforce. The FM operator will have to undertake a skills review and establish a training plan to ensure that its methods can be introduced effectively. Staff may be reticent to embrace new ideas and working methods. Where a large body of staff transfer the FM operator may be protected from recruitment risks at an early stage, but will still remain at risk for greater than expected staff turnover. Local market conditions for recruitment and retention and the type of operating environment will play a significant part in the staffing risk profile.

8.6.2 Monitoring the contract

The risks of performance and availability failures will have been well rehearsed by the FM operator, its funders and the client during the procurement process. Implementing the payment mechanism too early in the operational period can generate risks for the FM operator as mobilisation and familiarisation issues can lead to significant payment deductions. A bedding-in period of several months without payment deductions or limited payment deductions (e.g. performance only, not availability) is attractive to the FM operator and can have advantages to the client. This is also an issue for the operator's funders as it may be perceived that the FM operator is achieving unrealistically high penalty points over a short period. This can have an impact on contractual terms such as warning notices and step-in rights early in the contract without real cause. A penal environment for service delivery early in the contract also prevents the development of joint monitoring between the client and the FM operator and the building up of the relationship. A joint approach to monitoring can engender close working relationships so that objective measures can be applied when disputes arise. Table 8.11 shows an example of a joint approach to monitoring the contract.

During the early stages of the contract, the detail of the output specification is tested, especially when it is written to cover a wide range of buildings in portfolio projects. Where differences in interpretation arise, difficulties can develop which can have an impact on price and require the dispute resolution mechanism to be invoked. A well-designed dispute resolution procedure will have a number of checks and balances to ensure that small issues do not become major stumbling blocks to the delivery of the contract. This is especially important in long-term contracts such as PPP projects.

Both the client and the FM operator can underestimate the time and effort

Table 8.11 *Example of monitoring of compliance with output specification.*

Output specification requirement: opening buildings	Performance standard	Service delivery methodology	Operator monitoring methodology	Client monitoring methodology
The contractor must open the buildings in accordance with the Authority's requirements	No occasion of failing to meet the authority's requirements	Security guard arrives on site 10 minutes prior to agreed opening times and unlocks the building	*Pro-active:* • Spot checks by manager • Security guard reports to help desk on arrival at site • Manager monitors trends in calls to Help Desk *Reactive:* • Help desk calls back-up if security guard fails to arrive • Help desk receives call from Authority reporting performance failure. Help desk despatches back-up to open building	• Monitor help desk calls daily/weekly • Review monthly report and trends • Spot check on a predetermined programme • Respond to escalation in performance failures by increasing frequency of spot checks • Monthly meeting with contractor

required to monitor projects. While many contracts are encouraged to be 'self-monitoring' to reduce the requirement for client monitoring, especially on the larger portfolio schemes, a strong client role, with senior personnel involved over the long term, is an essential requirement. In the majority of projects, quality management techniques are used as a primary tool to monitor compliance with the output specification (e.g. spot checks, audits, statistical analysis). Where non-compliances are found, the quality management process is used to escalate issues, the payment mechanism is used to attract penalties and financial deductions.

Seldom are risk management techniques introduced to assess the impact of failure and the need to establish remedial action until a warning notice is issued. When this occurs, a remedial plan is implemented to allow opportunity for the FM operator to rectify substantial issues in its service delivery plan within an agreed timescale. It is seldom in either party's interests for the FM operator to be replaced although the SPV will have some rights within the FM contract and the Project Agreement to rectify problems, e.g. replace management, replace service sub-contractors.

8.6.3 Managing relationships

The nature of commercial negotiations can lead to the breakdown of relationships between the procurement process and the operational phase of projects. Users can become isolated from the commercial process, resulting in them having to live with the deal, not contribute to it. The technical team involved in the project's procurement may not have an active operational role. Lack of information about the scope and delivery methods of the FM service in the early stages of operation can also generate risks in client confidence and affordability issues where users contribute to the cost of services.

If the client is unable to provide strong leadership during the early operational phase, relationships between users and the FM operator can develop that are stronger than the relationship with the client, making enforcement of some contractual terms and management of change difficult for both sides. A well-developed change management process is essential to capture potential changes formally so that the impact on overall project risk can be taken into account.

8.6.4 Developing operator capability

Early PPP projects suffered in the initial stages of operation because the FM operator had a lack of market-sector specific knowledge. In many cases, the private sector had not had the opportunity to bid for the types of project brought about by PPP. This has been a short-term issue as commercially aware FM operators have climbed a steep learning curve and then successfully disseminated the knowledge to other projects. It has become common practice for bidders to employ sector-specific specialists to provide the necessary experience in new market sectors.

8.7 Conclusions

Within PPP projects there are a number of principles set out in government guidance that control the treatment of risk throughout the life of the contract, but risk is generally incumbent on the party best placed to manage it. In many market sectors industry guidance has been established to offer models that are intended to reduce time spent on the procurement process, reduce the potential risk premiums and introduce consistency to the procurement of similar projects. Current guidance offers high level advice that must be adapted to be applied to specific projects.

For individual projects the risk matrix developed at the OBC stage becomes the benchmark for reviewing risk and its allocation between the parties throughout the procurement process until the deal is signed. There is limited development of the risk matrix post contract close to establish the extent to which perceptions of risks have been accurate or whether the allocation of risk has brought financial benefits. There is limited feedforward to the business cases for other projects based on practical experience.

Industry guidance on the development of documentation makes little reference to the use of documents to make risk transfer explicit. Greater use of the key documents, such as the output specification, to highlight where risk is being transferred, e.g. free meal uptake, availability risk, would pre-empt bidders' use of the output specification to dilute their risks during commercial negotiations.

The nature of PPP projects, especially high-profile ones, often leads to unique partnerships between FM, design and construction companies. In such circumstances the opportunity to develop an optimised 'whole-life cost' model for hard FM services are limited. The risks of the premature failure of key building elements and their mitigation often remains unexplored during the design stages, reducing the overall effectiveness of a PPP procurement route. The unique factors in these types of project reduce the potential to replicate risk models for future projects.

The provisions for market testing and benchmarking of soft FM services during the life of the contract only make allowances for testing cost fluctuations. It does not make allowances for reprofiling the risk matrix between client and operator to offer better value for money, unless a 'change' to the project is invoked.

During the negotiations between client and bidders during procurement, commercial and technical issues may be treated separately. Few generalists are able to think across traditional professional boundaries and challenge bidders' assumptions about the links between design, construction and FM.

Little thought has been given to the handover requirements at the end of contracts and how risks will be transferred back to the client. As many of the contracts have at least 20 years before the handover requirements in the contract are invoked, there is no practical experience to feed forward into new deals. The robustness of these terms and conditions will remain largely untested unless they are applied to projects which terminate prematurely.

Badly written contracts will provide a steady work stream for legal advisers and risk managers in future years to address handover issues.

The operator of the contract puts great emphasis on the use of quality management techniques to manage the risks of non-compliance with the output specification. It is unclear whether the diagnosis element of reviewing non-conformances with the quality system are sufficient to highlight fundamental flaws in the service delivery and management process adopted by the operator in a holistic way.

References

4Ps (2000a) *Output Specifications for PFI Projects – 4Ps Guide*. Public Private Partnerships Programme, London.

4Ps (2000b) *Payment Mechanisms for Local Authority PFI Schemes*. Public Private Partnerships Programme, London.

4Ps (2000c) *Option Appraisal and the Outline Business Case*. Public Private Partnerships Programme, London.

4Ps (2001) *PFI in Schools – Update from the First Operational Schools Contracts*. Public Private Partnerships Programme, London.

Arthur Andersen and Enterprise LSE (2000) *Value for Money Drivers in the Private Finance Initiative*. Office of Government and Commerce, London.

McDowall E. (2000a) Delivering FM services during transition. In: *Facilities Management*, February, **7**(4), 8–9. Eclipse Group, London.

McDowall E. (2000b) Informing the PFI process. In: *Facilities Management*, April, **7**(6), 8–9. Eclipse Group, London.

McDowall E. (2001) *Public Private Partnerships*. Eclipse Group, London.

Treasury (1995) *Private Opportunity, Public Benefit – Progressing the Private Finance Initiative*. HMSO, London.

Treasury (1996) *How to Write an Output Specification*. HMSO, London.

Treasury (1998) *Appraisal and Evaluation in Central Government – Treasury Guidance*. HMSO, London.

Treasury Taskforce (1998a) *How to Construct a Public Sector Comparator – Technical Note No. 5*. Office of Government and Commerce, London.

Treasury Taskforce (1998b) *PFI Projects: Disclosure of Information and Consultation with Staff and other Interested Parties*. Office of Government and Commerce, London.

Treasury Taskforce (1998c) *How to Manage the Delivery of Long Term PFI Contracts*. Office of Government and Commerce, London.

Treasury Taskforce (1999a) *A Step-by-Step Guide to the PFI Procurement Process*. Office of Government and Commerce, London.

Treasury Taskforce (1999b) *Standardisation of PFI Contracts*. Butterworths, London.

9 | Developments in UK public sector risk management: the implications for PPP/PFI projects

John Hood, Andrew Mills and William Stein

9.1 Introduction

The election of the Conservative Government in 1979 had many radical effects on the UK public sector. One of these was the extent to which academic interest in the sector increased. Whilst academics have long been interested in this particular area of operation, the post-1979 period saw such radical changes in approaches to the public sector that the level of interest, and subsequent literature, moved to a new level (Isaac-Henry, 1997). Indeed, the very terminology surrounding the discipline changed from one of public *administration* to one of public *management* (Osborne & Gaebler, 1992). The whole raft of changes initiated by the Conservatives were synthesised by Hood (1991) into the generic term of 'New Public Management' (NPM). The seven 'doctrinal components' of NPM outlined below can be clearly related to traditional models of private sector risk management:

- Pro-active and professional management
- Explicit standards and measures of performance
- Emphasis on outcomes rather than inputs
- Disaggregation of large units
- Greater competition
- Adoption of a private sector management style
- Greater financial rigour.

As can be seen, the move towards NPM resulted in the public sector being forced to adopt more private sector management principles and practices. It also became exposed to market forces, deregulation and privatisation and, in reality, was considered by central government to be inherently inferior to the private sector. As we will discuss in this chapter, this move towards 'managerialism' has had a significant effect on the development of public sector risk management (PSRM).

Another effect of post-1979 policy and the principles of NPM is the fact that the discrete boundaries of the public and private sectors have become increasingly blurred (Lawton & Rose, 1994:2–4). It would be wrong to assume, however, that prior to 1979 the public and private sectors operated in total isolation from each other. The public sector has always relied on private sector providers for a whole range of goods and services, but traditionally the role which the private sector played tended not to impact directly on public service users. The post-1979 increased role of the private sector, through policies of privatisation and market testing has, however, led to a situation where 'front-line' services such as refuse collection and hospital catering have been provided by private sector companies. In effect, therefore, in public sector environments such as the National Health Service (NHS) and local authorities, aspects of service delivery are provided by private sector organisations. Public-private partnerships/private finance initiatives (PPP/PFI) are the latest manifestation of this and, arguably, blur the boundaries to the extent that it is not clear to 'consumers' of public services who is actually providing the service. For the purposes of clarity, however, this chapter will look at the development of risk management in three main areas of the public sector, although the main focus will be on the latter two:

- Central government departments
- Local authorities
- The NHS.

9.2 Why has PSRM developed?

High profile events involving serious loss of life such as King's Cross, Clapham and Dunblane, and their resulting enquiries (see Fennel, 1988; Hidden, 1989; Cullen, 1996), have played a major role in focusing the public spotlight on the underlying practices of risk management in public sector organisations. The prominence of PSRM has been further strengthened by a series of government publications and guidance documents. Organisations such as HM Treasury (1991, 1994, 1997), the Audit Commission (1997, 2001), the Accounts Commission for Scotland (1999), the National Audit Office (1996, 2000), the NHS Executive (1993, 1997, 1999), and Scottish Executive (2000) have sought to raise the profile of risk management across the public sector. This has involved the publication of a number of reports which have emphasised the importance of a systematic risk management structure in terms of reducing operational costs. The main reasons for, and the present state of, their development in the three areas of the public sector identified above will now be discussed.

9.2.1 Central government

Central government departments can be distinguished from local authorities and the NHS in that they are the providers, as well as the consumers, of risk management guidance. It is also the case that what are known as 'quangos' (quasi-autonomous non-governmental organisations) also play a role, on central government's behalf, in the provision of risk management information. Such quangos as the Accounts Commission for Scotland (now Audit Scotland) and the Audit Commission, both mentioned above, have been instrumental in raising risk management awareness in both local authorities and the NHS. The reasons for this awareness-raising are not entirely clear, but it would be reasonable to speculate that the much greater commercial awareness consequent to NPM and the increasing convergence between business practice and aspects of government have been instrumental.

There are numerous examples of central government departments publishing risk management guidance, policies, objectives etc., for use in their own departments (see for example such diverse government departments as the National Archives, www.pro.gov.uk/about/conservation/risk.htm, the Lord Chancellor's department, www.lcd.gov.uk/risk.htm and Higher Education, www.hefce.ac.uk/GoodPrac/risk). The office of the Prime Minister (10 Downing Street Newsroom, 2001) has also recently reinforced the Government's apparent commitment to a wide range of risk management measures:

'Effective management of a wide range of risks is essential both for the delivery of improved public services and for the achievement of the Government's wider goals. Government is concerned with managing risks to the public (including public health, social, environmental and safety risks) and also risks to the delivery of specific objectives and programmes (including financial, operational and technological risks).'

A detailed analysis of central government departments' utilisation of risk management is clearly, therefore, highly problematic in the context of this brief chapter. Such government departments (and agencies operating on their behalf) acting as advisers and/or regulators of other arms of the public sector have published a panoply of risk management papers. These have ranged from the basic and general to the highly technical and specific. Added to this, areas of central government, such as the Ministry of Defence and the Department of the Environment, Transport and the Regions, have produced PPP/PFI-specific risk management guidance.

Although local authorities and the NHS across the UK are not entirely homogenous, their risk management development has had a much greater level of homogeneity than that of central government. Consequently, therefore, although we recognise that central government risk management policies, practice and guidance have had important implications for PPP/PFI, the main focus of this chapter will be on local authorities and the NHS.

9.2.2 Local authorities

There are a number of reasons for the development of risk management in local authorities, e.g. central government pressure, greater exposure to private sector practices and moves towards risk-based health and safety legislation, but arguably one is dominant – the change in the local authority insurance market which occurred in the early 1990s. Prior to 1992 if risk management in UK local authorities, as a distinct management practice, existed at all, it did not exist to any great extent. Fone and Young (2000:37) identify the main reason for this: 'The lag in adoption more likely was due to the particular characteristics of the insurance market formed to serve local authorities in the UK'. This insurance market was dominated by the Municipal Mutual Insurance (MMI), which until the late 1980s provided insurance cover for over 90% of local authorities. The MMI was owned by its policyholders – primarily the local authorities – and was not therefore answerable to the stock market and to a wider body of shareholders. As a consequence, the MMI's prime objective was the provision of the widest form of insurance cover at the lowest possible rates. Local authorities were therefore substantially immune from the commercial realities of the wider insurance market. These realities had long forced large private sector organisations to use insurance as part of a wider risk management strategy, and not to look upon it as the only response to risk. The combination of increased numbers of claims, higher settlements for personal injury claims and an ambitious, but ultimately flawed, diversification strategy seriously affected the viability of the MMI. By 1992 the company was insolvent and was purchased by the Zurich Insurance Group. It would not be appropriate here to compare the attitudes of the MMI and the Zurich to local authority insurance business, but suffice to say the Zurich took a much more rigorous and commercial approach to both underwriting and claims handling.

It was the failure of MMI that led many local authorities to the realisation that they had neither strategy nor operational procedures for risk management. They had been protected from commercial realities by the MMI and were now on a very steep learning curve. This, along with the other influencing factors outlined above, forced local authorities into a position where they had to quickly develop risk management strategies and practices. It is not at all clear (Hood & Kelly, 1999) as to how successful they have been, but there is some evidence that they have come some way in a relatively short time. Fone and Young (2000:64) suggest, however, that in general they have developed broader risk management skills for those risks which they would have previously insured with the MMI. Therefore whilst local authority risk managers may be involved in such diverse areas as financial risk management, crisis management and aspects of human rights legislation, the evidence would suggest that their 'insurance' background means that their primary function remains rooted in the management of 'insurable' risks.

Hood and Kelly (1999:274–5) have summarised the post-1992 growth in local authority risk management (LARM). This growth has, to date, not been accompanied by a large amount of academic interest. However, there is now evidence that the risk management function in local authorities is developing a higher profile in terms of academic research, central government guidance and in practitioner guidance devised by a variety of professional bodies. There remains, however, little attempt to address any conceptual conflict which may arise between the corporate managerialism implicit in a LARM system, which has its roots in a private sector model, and the predominant professionalism of specific local authority services. Clearly this conflict between the traditional professionalism of local authorities, e.g. education departments being run by professional educationalists and not professional managers, and the managerialism implicit in NPM has ramifications for PPP/PFI projects. Vincent (1996) suggests that the professional and practical application of risk management may be underdeveloped due to the public sector professions failing to address adequately the management of risk in their own services. Vincent's hypothesis is further developed to include the theory that the fragmented nature of PSRM results in an over-reliance on legal compliance and in financial accountability.

The formal organisation of local authorities has changed greatly since the Bains (1972) and Paterson (1973) reports. Pre-dating NPM, these organisational changes were intended, amongst other things, to move away from a purely professional approach to service delivery. In many respects senior officers in the whole range of local authority services have been required to develop wider management skills. Kogan (quoted in Kerley 1994:52) succinctly summarises this development:

> '... and so an Education officer finds himself looking after the education of children and the welfare of the community but to some extent he's also a transport manager, and a catering manager, and he has all manner of other such sidelines that he must attend to.'

In the example of schools, this statement would suggest that education officers would be perfectly able to deal with the multiplicity of non-educational factors inherent in PPP/PFI projects. On the other hand, Kerley (1994:55) also refers to managers in local authorities seeing the professional task as their 'real work' and, presumably, other tasks associated with corporate or managerial issues as peripheral distractions. It is not practical here to analyse in detail the level of acceptance by local authority professionals of wider managerial tasks, but it would seem reasonable to speculate that many mangers in local government still view themselves predominately as professionals and relegate wider corporate or managerial issues to a lower order of priority. If this hypothesis is valid, it could be argued that any corporate risk management strategy, which requires a wider strategic view and inter-service co-operation, would be severely undermined. Case study research by Harrow (1997), whilst concerned mainly with the 'professional' risks, does connect normative and prescriptive risk choices which Social Services professionals must make, with the pursuit of quality. This emphasis on quality is core to the private sector influence on contemporary public sector management. Whilst quality is at the heart of the NPM, Harrow recognises the dangers of over-simplifying the generic applicability of private sector management models to the public sector. This view has been echoed by others critical of the 'New Right' (Isaac-Henry, 1997). Indeed, Christopher Hood, who has been at the forefront of NPM thinking, has questioned whether 'business' risk management is entirely applicable to the public sector (Hood & Rothstein, 2000).

These conceptual difficulties surrounding LARM have not, however, prevented many aspects of its practical development. The period post-1992 has seen a number of initiatives aimed at furthering the practice of risk management in local authorities, many of which have their roots in the demise of the MMI. Paradoxically, therefore, the insurance-related crisis that was the collapse of the MMI has led to improvements in the way that local authorities view, and subsequently manage, risk. As indicated above, however, these improvements have tended to be restricted to the insurable-type operational risks, and local authority risk managers have only seldom become involved in the speculative or strategic risk arena.

Amongst the major drivers behind LARM is the Association of Local Authority Risk Managers (ALARM) and its Scottish equivalent (ALARMS). Both were founded primarily in response to the MMI collapse. The aims of ALARM and ALARMS, as outlined in their constitution (ALARM, 1997), are: 'to enable local authorities and the public sector generally to develop risk management strategies which address the incidence and consequence of injury, crime, loss and damage'. In line with comparable private sector approaches, ALARM and ALARMS emphasise that good risk management practice involves a staged process of risk identification, evaluation and the implementation of measures to control the risk. They have, however, said little about how this process is to be financially planned and controlled. Risk management as a function, especially in local authorities, cannot be actively pursued unless there is sufficient financial backing. In the ongoing financial climate of public expenditure constraints, the planning and targeting of the risk spending has to play a central role in local authorities.

Other organisations have contributed to the development of risk management initiatives in local authorities (Accounts Commission for Scotland, 1999; NAO, 2000; Audit Commission, 2001; CIPFA, 2001; EVH, 2001). For example, it is the belief of the Accounts Commission that significant 'value for money' savings can be achieved by local authorities if there is successful planning of, and a corporate dedication to, risk management. The Commission calls for several measures which can be undertaken by local authorities to develop an effective approach to corporate risk management. Key recommendations of this report include:

- The adoption of an organisational commitment to risk management
- The improvement in risk management education, training and awareness
- The identification of specific risk areas
- The development of effective information systems.

Like ALARMS' recommendations, those of the Accounts Commission give little information on the financing of PSRM units. Indeed, little reference at all is made to the financial implications of, and budgetary procedures required when, adopting better risk management practices.

Many of the other guidance documents adopt a similar approach, so clearly there is no shortage of guidance on what local authorities should be doing as regards risk management. There is a wealth of guidance on such areas as:

- The benefits of risk management
- How to identify the major risks
- The roles of members, i.e. politicians and officers, i.e. employees
- Development of a risk management strategy.

Arguably, therefore, there is little reason for local authorities to lack knowledge of risk management, although as indicated above, issues of financing risk-related initiatives may be problematic.

Despite, however, this plethora of guidance on LARM and clear development of it as a distinct management function, there is some evidence that its actual practice is underdeveloped in key areas. Based on empirical research of Scottish local authorities, Hood and Kelly (1999) concluded that the lack of a corporate approach to risk management and the non-systematic and arbitrary methods of risk financing, exacerbated by the diffusion of risk management activities and expenditures across service departments, prevented a strategic cost–benefit comparison of alternative measures which could have been employed to reduce risks. Survey data indicated that no standard existed for spending, control or budget setting for the risk management function within Scottish local authorities. Furthermore, the risk management budget setting process did not follow any systematic process and, in all cases examined, budget allocation was based on historic data which was calculated prior to local government re-organisation in 1996, and/or determined through general negotiations within the authority as to what would be felt appropriate. These arbitrary and *ad-hoc* methods employed in budgeting for risk management were likely to be perpetuated year after year. In an era of Best Value, it was difficult to envisage a situation where such a fragmented risk management budgeting system would satisfactorily meet the criteria laid down by central government. Irrespective, however, of any need to satisfy central government requirements on value for money, failure to adopt a more co-ordinated and corporate approach to risk management will leave authorities in a situation where they are failing to benefit fully from it.

Further research (Hood & Allison, 2001), also suggested that a number of barriers still existed which militated against good risk management practice within local authorities. This research concluded that there appeared to be two main obstacles to achieving effective risk management strategies across local authorities. Firstly, lack of knowledge, and in some cases lack of willingness, of senior members of staff to endorse risk management. Secondly, the responsibilities of the risk manager and their position within the organisational structure. In effect, there was a clear dichotomy between the managerialism implicit in any corporate risk management strategy and the exclusively professional risks with which service department staff felt more comfortable. This situation has its origins in, and was exacerbated by, the failure of local authorities to separate risk management from the insurance function. It would be flawed to claim that the situation could be improved by service department staff relegating the professional risks to some lower order of priority. Fundamentally, and understandably, the professional risk and its management will always be at the heart of the approach of, for example, educationalists or social workers. What is needed, however, is a wider understanding of how non-professional risks can impact on both the delivery of their service and across the authority as a whole. Clearly this whole area has implications for PPP/PFI projects in such services as education.

Finally, as yet unpublished research (Hood & McGarvey, 2001) would suggest that the involvement of local authority risk managers in the PPP/PFI process is marginal. Survey evidence suggests that scope exists for poor risk management decisions to be taken in Scottish local authorities in relation to PPP/PFI projects. The degree of involvement of risk management departments is variable, and where it does exist its extent is frequently limited in scope, primarily to the risk financing aspects, i.e. insurance of residual risks. This finding tends to support the view (see for example Gaffney & Pollock, 1997; Broadbent & McLaughlin, 1999) that risk transfer is unclear, poorly understood and weighted in favour of the private sector. The research did not explore whether some officer or department, other than the risk manager/ management function, was involved in the fine detail of risk transfer in PPP/ PFI projects, although we must accept that this is likely. Even if others were involved (e.g. lawyers or finance staff) it is arguable as to whether they would have the breadth of knowledge of risk management, in a corporate sense, which is likely to exist within the private sector partners. Little evidence therefore exists that Grout's (1997) view on bias against the private sector is valid as regards PPP/PFI risk management in Scottish local authorities.

9.3 National Health Service

To make sense of where the NHS stands today on risk management it is necessary to look back over a period of 25 years and more (Price *et al.*, 2000). The term 'risk management' was then little used in UK healthcare but, fundamentally, the risks faced by the NHS were not so different from today. Risks were responded to, but not in an integrated way. There were two clearly separate strands of response to risk: 'health and safety' and 'clinical'. Measures to ensure the health and safety of staff, patients and visitors to hospitals were essentially a matter for estates management; and measures to ensure standards of clinical care and treatment were very much a matter for the medical and related professions rather than the hospital administrators. Current risk management practice in the NHS has evolved along these two quite distinct strands.

9.3.1 Non-clinical risks

The Health and Safety at Work Act 1974 and related regulations hold negligent employers accountable to the State or individuals. In 1974 hospitals were exempt from prosecution, as they were 'Crown' property, meaning that they belonged to the State. Health authorities became accountable for the management of these risks as a result of the removal of Crown Immunity by the NHS (Amendment) Act 1986. This situation prevailed until the establishment of new entities called 'NHS Trusts' in the NHS and Community Care Act 1990 at which point individual Trusts became accountable for their own management, including that of managing risk, health and safety. Fire prevention had always maintained a high profile in hospitals but it too received greater emphasis with the lifting of Crown Immunity and with the transfer of responsibility for fire safety inspection from the Home Office to local fire authorities. European Directives dealing with health and safety resulted, in 1993, in new UK regulations dealing with risk assessments, preventive and protective measures and identification of those responsible for health and safety procedures. In line with such developments, accountability for risk, health and safety matters in the NHS has moved downwards from the macro to micro level. Accountability for the strategic and operational management of risks now rests at local level.

At present, NHS employers' statutory accountability for managing risks only extends to breaches in the Health and Safety at Work Act and related regulations, but they are being stimulated to consider the need to promote pro-actively the health of the workforce. Several recent policy documents have focused on the health of the workforce. The *Health at Work: in the NHS* initiative (Health Education Authority, 1992) *Improving the Health of the NHS Workforce* (Williams *et al.*, 1998), and *Working Together* (Department of Health, 1998) have urged NHS employers to promote health and fitness as well as prevent disease. These documents have encouraged NHS employers to consider a wider range of risks to health and include the conditions of work (rather than just the systems of work) and softer risks such as stress, hours of work, workload and management style.

9.3.2 Clinical risks

A recent development in the accountability of NHS employers is the focus on the control of clinical risks. From the 1980s onwards, the UK Government has issued a number of directives specifically related to the quality of clinical care. These have stimulated examination of the process of care and included initiatives such as standard setting, quality assurance, clinical audit, clinical effectiveness and evidence-based practice. The NHS response to clinical risks took longer to develop than its response to health and safety risks, but has quickly assumed even greater significance. Until the late 1980s, managing clinical risk was largely left up to individual clinicians and their peers. Clinical risk management was not a recognised function. Problems with clinical practice and performance were often avoided or tolerated rather than being addressed (Walshe & Dineen, 1998). In around ten years, clinical risk management has become a key concern for all NHS Trusts and health authorities, and especially for board members, managers and senior clinicians. This change has been brought about by a combination of developments: spiralling costs of litigation for clinical negligence, the rise of healthcare quality improvement, increasing public expectations and awareness of clinical risk, and the growth of regulation by the professional bodies.

For NHS Trusts in England, additional impetus came from the establishment in 1995 of a mutual fund to meet the costs of clinical negligence. NHS Trusts joining the Clinical Negligence Scheme for Trusts (CNST) were required to meet a set of minimum standards for risk management, both clinical and non-clinical. The emergence of the concept of Clinical Governance (e.g. Department of Health, 1997; Scottish Office, 1997) has renewed the emphasis on consistent high quality care and clinical effectiveness by placing Trusts under a statutory duty to control clinical risks. In England this was translated into action in the shape of the Controls Assurance Project (NHS Executive, 1997, 1999) which set a wide range of risk-related standards and benchmarks. There are now 19 separate standards including, for example, a risk management system standard (Standards Australia, 1999), a medical devices standard (Medical Devices Agency, 1998) and a waste management standard (Health Services Advisory Committee, 1999). The NHS in Scotland has adopted a simpler approach that could be described as an amalgam of (1) a *risk pool* – like the English CNST; and (2) a *set of standards or benchmarks* – like the English Controls Assurance Project. In Scotland, the Clinical Negligence and Other Risks Indemnity Scheme (CNORIS) (Scottish Executive, 1999) was launched on 1st April 2000. Membership is mandatory for all publicly funded health bodies in Scotland. CNORIS sees itself as central to risk management, summed up in its slogan 'Delivering an integrated risk solution to NHS Scotland'. Wales and Northern Ireland are expected to follow routes similar to that taken by the NHS in England and in Scotland.

9.3.3 Integrating clinical and non-clinical risk management

The expansion of NHS Trusts' statutory requirements to account for their management of health and safety *and* clinical risks has driven them to work towards a more integrated way of controlling these risks. Bringing the management of both clinical and non-clinical risks together is intended to result in a holistic approach. The perceived benefits of a holistic approach are to improve the quality of care given by incorporating procedures such as clinical audit, incident reporting, claims management, benchmarking, variance tracking and analysis, clinical competence and record keeping which are integral parts of the whole process of establishing quality care (Moss, 1995). The co-ordination of risk management and clinical and financial governance is seen as essential to promote maximum effectiveness, efficiency and quality of care. There are a number of organisational characteristics evident in the NHS that have some influence on progress towards an integrated model of risk management:

(1) A limited culture of risk transfer: risk transfer, either by conventional insurance or by some form of 'alternative' making use of the mechanism

of the capital markets, is not an option for the NHS which is now mostly prevented from transferring risks for payment or from contracting out of liabilities to its patients.

(2) No threat of 'going out of business': the presence of obvious threats to the survival of a business may ensure that commercial and industrial risk managers are taken seriously and given the resources they need to do their job but if a hospital burns to the ground, or is sued by many of its patients, the delivery of essential healthcare will have to continue. Political pressure is expected to see to that. The survival motive to practice good risk management is in this respect thus diminished.

(3) Healthcare staff serve many masters: the organisation or command structure in healthcare is vastly different from most non-health organisations. Individual healthcare workers owe their loyalty to many masters: the NHS Trust management representing the direct employer; clinical supervisors regarding day-to-day aspects of care; registration bodies that determine fitness to practice; professional bodies that develop the knowledge of their discipline; trade unions; and the peer pressure of fellow colleagues.

(4) The existence of a wide variety of organisational models: there is considerable variation in both the size and function of NHS Trusts and in the way that risk management has developed in NHS Trusts in the UK (Walshe & Dineen, 1998). The existence of a risk management department or a jobholder with the title of risk manager or risk director cannot be taken as any indication of the scope or effectiveness of that function in any particular Trust.

(5) Dependence on partner organisations: individual NHS Trusts are very dependent on external partners for the successful delivery of their end product, e.g. the quality of staff being trained in the universities and teaching hospitals, the co-operation of general practitioners as independent contractors and of local authority social service departments. These 'partners' cannot easily be sacked or replaced. Hospitals must find a way to work with what they have got.

(6) Political pressure and morale: political pressure and media scrutiny is intense. Several prominent scandals have produced a situation where it is 'open season' on doctors and hospital managers, including those that manage risk. The media portray an organisation where other health workers are struggling to deliver good healthcare in spite of the system they are forced to work within.

9.3.4 NHS guidance to Trusts on PPP/PFI

The first major hospital development under the PFI was given the go-ahead in May 1997. The current guidance to NHS Trusts was published in December 1999 (Department of Health, 1999a). The guidance is based on best practice and lessons learnt from PFI schemes completed successfully. The guidance is supplemented by a standard form contract for major PFI schemes. All the major bodies representing those involved in and affected by

PFI projects were consulted on the drafting of the contract – the construction industry, facilities management providers, Community Health Councils, health sector trade unions, doctors' leaders and other healthcare professionals. A stated objective of the 1999 Guidance and its associated standard contract is to bring standardisation and consistency to the PFI procurement process, bringing reductions in time taken and in transaction costs (Department of Health, 1999b). The introductory notes to the Guidance emphasise that its aim is to offer practical advice based on experience but warn that NHS bodies and private sector parties should seek their own legal advice before and during any PFI procurement.

The NHS Guidance notes suggest that major PFI schemes will typically be DBFO (design, build, finance and operate) meaning that the private sector partner is responsible for designing the facilities (based on the requirements specified by the NHS); building the facilities (to time and at a fixed cost); financing the capital cost (with the return to be recovered through continuing to make the facilities available and meeting the NHS's requirements); and operating the facilities (providing facilities management and other support services). The Guidance goes on to state:

> 'Risks in each of these areas will be assumed by the PFI partner, if best placed to bear them, in such a way that overall the risks associated with procuring new assets and services for the NHS will be reduced. Moreover, because the PFI partner's capital is at risk, they will have strong incentives to continue to perform well throughout the life of the contract.'

Specific guidance on risk analysis is provided under 'Technical issues'. It warns that throughout a PFI process, the NHS Trust should undertake extensive analysis of risks in order to ensure that it makes the right decisions at the appropriate stages. The risk analysis approach is outlined in some detail, including:

- The types of risk that need to be analysed (categories of risk suggested are design, construction and development, performance, operating cost, variability of revenue, termination, technology and obsolescence, control, residual value, and other project risks)
- The extent of the analysis required. The use of sophisticated techniques such as multi-point probability analysis or Monte Carlo sampling is advocated for those risks that are inherently quantifiable. The use of weighting and scoring approaches is advocated for analysis of all risks that are inherently non-quantifiable (e.g. the risk of changes in government legislation)
- The way in which the analysis should be presented in the relevant business case. This includes a clear demonstration of the robustness of assumptions in risk assessments and demonstration of management of the assessed risk. This will include setting out how any potential risks will be monitored in order that their materialisation can be identified at an early stage. The private partner should have a plausible strategy for managing the risk it bears.

The Guidance commends the development of a risk matrix as a useful tool for ensuring that all the individual risks over the life of the project – from design through to residual value – are properly considered. The thrust of the guidance to the NHS is not to identify and transfer all risks to the private sector. Rather it is to be very clear about the nature and extent of risks and to allocate responsibility either to the NHS or to the private partner as appropriate. Implicit in this view is a recognition that the private sector will levy a charge for risks that are transferred to them.

9.4 Conclusions

Before we draw together our picture of risk management in the public sector and consider the implications for handling risk in PPP/PFI contracts, it may be helpful to begin with the premise that the private sector partner will, by its very nature, act in such a way as to either minimise its acceptance of risks or, if all else fails, to maximise the compensation it receives for the assumption of risks. Put simply, the types of private sector companies operating in the PPP/PFI arena have had a much longer and more wide-spread experience of complex negotiations and contracts. Secondly, we can acknowledge both the novelty and the complexity of PPP/PFI and the likely vulnerability of both local authority and NHS trust negotiators in the early days of dealing with sophisticated private sector operators. However, almost ten years have elapsed since the first contracts and a knowledge base has now been built up in the public sector. This suggests that the public sector ought to be getting more and more skilled in both understanding the obligations of PPP/PFI deals and also in driving a hard bargain in negotiations with potential private sector providers. As more PPP/PFI contracts are agreed and implemented, it will also help to grow the expertise in assessing risks and in managing them. As explained above, we can see evidence of that in the timescale from the first contracts and the subsequent appearance of detailed guidance. Despite, however, any build up of expertise, in individual NHS Trusts, government departments or local authorities, PPP/PFI contracts are relatively few and far between and many officials will yet have no practical experience of negotiating a PPP/PFI contract. Given central government's enthusiasm for it, PPP/PFI is, however, a fact of life for local authorities and the NHS. Although the Public Sector Comparator exists as a mechanism for comparing the cost of totally public provision of capital projects, the prevailing view is that government will tend towards PPP/PFI wherever possible. This is likely to lead to an ever-increasing number of PPP/PFI projects and a concomitant rise in risk transfer considerations. The work of Hood and McGarvey (2001) suggests that many Scottish local authorities are ill-prepared to address these considerations, and unless radical action is taken the private sector will retain a substantial advantage over them in the arena of risk transfer negotiations. Anecdotal evidence would suggest that the situation in the rest of the UK is unlikely to be substantially different. Local authorities and the NHS have, for a variety of reasons, adopted the principles of risk management from the private sector, but, certainly in the context of PPP/PFI projects, they have much to do on operationalising its practices.

We might question where public sector expertise in PPP/PFI now lies. It would appear that assessment and management of PPP/PFI risks in public sector organisations is largely in the hands of legal advisers, contracts managers, buildings and facilities managers and general managers, or, indeed, is being outsourced to consultants. The public sector are still in the process of embedding risk management into their organisations and in determining the exact nature of the specialist advisers (either in-house or outsourced) that will be used to provide expert guidance on risk management to functional managers. Public sector risk managers already have much on their hands – the full panoply of risks relating to the facilities occupied by the organisation, the people they employ and come into contact with, and the services and processes they provide. By and large they are unlikely to have knowledge and familiarity with PPP/PFI but they do have a culture of thinking about risk and of responding to risk. However, it would appear that the risk assessment expertise and structures that do exist in the public sector contribute only in a very minor way, if at all, to PPP/PFI. There is an obvious case to be made that PSRM expertise should be an integral part of the process.

There may be a fundamental mismatch in some aspects of the risk management expertise of the private sector and the public sector. Some public authority risk managers, in particular NHS risk managers, are likely to have a limited involvement and thus limited knowledge of the insurance and alternative risk transfer market. We could question whether they really are able reliably to assess and monitor (on an ongoing basis) the risk strategy of the private sector partner. Admittedly, this may be a skill that can be bought in by the public sector for PPP/PFI purposes.

No matter what good guidance is given on procedures, risk analysis will remain far from an exact science. Many assumptions will have to be made in the quantitative and qualitative data underlying assessments. Some of those assumptions are likely to be wrong and the chances of this happening are exacerbated by the typical length of a PPP/PFI contract. Also, regardless of any lingering doubts about the accuracy of the risk analysis, there may be yet more problems with what happens next i.e. after the responsibility for carrying risks is divided. When risks are transferred to the private partner, their strategy to bear them is likely to be gauged in terms of a financial capacity (by insurance or some other funding mechanism). This may not reflect the cost to the Trust or local authority in terms of damage to reputation. For an NHS Trust, a local authority or a government department to be able to say that a risk was anticipated and that the private partner is now footing the bill for it may be of little comfort to patients or service users suffering from the absence, delay or reduction in a service. It is true that the private sector are not entirely immune from reputational risk but those that govern the public sector are increasingly subject to extraordinary pressures of this kind. They might do well to take full advantage of their organisation's own risk professionals.

References

10 Downing Street Newsroom (2001) *Prime Minister Announces Review of Government's Management of Risk*, www.number-10.gov.uk/news, 19 July 2001.

Accounts Commission for Scotland (1999) *Shorten the Odds: a Guide to Understanding and Managing Risk*. ACS, Edinburgh.

ALARM (1997) *Membership Directory and Handbook*. ALARM, Exmouth.

Audit Commission (1997) *Insurance Arrangements and Risk Management*. Technical release TR/26/97, Audit Commission, London.

Audit Commission (2001) *Worth the Risk: Improving Risk Management in Local Government*. Audit Commission, London.

Bains M. (1972) *The New Local Authorities: Management and Structure*. HMSO, London.

Broadbent J. & McLaughlin R. (1999) The Private Finance Initiative: clarification of a future research agenda. *Financial Accountability and Management*, **15**(2), 95–114.

CIPFA (2001) *Risk Management in the Public Services*. Chartered Institute of Public Finance and Accountacy, London.

Cullen, the Hon. Lord (1996) *The Public Enquiry into the Shooting at Dunblane Primary School on 13th March 1996*. HMSO, London.

Department of Health (1997) *The New NHS: Modern Dependable*. The Department of Health, London.

Department of Health (1998) *Working Together*. Department of Health, Wetherby.

Department of Health (1999a) *Public Private Partnerships in the National Health Service: the Private Finance Initiative*. Department of Health, London.

Department of Health (1999b) Press Release 1999/0760 Wednesday 15th December 1999. *New PFI Manual to Ensure Best Value for Taxpayers*. Department of Health, London.

EVH (2001) *Developing and Implementing an Effective Risk Management Strategy: a Guide for RSLs*. Employers in Voluntary Housing, Glasgow.

Fennel D. (1988) *Investigation into the King's Cross Underground Fire*. HMSO, London.

Fone M. & Young P. (2000) *Public Sector Risk Management*. Butterworth Heinemann, Oxford.

Gaffney D. & Pollock A. (1997) *Can the NHS Afford the Private Finance Initiative?* British Medical Association, London.

Grout P. (1997) The economics of the private finance initiative. *Oxford Review of Economic Policy*, **13**(4), 53–67.

Harrow J. (1997) Managing risk and delivering quality services: a case study perspective. *International Journal of Public Sector Management*, **10**(4/5), 331–352.

Health Education Authority (1992) *Health at Work in the NHS*. Health Education Authority, London.

Health Services Advisory Committee (1999) *Safe Disposal of Clinical Waste*. Department of Health, London.

Hidden A. (1989) *Investigation into the Clapham Junction Railway Accident*. HMSO, London.

HM Treasury (1991) *Appraisal in Central Government: a Technical Guide for Government Departments*. HMSO, London.

HM Treasury (1994) *Risk Management Guidance Note*. HMSO, London.

HM Treasury (1997) *Appraisal and Evaluation in Central Government*. HMSO, London.

Hood C. (1991) A public management for all seasons? *Public Administration*, **69**(4), 3–19.

Hood C. & Rothstein H. (2000) Business risk management in government: pitfalls and possibilities. In: *Supporting Innovation: Managing Risk in Government Departments*, pp. 21–32. Report by the Comptroller and Auditor General. National Audit Office, London.

Hood J. & Allison J. (2001) Local authority corporate risk management: a social work case study. *Local Governance*, **27**(1), 3–17.

Hood J. & Kelly S. (1999) The emergence of public sector risk management: the case of local authorities in Scotland. *Policy Studies*, **20**(4), 273–283.

Hood J. & McGarvey N. (2001) *Managing the Risks of Public-Private Partnerships in Scottish Local Government*. Unpublished manuscript.

Isaac-Henry K. (1997) Development and change in the public sector. In: *Management in the Public Sector: Challenge and Change*, 2nd edn (eds K. Isaac-Henry, C. Painter & C. Barnes), pp. 1–25. International Thomson Business Press, London.

Kerley R. (1994) *Managing in Local Government*. Macmillan, London.

Lawton A. & Rose A. (1994) *Organisation and Management in the Public Sector*, 2nd edn. Pitman, London.

Medical Devices Agency (1998) *Medical Device and Equipment Management for Hospitals and Community-based Organisations MDA DB 9801*. Medical Devices Agency, London.

Moss F. (1995) Risk management and quality of care. *Quality in Health Care*, **4**, 102–107.

National Audit Office (1996) *Health and Safety in NHS Acute Hospital Trusts in England*. Report by the Comptroller and Auditor General. HMSO, London.

National Audit Office (2000) *Supporting Innovation: Managing Risk in Government Departments*. HMSO, London.

National Audit Office (2001) *Supporting Innovation: Managing Risk in Government Departments*. HMSO, London.

NHS Executive (1997) *Controls Assurance Project Guidance: First Principles*. Department of Health, Leeds.

NHS Executive (1999) *Controls Assurance Statements 1999/2000: Risk*

Management and Organisational Controls. Health Service Circular (99) 123, issued 21 May 1999. Department of Health, London.

NHS Management Executive (1993) *Risk Management in the NHS*. NHSME, Leeds.

Osborne D. & Gaebler T. (1992) *Reinventing Government*. Addison Wesley, Reading, Mass.

Paterson I. (1973) *The New Scottish Local Authorities: Organisations and Management Structures*. Scottish Office Development Department, Edinburgh.

Price L., Maclaren W., Stein W. & Dickson G. (2000) *A Comparative Study of Accident and Emergency Nurse and NHS Trust Management Perceptions of Risks*. Zurich Municipal, Farnborough.

Scottish Executive (1999) *Clinical Negligence and Other Risks (Non-Clinical) Indemnity Scheme (CNORIS)*. Health Department, Directorate of Finance. Management Executive Letter NHS MEL (1999) 86, Issued 21 December, 1999. Scottish Executive, Health Department, Edinburgh.

Scottish Executive (2000) *Mental Health Reference Group – Risk Management*. Scottish Executive, Health Department, Edinburgh.

Scottish Office (1997) *Designed to Care: Reviewing the National Health Service in Scotland*. HMSO, London.

Standards Australia (1999) *Risk Management: AS/NZ 4360: 1999*. Standards Association of Australia, Strathsfield.

Vincent J. (1996) Managing risk in public services: a review of the international literature. *International Journal of Public Sector Management*, **9**(2), 57–64.

Walshe K. & Dineen M. (1998) *Clinical Risk Management: Making a Difference?* The NHS Confederation, London.

Williams S., Michie S. & Pattani S. (1998) *Improving the Health of the NHS Workforce: Report of the Partnership on the Health of the NHS Workforce*. The Nuffield Trust, London.

PART THREE

International perspectives on public-private partnership risks and opportunities

10 Public-private partnership risk assessment and management process: the Asian dimension

Robert Tiong and Joseph A. Anderson

10.1 Introduction

Over the past decade, international investors have sought a meaningful role in the infrastructure sectors of key Asian markets such as China, India, Thailand, the Philippines, Vietnam and other countries. Governments in the region have faced a continuing need to upgrade and expand their electricity generation and distribution systems, water treatment and distribution facilities, road and rail transportation systems and telecommunications facilities. Asian governments have sought private investment in infrastructure development, primarily through build-operate-transfer, build-own-operate, and concession arrangements, as a means of tapping new sources of capital. Even in the current investment climate, some observers have claimed that 54 800 megawatts (MW) of capacity may be open to development in India's electricity sector by the private investors through 2005; and 50 000 MW of capacity needed in China by 2010 could be open to private investment (*International Private Power Quarterly*, 2001).

The last decade was witness to a host of foreign investors establishing operations in Asia in order to pursue opportunities in the infrastructure sector. However, most large-scale infrastructure projects are financed using non- or limited recourse project financing, a form of debt financing in which lenders rely exclusively on the revenue stream generated by an infrastructure project as the source of loan repayment. Since lenders assume most of the risk in such projects, providing 70–75% of the capital costs in the form of debt financing, lenders have developed risk allocation requirements which must be met in any international project financing. The success of privately developed infrastructure projects in Asia is therefore highly dependent on the ability of investors to meet the risk allocation requirements of international non-recourse lenders and to assure that legal commitments made in connection with their projects are upheld once funds are invested. Although a number of high-profile projects have been successfully financed in markets such as China, India, Thailand, the Philippines, Pakistan and Indonesia, each of these countries has also hosted projects developed by international investors which are either in default or have faced other difficulties, such as the inability to achieve financial closure because of bureaucratic inefficiency or political opposition. Asia's track record with respect to effective public-private partnerships (PPP) in the infrastructure sector is therefore mixed. Lessons have been learned, however, which have the potential to provide a strong foundation for the future of private sector participation in the Asian infrastructure market.

10.2 PPPs and the Asian infrastructure sector

No matter their ideological inclinations, Asian governments have turned to private sector investors due to the unavailability of public financing to meet the demand for new infrastructure within their countries. In the power sector, the typical cost of building a new power station is approximately $1 million per megawatt. For a government to finance a new 1000-MW power station at a cost of $1 billion is prohibitive, especially in cases where the public sector is subject to tight credit, as with state-level governments in India or national treasuries such as in Vietnam or Pakistan. Even in countries such as Thailand, the Philippines and Indonesia, which experienced strong growth during the 1980s, the demands for spending in areas such as healthcare, education, and general government operations competed with the need for infrastructure spending. These circumstances created the opportunity for private sector involvement in infrastructure development.

The form of private sector participation for major projects has typically been either build-own-transfer, build-own-operate, or an operating concession for a fixed period. Build-own-transfer arrangements, also known as BOT, and build-own-operate or BOO, are typically used for the construction and operation of discrete facilities. Facilities such as water treatment plants or independent power projects are usually developed in most Asian markets under BOO or BOT arrangements. Under a BOO arrangement, a private investor will develop, finance, construct and operate a project on the basis of an offtake contract under which a government-owned utility will agree to purchase the output of the facility. In the power sector, private investors enter into power purchase agreements with utilities for a period of 20–30 years. The utility agrees to purchase electricity generated by the power station on a take-or-pay basis, producing a fixed revenue stream which the investor can utilise as the basis for non-recourse financing. The investor owns and operates the power station, and is free to mortgage its assets to lenders as part of the long-term debt financing provided by international financial institutions. BOT operates in much the same manner as BOO, with the only difference being the obligation of the investors to transfer the project to the ownership of the government upon the expiration of the offtake contract. BOO structures have been more common than BOT, although notable projects such as the Shajiao B and Shajiao C power stations built by Hopewell Holdings in China have been developed on a BOT basis.

Concession arrangements are typically utilised in projects which sell their output directly to consumers on a retail basis, such as fixed line tele-communication facilities or water distribution facilities. Investors are given the exclusive right to operate such facilities during the term of a concession contract in exchange for a payment or series of payments to be made to the government agency which grants the concession. The investor is usually obligated to invest in the expansion and upgrading of the facility, with the ability to pass along the investment costs to consumers. Tariffs charged to consumers, however, are usually regulated, and approval must be obtained by the investors from the appropriate government regulatory agency prior to increasing tariffs. Projects such as the East and West water concessions in Manila, the Philippines, and the fixed line telecommunications operating concessions in Indonesia were organised using concession arrangements.

Relatively few infrastructure projects in Asia are organised in the form of a direct partnership or joint venture with the host country government. The nature of the PPP is typically more indirect, with the government agreeing to delegate the right to build and operate infrastructure facilities to private investors and in some cases, as with a power station or water treatment plant, to purchase output generated by the facility. The reasons that governments agree to private sector involvement typically involves the need to tap into private sector capital and expertise due to scarce government resources, rather than any ideological imperative to privatise. In fact, private sector infrastructure development has sometimes experienced delays due to the tension between a government's desire to attract private (usually foreign) capital and its ideological objections to a wide-ranging role for the private sector in the infrastructure sector. Vietnam is a good example of a country which announced during the early 1990s its desire for foreign, private investment in the power sector, even passing a BOT law. To date, however, there has not been a major power project in Vietnam which has achieved financial closing. This has been due in part to the difficulty of negotiating power purchase agreements and related documents with the applicable government agencies and the reluctance of government officials to furnish investors with appropriate assurances regarding the long-term security of each project's revenue stream. Project finance professionals involved in Vietnam's infrastructure sector have remarked that one of the reasons for the delay in the development of projects is the continued distrust of the private sector held by many in government.

Vietnam, however, is an exception. Most countries in Asia have a history of state control in sectors such as electricity, water, telecommunications, and transportation. However, most governments have been very pragmatic regarding the need to utilise private capital in the infrastructure sector. Rarely has there been extensive ideological opposition to a role for the private sector in infrastructure development. To the extent opposition has arisen to particular projects, the controversy has usually centered around the economics of the project, as with Enron's Dabhol power project in India, or environmental considerations, as with the opposition by non-governmental organisations to cross-border hydroelectric facilities under development in Laos.

10.3 Role of non-recourse financing

The extent of successful private sector-led projects in Asia's infrastructure sector is dependent on the availability of non-recourse project financing to fund most of the capital costs incurred in connection with most major projects. Non-recourse debt financing is usually not recorded as a liability on the balance sheet of a project sponsor. Since many major infrastructure projects incur capital costs in the hundreds of millions of dollars, most international investors will proceed with such projects only to the extent non-recourse financing is available. The success of private sector involvement in Asia's infrastructure sector is therefore reliant on the ability of major projects to meet the risk allocation requirements of international non-recourse lenders.

Non-recourse risk allocation requirements are based on certain principles that exist in any international market, whether the country is a developed economy or an emerging market. As a threshold matter, lenders will want assurances that the project will have the ability to generate sufficient revenue to meet debt service obligations for the term of the loans. The non-recourse financing market grew significantly in the US following the passage of the Public Utilities Regulatory Policy Act of 1978, which required US utilities to purchase electricity from independently-owned electricity generators using cogeneration or alternative fuels. Many of these facilities were financed with non-recourse project financing. Lenders were comfortable that publicly listed US utilities with strong balance sheets would have the financial resources required to meet payment commitments to independent power projects under power purchase agreements that typically ran for a term of 20 years. On this basis, numerous projects were financed with an 80/20 or even a 90/10 debt–equity ratio. A few projects were even financed with 100% debt. When the Asian infrastructure market opened for private and foreign investment during the early 1990s, especially the electricity sector, many investors hoped to duplicate in Asia the successes that had been achieved in the US market.

10.4 Public sector credit risks

A number of key conditions which contributed to the success of the US project finance market were absent from most Asian markets. Perhaps the most significant consideration in the electricity sector was the absence of creditworthy utilities operating under transparent legal and regulatory regimes. Virtually all the utilities in Asia were state-owned, and many were in poor financial condition. Lenders were therefore not confident that such utilities would have the financial capability to meet their payment obligations under long-term power purchase agreements. In most Asian markets, the primary issue which impeded financing was whether the government would step in and provide guarantees to investors to assure that payment would be made under power purchase agreements in the event that the utility defaulted on its payment obligations. In India, where most of the state electricity boards were subject to severe financial difficulties, the central Government agreed to provide counter-guarantees to seven 'fast-track' projects to support guarantees issued by state governments. In the Philippines, the Government provided guarantees to support the National Power Corporation's payment obligations under energy conversion agreements entered into with independent power projects. The Indonesian Government generally resisted providing guarantees, although the Ministry of Finance did issue 'comfort letters' in support of power purchase obligations in a number of projects. The enforceability of such comfort letters, however, is considered highly doubtful by many. In Thailand, projects were financed based on an understanding of Thai law which held that the Government was obligated to stand behind the third party payment obligations of state-owned companies and agencies, including the Electricity Generating Authority of Thailand.

The provision of state support enabled non-recourse financing to be successfully obtained by a number of high-profile projects throughout Asia. In Thailand, the Rayong electricity generating station was privatised in 1994 through the use of $770 million in non-recourse financing, including a tranche of debt provided by US institutional investors. This was followed the next year by the privatisation of another major power station in Thailand, the Khanom project, and by a number of independent power projects developed under EGAT's 1994 tender for independent power projects. In India, Enron successfully financed the Dabhol gas-fired power project using international non-recourse debt, and Mission Energy and its partners achieved financial closing for the Paiton coal-fired project in Indonesia. Similar success stories were seen in markets such as China, Pakistan, and the Philippines, as developers succeeded in meeting the requirements of international non-recourse lenders. Successes were not restricted to the electricity sector. Telecommunications projects such as the Aria West concession in Indonesia were able to achieve financial closing, and financing was procured for a small number of transportation projects, such as the Bangkok Light Rail Project in Thailand and the EDSA Light Rail Project in the Philippines.

Lenders were willing to provide non-recourse financing to Asia's independent power sector due in part to the strong economic conditions which existed in 1993–1997, when most of the financings were closed. In instances where governments provided direct guarantees of offtake obligations, as in India's Dabhol Project, lenders felt confident that any concerns about the credit of the utility offtaker were satisfactorily addressed through explicit legal recourse to the government's own balance sheet. In projects which did not have explicit guarantees, such as the Indonesian power projects in which comfort letters were issued by the Ministry of Finance, lenders were confident that the country's strong economic conditions would enable the utility to meet its long-term payment obligations, thereby enabling the borrower to pay debt service.

10.5 Risk management

Periods of heightened volatility bring the issue of risk management to the top of every investor's and lender's agenda. As recent events have shown, lenders will be just as concerned with foreign exchange risks, in particular the convertibility and availability, as they are with political risks (Boey, 1998). In a 1997 survey of 188 Japanese companies by the Nikkei weekly on risks that they face in doing business in Asia, foreign exchange risk was cited as the top problem. Lenders and investors become more conservative in difficult economic times regarding their requirements for political and commercial risk coverage. However, governments' ability to provide guarantees is constrained by the plunge in the value of their currencies and drop in sovereign and bond ratings. A range of risk management measures will therefore need to be explored if projects are to proceed to financial closing in the current environment.

The issue of offtake credit was only the initial, threshold risk management concern faced by international non-recourse lenders. Since power purchase agreements were entered into with state-owned utilities under common control with the agencies which regulated the projects, lenders were also concerned that the government could take actions by changing laws or regulations in a manner adverse to a project. In the power purchase agreement, such changes in law were classified as 'government *force majeure* events' (beyond one's control), along with other acts or omissions by government agencies which might impair the successful operation of a project. Upon the occurrence of a government *force majeure* event, the utility would still be obligated to make the fixed payments, or capacity payments, required under the power purchase agreement so that the project could continue to meet its debt service obligations. If a utility default or a government *force majeure* event continued beyond a set period, the project's investors and lenders were given the option of selling the project back to the utility, at a price sufficient to pay down all outstanding debt and provide compensation to the equity investors. These contractual measures were intended to insulate a project's revenue stream from any political risks which might occur in a particular market.

Lenders and investors had other means of mitigating political risks. In many markets, it was felt that the involvement of local partners with strong political connections and a high profile in the domestic market would help provide protection against changes in the political climate. The involvement of public sector lenders was also viewed by private commercial bank lenders as a useful risk management strategy. Lenders surmised that if a state-owned utility or a government guarantor defaulted on its payment obligations, an export credit agency or a multilateral lender would be in a better position to compel the government to meet its obligations to the project investors and lenders. A number of major power projects, such as Hub River in Pakistan, which was supported by the World Bank, Meizhouwan in China, which included the Asian Development Bank as both a lender and an equity investor, and Paiton in Indonesia, which included public sector financial institutions such as the US Export-Import Bank, Japan Bank for International Cooperation, and US Overseas Private Investment Corporation, arguably would not have been financed without the participation of public sector lenders.

10.5.1 Political and commercial cover

The support of multilateral lending agencies and ECAs is perhaps even more crucial in the current project finance market. According to a recent report by Project Finance International, lenders' credit committees are cutting back uncovered exposure and are asking for political risk and commercial risk cover for loans to Asian countries. The $2.2 billion 3000 MW Shandong power plant in China (Table 10.1), whose financing was closed recently, proved that such deals can work. It is the first time ECGD of UK provided 100% political and commercial risk cover for a project in China. The participation of Hermes and CESCE in the Rizhao and Hanfeng financing and of COFACE in Laibin has demonstrated that substantial progress has been achieved on limited-recourse export credits for China, in particular with regard to regulatory and tariff approval issues and foreign exchange concerns (Edwards and Kuan, 1998). In the case of the PT Jawa power project in Indonesia (Table 10.2), the political risk cover was covered by US Exim Bank and Germany's Hermes.

Table 10.1 *Shandong Zhonghua power project.*

Total cost	$2.2 billion
Borrower	Shandong Zhonghua Power, comprising Shandong Electric Power (36.6%), China Energy Investment (29.4%), Electricité de France International (19.6%) and Shandong International Trade & Investment (14.4%)
Debt	$1.484 billion, lead arranged by Greenwich NatWest IBJ Asia and Société Générale
Offshore commercial loan	$350 million, 12-year loan with a margin of 185 basis points over Libor pre-completion and 175bp post construction
ECGD-backed loan	$312 million, 17.5-year loan at 57.5bp
Renminbi loan	$822 million provided by China Construction Bank and Sitic
Sponsor equity and interest	16 million
Equipment suppliers	Mitsui Babcock Energy (providing boiler units) and Shanghai Turbines (a subsidiary of Westinghouse)

Table 10.2 *PT Jawa power.*

Sponsors	Siemens, PowerGen, Bumipertiwi Tatapradipta
Arrangers	Toronto Dominion, Sanwa, Credit Suisse, Dresdner
Tranche 1	
Amount	$389 million
Tenor	Construction plus 12 years
Margin	137.5 bp (construction)
	155 bp (years 1–4)
	165 bp (years 5–8)
	180 bp (years 9–12)
Cover	Political risk from US Ex-Im
Agent	Credit Suisse
Tranche 2	
Amount	$444 million
Tenor	Construction plus 12 years
Margin	50 bp
Cover	95% political and commercial from Hermes
Agent	Dresdner
Tranche 3	
Amount	$250 million
Tenor	Construction plus 15 years
Margin	110 bp
Cover	100% political and certain commercial risk cover from the German Government under the BKA Instrument
Fund provider	KfW
Tranche 4	
Amount	$82 million
Final maturity	15 years
Margin	150 bp (construction)
	165 bp (years 5–8)
	175 bp (years 9–12)
	190 bp (years 13–15)
Cover	Uncovered
Agent	Credit Suisse
Tranche 5	
Amount	$200 million
Final maturity	15 years
Cover	Uncovered
Fund providers	US institutional investors
Agent	Credit Suisse
Adviser	JP Morgan
Lead institutional investors	TIAA, American Prudential

10.5.2 Managing foreign exchange risks

Foreign exchange risks result from the mismatch between the currency of payment under the offtake contract and payment obligations for taxes (usually in local currency), operating expenses (sometimes in hard currency), debt service payments (mainly hard currency) and dividend payments and profit repatriation (mainly hard currency). For example, in the electricity industry, most power purchase agreements in the Asian independent power market require payments in local currency. However, most equipment supply and construction costs, together with the cost of offshore financing and the equity returns of international investors, are payable in US dollars. Most power purchase agreements addressed this issue through indexing the local currency payments to the value of the US dollar. This approach has proved workable in countries such as Thailand and the Philippines, where no known defaults have occurred notwithstanding the collapse in value of the Baht and the Peso, respectively, during the 1997 Asian currency crisis. However, payment defaults did occur in markets such as Indonesia where currency devaluation was accompanied by political collapse. The Indonesian Government was not in a position to support the payment obligations of PLN, Indonesia's utility, as it was preoccupied with the systemic upheaval which accompanied President Suharto's sudden departure from office following approximately 30 years of strongman rule.

10.5.3 Domestic equipment and finance

For infrastructure project developers, obtaining debt financing in local currency helps to mitigate the risk of local currency devaluation. In China, there is a trend towards the use of domestic equipment and finance. In the Shandong power project, a sizeable local Renminbi tranche (Renminbi loan equivalent to US$822 million was provided by China Construction Bank and Sitic) was combined for the first time with a large US dollar tranche. Due to the cost advantages of manufacturing in Asian markets, a number of major equipment suppliers are exploring the establishment and/or expansion of manufacturing capacity in key Asian countries. If they prove successful, these manufacturers could offer an important contribution in reducing the currency risks faced by governments and investors in regional BOT projects. Alternative collateral packages could also be explored such as the right to receive royalty revenues from a resource project with hard currency revenues.

10.5.4 Survey on mitigating measures

A recent survey was conducted (Wang *et al.*, 2000) involving 40 project financiers and developers on effectiveness of mitigation measures for exchange rate and convertibility risks in China's power projects. The options that are presently available to foreign companies in China wishing to convert

RMB into foreign currency include: direct conversion through Chinese banks, currency swaps with other companies, balancing between projects, dual currency contracts, and special hedging measures for RMB. However, from the survey, obtaining government guarantees on exchange rate and convertibility, e.g. fixed rate or tariff adjustment or concession extension to cover the cost, was ranked as the most effective measure for mitigating the risks. As shown in Table 10.3, this measure is regarded as much more effective than the other two measures. The second effective measure is to use dual-currency contracts with certain portions of tariff to be paid in RMB and other transactions denominated in foreign currency. (These measures were partially adopted in the Laibin B project.)

The use of hedging tools that are specifically tailored to China's partially convertible currency by foreign banks is the third but less effective measure, according to the survey. Foreign banks are strictly limited in the scope of their transactions allowed within China using RMB. However, banks in China and Singapore have come up with ways around these limitations to respond to demand by foreign companies in China for ways to protect against devaluation of the RMB. One example is the 'Non-Delivery Forward', a hedging instrument where no actual RMB cash is required for the transaction. Instead the entire transaction is conducted in foreign currency outside China, with the payout tied to the official closing RMB rate posted each day by the State Administration for Exchange Control. Other measures suggested by the respondents for this risk include: (1) linking tariff to exchange rate change; (2) using an experienced-in-China bank or financial institution to arrange financing; (3) obtaining SAEC guarantees plus foreign exchange adjustment formula in PPA with sufficient support from a government authority; or (4) using an offshore branch of a Chinese bank to support foreign currency requirements.

Table 10.3 *Effectiveness of mitigating measures for exchange rate and convertibility risk.*

Mitigating measure	Effectiveness	
	Mean score	Ranking
Obtain government's guarantee of exchange rate and convertibility, e.g. fixed rate or to adjust tariff or extend concession to cover the cost	3.97	1
Use dual-currency contracts with certain portions to be paid in RMB and other transactions denominated in foreign currency	2.94	2
Use hedging tools, e.g. forward, swap	2.50	3

0, not applicable; 1, not effective; 2, fairly effective; 3, effective; 4, very effective; 5, extremely effective.

10.6 General risk mitigation requirements

Non-recourse lenders' participation in Asian projects was based primarily on contractual protection intended to address risks such as changes in law and regulation, negative exchange rate fluctuations, and the availability of adequate financial resources to meet the payment obligations of state-owned utilities under offtake agreements. In addition, lenders relied on practical measures such as the involvement of well-connected lenders and investors as a strategy for helping to mitigate political risks. The credit and political risks which lenders sought to address in Asian projects presented the most significant challenge to the wide-scale use of non-recourse project finance in Asia. However, investors seeking to tap the non-recourse financing market must meet other risk allocation standards typically required by international lenders. Such requirements are not unique to Asia or to other emerging markets, and must usually be met in any market where a project is financed with non-recourse debt.

Construction period risk is a major concern of non-recourse lenders. During construction periods, which can last anywhere from 1 to 2 years, projects have not commenced commercial operation and are not typically generating any revenue. Furthermore, under most power purchase agreements and similar offtake or concession contracts, a project must achieve commercial operation by a certain date or face termination by the government. If significant construction delays occur which are not excused by *force majeure*, the lenders could be placed at the risk of losing most of the funds invested in a project if construction is not completed in accordance with the required schedule. Non-recourse lenders mitigate such construction risks through requiring project borrowers to enter into lump-sum turnkey construction contracts. Such contracts require a single contractor to assume full responsibility for the performance of all sub-contractors and equipment vendors, and to provide performance and schedule guarantees backed by liquidated damages to assure that a project is completed on time and in accordance with required specifications. The turnkey contractor will typically receive a premium incorporated into the lump-sum price as compensation for assuming such a broad range of risks.

Lenders sometimes prefer that a third party operator be retained to manage a project's operations following the completion of construction. The project borrower will enter into an operations and maintenance contract which requires the operator to comply with annual budget restrictions and to pay damages if specified performance guarantees are not met. However, due to the limited fee income which operators are eligible to earn on an annual basis, most operators seek to impose stringent limits on their annual liability under operations and maintenance contracts. Despite the relatively small amount of damages payable by an operator, lenders often prefer to involve third party operators as part of their strategy of allocating as many risks as possible to third parties such as construction contractors, operators, and suppliers. As noted above, this principle is observed regardless of whether a project is located in an emerging market region such as Asia or in a developed country.

Another major risk mitigation strategy observed by international non-recourse lenders is the control over a project's operations and cashflow provided in the credit and security documentation entered into by project lenders and borrowers. The project borrower is organized as a single-purpose company with no other business aside from the construction and operation of the project. The project's cashflow is placed under the effective control of the lenders through the provisions of a disbursement agreement which allocates project revenues to a series of 'waterfall' accounts, with payment of dividends to the project's sponsors falling last in priority. When revenue is received by a project borrower, the funds are typically deposited into a receipt account, and subsequently into an operating account in order to meet monthly operating expenses, debt service accrual accounts in order to meet debt service payment obligations which might arise on a quarterly or semi-annual basis, and reserve accounts maintained to address various risks. Only after all these accounts are fully funded may revenues be deposited into the borrower's account from which dividends are paid to project sponsors. Dividends are usually only payable following a debt service repayment date, provided that no events of default have occurred and the project has maintained a required debt service coverage ratio or met similar financial tests.

Project sponsors are heavily restricted by the terms and conditions of the credit and security documentation in actions they might take with respect to a project's operations. The sponsors are typically prohibited from transferring any shares in the project company without the lenders' consent during the early years of operation and must maintain compliance with a long list of covenants set forth in the main credit agreement throughout the repayment period of the loans. Such credit agreements also typically contain a detailed list of events of default, upon which lenders may take control of a project due to the security interest maintained over all project assets, including assignment of project contracts, mortgaging of real property assets, and pledging of shares in the project borrower. Lenders therefore have the ability to take control of a project should threats emerge to the integrity of its revenue stream, and to remove the sponsors if they have not been successful in managing the construction or operation of a project.

10.7 Case study: Hub power project

Hub power plant is the first BOO-type power project in Pakistan. It was initiated in the late 1980s and developed in the 1990s. The project is an oil-fired power plant comprising four turbine-generator units with a gross installed capacity of 4×323 MW at a new site. The size of the project and its investment environment at that time resulted in one of the more complex financing structures to date. Negotiations occurred over 8 years, and 59 lenders participated in the external financings (*Power in Asia*, 2001). Although the project achieved commercial operation, its sponsors engaged in a drawn-out legal battle with the state-owned Water and Power Development Authority (WAPDA) after Benazir Bhutto's Government was thrown out of power. The independent power project crisis in Pakistan is a good case to examine the effectiveness of risk management in privately financed projects and Hub power project is worth studying in detail to illustrate the role of the importance of integrated risk management.

10.7.1 Financial structure

The overall financing required for the project was over US$1.7 billion, including standby funding. Debt financing was required and had to be raised from a large number of financial sources. However, at the outset of the project, international banks had modest credit limits for Pakistan, and domestic banks had limited lending capacity.

With the support of the World Bank and export credit agencies of Japan, France, and Italy (i.e. JEXIM, MITI, COFACE, SACE) in the form of political risk guarantees and insurance, the project successfully reached its financial closure after many years of negotiation and structuring. The financial structure included equity equal to US$371 million (some 25% of the project funding), subordinated debt equal to US$572 million, and senior debt totalling US$823 million.

The project financing introduced a number of innovative features: (1) utilisation of guarantees for political risk on the offshore commercial debt by the World Bank and JEXIM; (2) utilisation of multilateral and bilateral funds for project financing through the Private Sector Energy Development Fund (PSEDF); and (3) utilisation of Islamic financing principles.

10.7.2 Developing contractual structure

The project's participants included two sponsors, five arrangers, five guarantors, four contractors and several different Pakistani government agencies. This resulted in very complicated contractual arrangements, which reflected the requirements of risk management.

Following an approach by the Government of Pakistan to the private sector, Xenel, a Saudi company, identified the opportunity to develop Hub power project. Since it was not an electrical power generator or a construction contractor, Xenel sought to include a construction consortium led by Mistui & Co. Ltd in the promoting sponsor group in order to make proposals bankable. Furthermore, to ensure that the project structure and arrangements would be accepted by international financial institutions, National Power plc was invited to join the promoting sponsors as the operator for the power plant. The promoting sponsors provided mobilisation finance, and financed 20–25% of project in the form of equity. In addition, they agreed to provide standby equity of up to US$50 million.

The project was constructed by using a date-certain, fixed lump-sum price turnkey contract with a consortium led by Mitsui & Co. Ltd of Japan. The obligations of the consortium were supported by bank bonds and guarantees. Furthermore, to reduce the construction risk, the contractors' work was supervised by K&M/Raytheon Ebasco, the project company's consultant, under a Consultancy Service Agreement. The agreement was guaranteed by Raytheon Engineers & Constructors, Inc., its parent company.

The project is operated by National Power International, whose obligations are guaranteed by National Power plc. Pakistan State Oil Co. Ltd (PSO) agreed to supply the project's fuel under a 30-year fuel supply contract. WAPDA agreed to purchase the output of the project under a 'take-or-pay' Power Purchase Agreement. Both the fuel supplier's and the power purchaser's obligations are guaranteed by the Pakistani Government. Moreover, the project company entered into an exchange risk insurance scheme with the State Bank of Pakistan (SBP) to cover the currency-related risk.

10.7.3 Government support

To facilitate the financing of independent power projects, the Government of Pakistan established a fund known as PSEDF with the support of international donor agencies, particularly the World Bank and the United States Agency for International Development. The fund aimed at providing financial support for such projects in the form of long-term subordinated debt. As the first IPP, Hubco received over US$500 million from the Fund through the National Development Fund Corporation, the administrator of the Fund.

To attract foreign investments, the Government also established an exchange risk insurance scheme. The scheme covers currency risk such as availability of foreign exchange, free transfer of funds, exchange rate risk, and convertibility of local currency.

Finally, to enhance the creditworthiness of local participants, the Government agreed to guarantee the obligations of WAPDA under the power purchase agreement, PSO under the fuel supply agreement, and SBP under the foreign exchange risk insurance scheme.

Figure 10.1 provides a detailed overview of the contractual arrangements and the key measures of mitigating risks such as political risk guarantees/insurance from the World Bank and export credit agencies and government guarantees. The integrated, interdependent contractual structure enabled the project to be financed by limited recourse project facilities (Ye, 2001).

10.7.4 Lessons learned from Hub power project

The Hub River project became subject to one of the world's highest profile independent power disputes when the Government of former Prime Minister Nawaz Sharif cancelled the project's power purchase agreement, alleging that bribery had taken place when the contract was awarded. Following substantial pressure from the World Bank and the other lenders to the project, the current Government settled the dispute in December 2000. As

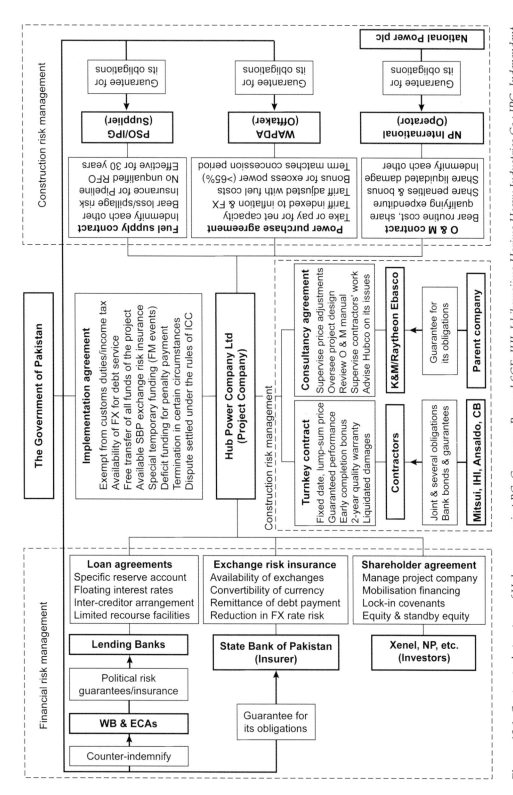

Figure 10.1 *Contractual structure of Hub power project.BC, Compenon Bernard SGE; IHI, Ishikawajima-Harima Heavy Industries Co.; IPG, Independent Petroleum Group; NP, National Power plc (UK); PSO, Pakistan State Oil Co.; WAPDA, Water and Power Development Authority.*

part of the settlement, the project's lenders and investors agreed to tariff reductions which will save WAPDA an estimated US$3 billion over the 27-year life of the project (*World Gas Intelligence*, 2001).

The Hub River dispute, which lasted approximately four years, demonstrates the value of public lender involvement in major emerging market infrastructure projects. The World Bank's leverage with the Pakistani Government was probably a key factor in the successful resolution of the dispute. It is doubtful that private commercial bank lenders would have been able to bring the same pressure to bear in negotiating a settlement. In the current market environment in Asia, most major infrastructure projects are proceeding only with political risk cover provided all or in part by public sector agencies or with indirect political cover through involving public agencies such as the Japan Bank for International Cooperation and the International Finance Corporation as part of the lender group.

The Hub River dispute also put lenders and investors on notice of the risk of undertaking projects in countries subject to political upheaval. Pakistan has had a history of instability since the death of its founder, Mohammed Ali Jinnah and the assasination of his successor, Liaqat Ali Khan. Since then, a series of military governments has presided over the country, with intermittent periods of democracy led by a small group of dominant families. When systemic collapse occurs in this type of political environment, the needs of international investors and lenders are likely to be low priorities for a government which is struggling to survive. Lenders and investors are therefore likely to participate in markets such as Pakistan only with the cover afforded by institutions such as the World Bank, if at all.

10.8 Testing of risk management measures

The 1997 Asian currency crisis, together with market-specific problems in countries such as India, challenged many of the risk management strategies developed by international investors and lenders in connection with Asian projects. In Pakistan, the Hub River project was the subject of an extended payment default even with the involvement of the World Bank in the project. Conventional wisdom had held that an emerging market government would be very reluctant to allow a project backed by key multilateral lenders to default on any of its obligations. In Indonesia, national utility PLN became insolvent and defaulted on most of its power purchase obligations with independent power producers. The collapse of the Suharto Government and the following political upheaval seemingly overwhelmed the ability of the Indonesian Government to negotiate a timely settlement with the investors and lenders involved in the different projects. In China, where the effects of the currency crisis were not as pronounced as in southeast Asia, many small and mid-sized power projects experienced payment defaults when municipal and provincial utilities, who often owned an interest in the independent power projects from which they were purchasing electricity, refused to make required tariff payments. The Indian market was also marginally affected by the Asian currency crisis. However, most infrastructure projects under development in that country proceeded at a very slow pace due to changes in regulatory policy, the difficulty of securing required approvals from multiple state and central government agencies, and financial difficulties faced by various state electricity boards. The disputes over Enron's Dabhol power project also impaired the confidence of international lenders in India's project finance market. That project had been backed by a central government counter-guarantee, and was the first 'fast-track' project to have achieved successful financial closing. However, upon a change of government in the State of Maharashtra, the new BJP/Shiv Sena Coalition Government charged that the tariff under the power purchase agreement was too expensive, and accused the predecessor Congress Party Government of engaging in fraud in connection with the negotiation of the power purchase agreement. Although the initial dispute was settled, following another change in government in Maharashtra, a new Congress-led Coalition Government has raised similar charges regarding the renegotiated power purchase agreement, resulting in payment defaults which have led the Dabhol sponsors to invoke their remedies under the power purchase agreement.

The impact of the Asian currency crisis on privately-backed infrastructure projects was not all negative, however. Notwithstanding the devaluation of the peso and substantial political upheaval, the Philippines has apparently honoured its outstanding payment commitments under energy conversion agreements entered into with private sector independent power producers. The Philippine Congress even managed to pass an omnibus electricity reform law during the first half of 2001 which will lay the basis for greater private sector participation in the industry's generation, transmission and distribution sectors. International lenders gave an important vote of confidence to the Philippines' infrastructure market through providing non-recourse financing to the 1200 MW Ilijan gas-fired project during 2000 and the CBK hydroelectric generating facility in 2001. The latter project was financed using privately furnished political risk insurance, without cover provided by public sector lenders. In Thailand, the Electricity Generating Authority of Thailand negotiated delayed commercial operations dates with independent power producers in an orderly manner, and otherwise observed its legal obligations to private sector investors and lenders. Notwithstanding the difficulties in China experienced by smaller-scale power projects, large-scale projects with substantial input from the central government, such as the Laibin B and Meizhouwan projects, proceeded without any disruption or defaults from state-owned counter-parties.

10.9 Effectiveness of mitigating measures

The effectiveness of risk mitigation measures employed by non-recourse lenders in Asia has been mixed. It has become evident that lenders and investors cannot rely solely on legal commitments to protect against political risks in Asia. Projects subject to default in countries such as Pakistan and Indonesia were very well structured from a contractual perspective, containing all of the risk mitigation provisions typically required by international non-recourse lenders in emerging markets. However, these projects were overwhelmed by the effective collapse of each country's political system, demonstrating that individual projects will not be accorded a high priority when a country is experiencing extreme political upheaval. On the other hand, the Philippines and Thailand experienced political turbulence, but commitments made to investors and lenders in the power sector were kept, perhaps due to a strong track record of successful development and financings in each country's independent power sector. In the future, investors and lenders are likely to approach analysis of political risk with even greater scrutiny, and probably proceed with their projects only with appropriate political risk insurance in place or with the involvement of public sector lenders such as export credit agencies and multilateral lenders. The involvement of public sector lenders and reliance on government-to-government relationships may revive a renewed emphasis based on recent experience. Despite initial difficulties, the Hub River project in Pakistan was able to negotiate a workout plan with the Government of Pakistan to address payment defaults faced by that project. The Pakistani Government's need to maintain a positive working relationship with the World Bank was probably a factor in achieving the settlement. Many believe that the power projects in Indonesia which have public lender involvement may be in a better position to negotiate a settlement with the Indonesian Government than projects which cannot call on intervention from other governments.

Recent experience has also taught both lenders and investors that they cannot rely solely on the legal commitments made in an offtake or concession agreement to assure that payments are made throughout its term. Many have concluded that the tariff must be priced on par or at a lower cost than a project's competitors in order to avoid the tariff becoming a political issue at some point in the future. In Asian countries with free presses and competing political groups, the costs charged to the public for basic needs such as water, electricity and transportation provide fertile ground for ambitious politicians seeking publicity. Investors and lenders must assure that even if a binding contract is in place calling for a specified tariff rate, the tariff should be defensible from the political attacks which are likely to be made in the future to the extent the public is made to feel it is overpaying for a project's output.

The experience developed by investors and lenders in Asia's infrastructure market, although mixed, is likely to provide a strong foundation for future projects. Market participants, after almost ten years of experience in both periods of growth and recession, have a good idea of what has worked and what has failed to effectively address market risks. New investors are beginning to emerge in the Asian infrastructure market, and lenders claim that financing is available for projects that are well structured and make good commercial sense. When investors and lenders proceed to apply their hard-won experience in pursuing new opportunities, projects likely to prove sustainable even in difficult political or economic times will hopefully result.

10.10 Conclusion

The Asian infrastructure market will continue to provide exciting challenges and opportunities. A lot can be learnt from the impact of the current financial crisis and the refinancing and restructuring of on-going projects. The future will require new innovative financing solutions, backed by rigorous and sophisticated risk management analysis and techniques, and tight contractual structures, in which the international debt capital markets, export credit agencies, multilateral lending agencies, international and domestic banks, and private investors will play an increasingly important role.

References

Boey K. (1998) A leaner and tougher PF market. *PFI Asia Pacific Review – News & Comment*, 4–6.

Edwards S. & Kuan O. (1998) China's power sector – trends and opportunities. *Project Finance*, 78–79.

International Private Power Quarterly (2001) Third Quarter, 19–37.

Power in Asia (2001) Hub Co's 1st Dividend in 4 years. *Power in Asia*, 30 October, Issue 340, p. 8.

Wang S., Tiong R., Ting K. & Ashley D. (2000) Evaluation and management of foreign exchange and revenue risks in China's BOT projects. *Construction Management and Economics*, **18**, 197–207.

World Gas Intelligence (2001) What's new around the world. *World Gas Intelligence*, January 11, **12**(1), 11.

Ye S. (2001) *A Study of Concession Design and Risk-return Trade-offs in Privately-Financed Infrastructure Projects*, unpublished doctoral dissertation, Nanyang Technological University, Singapore.

11 Risk management in an Austrian standardised public-private partnership model

Michaela M. Schaffhauser-Linzatti

11.1 Introduction

Public enterprises have always had an important share in the postwar Austrian economy. In 1985, 36.2% of all companies were publicly owned, of which 25.9% belonged to public federal institutions, 10.3% to municipalities and local corporations (AK, 1985). In 1987, Austria had the highest public sector share of all 22 OECD-countries (Schüssel, 1987). Policy makers realised that a higher private share would be necessary for a competitive Austrian economy in the long run. Furthermore, the high budgetary deficit had to be released if Austria wanted to join the European Union. Consequently, large-scale privatisation of nationalised industries started in the same year. Among others, important nationalised companies like OMV, VATech, Böhler-Uddeholm, Austrian Airlines, or the Vienna airport were sold through the stock exchange during the following 10 years. This step to privatisation started about 20 years later than in most other western European countries. Contrary to initial expectations, the Austrian sales of public companies have not been successful. The business performance of the privatised companies did not improve to a large extent, nor did their shares develop better than average at the Vienna Stock Exchange (Schaffhauser-Linzatti, 2000b).

The economic environment has changed significantly since privatisation has started. Austria has been a member of the European Union since 1996 and takes part in the Euro zone. It has to meet the Maastricht Criteria and observe the EC guidelines on competition and market liberalisation. These regulations also influence the organisation of European infrastructure. So far, high-volume investment in infrastructure has been made by public institutions or, in the rare case of private ownership, has at least been heavily subsidised. Now, policy makers must find alternatives as subsidies are forbidden and public institutions suffer from budgetary constraints. They face the problem that investment in infrastructure must be made to ensure the competitiveness of the overall Austrian economy, but not enough private capital is available to cover these necessary investments.

The concept of public-private partnership (PPP) fulfils the requirements of attracting private capital as well as ensuring a certain level of investment. Many countries already use this kind of co-operation for infrastructure projects in numerous fields, for example for transportation, education, or water supply. PPP is also slowly gaining recognition in Austria. Federal as well as local governments have already installed some PPP co-operations, for example for garages and terminals. They aim mainly at reducing subsidies; raising new equity and credit resources; and sharing risks with private partners (Gürtlich, 2000). These co-operations have been started by single initiatives. They are not co-ordinated on a national basis, nor do any general guidelines exist so far.

It is a challenge for politicians and academics at this stage of PPP implementation to develop such uniform definitions and legal frameworks. This chapter contributes to the discussion by developing an organisational model on how PPP projects could be structured in Austria. This so-called Austrian standardised model should help shorten the time for preparation and lead to reductions in planning costs. The term 'model' is defined as an organisation chart that includes ownership structure, contractual relations and fixed payments. It is based on a detailed analysis of existing Austrian experience, in-depth interviews with main representatives in charge of PPP projects, civil engineers, and lawyers. The Austrian standardised model integrates multiple aspects of a PPP project such as economic success or social services. Furthermore, it regards risk as one of the most important factors. Beside risks that occur in each project independent of its owner, the model must identify specific risks of PPP, i.e. political, legal, technical, and economic risks. It will be a challenge to allocate these risks to the participating partners who can best bear them in the Austrian standardised model.

This chapter is structured as follows: Section two explains the legal framework of PPP in Austria. Section three analyses two federal PPP projects, Climate-Wind-Channel Vienna and Cargo-Terminal Werndorf. Based on these cases, a so-called Austrian standardised model is developed in section four. Section five includes a classification and brief discussion of risks that are typical for PPP projects. These risks are further allocated to the participating public and private partners of the Austrian standardised model. Section six concludes the chapter and summarises the results.

11.2 Legal framework

The State as a territorial sovereign has two functions. First, it acts as a public authority towards its citizens. The Austrian State consists of the public bodies federal government, nine federal states, municipalities, and autonomous corporations such as legally obligated lobbies. Second, it acts as a private entrepreneur. Hereby, it abandons all rights as an authority and faces the identical duties and rights as its citizens. Within this private sector administration, the State has a legal capacity. While an individual person has legal status *per se*, artificial persons have to be founded by law. They may either be founded under civil law or as public corporations. The federal constitution (B-VG) is the most important legal basis on which the Austrian State is allowed to act as a public corporation. In contrast to the *ultra vires* doctrine in Anglo-Saxon countries, the Austrian legal system equates individual with artificial persons. The legal capacity of the federal government and the federal states is codified in para. 17 B-VG while para. 10 to 16 B-VG split all functions between federal government and the federal states according to Austrian federalism. Rights and duties of the municipalities are codified in para. 116 B-VG.

PPP belongs to the private sector administration. Within certain restrictions, the federal government and the federal states have the right to privatisation, municipalities depend on their corresponding federal or state laws. PPP must be seen as a special form of privatisation that can be split into three different categories (for a detailed discussion see Schaffhauser-Linzatti, 2000b). First, organisational privatisation, also called formal privatisation, is a specific kind of spin-off; second, task privatisation allows a private firm to fulfil public tasks (Gutknecht, 1994); and third, finance privatisation only transfers the funding for public tasks to a private entity. If public institutions also contribute financial means, PPP can be classified as a sub-category of finance privatisation (Kühteubl, 1998). The few legal definitions of PPP are not very precise. Para. 94 of the Schieneninfrastrukturfinanzierungsgesetz 1996 only mentions the term when it describes the conclusion of a contract about co-financing and profit realisation of rail links as one task of SCHIG (see next section). The government bill to the Strukturanpassungsgesetz 1996 states that PPP is a form of contracted co-funding for investments in private railroads and for the insourcing of new financing methods through third parties. These explanations are not satisfying. Therefore, this chapter summarises statements from political discussions and unpublished working papers and defines PPP as a tight co-operation between public and private firms, independent of the degree to which each partner contributes to the project.

The allocation of rights and duties in the B-VG gives the federal government greater possibilities for privatisation and consequently for PPP. Para. 10 sect. 6 B-VG allows the government to enact a special law for PPPs. The federal states are limited by para. 15 sect. 9 B-VG. In general, laws on budgeting, competition, data protection, or labour legislation further pose specific restrictions on privatisation and spin-offs. General limits are defined in para. 77 B-VG. It permits only single tasks to be transferred to individual companies out of public corporations and forbids spin-offs for whole sectors, but does not state clearly how to define those single tasks and whole sectors. This restriction will lead to discussions about the way in which PPP projects are installed now because the government tends to transfer whole sectors (Korinek, 1998).

Austria, as a member state of the European Union, also has to observe European laws. Among others, para. 5 states that the rules of competition have to be applied to public enterprises, and according to Directive 80/723/

EWG, public and private enterprises have to be treated equally. Para. 3a, 20, 30, 86 and 92 EG-V enforce privatisation, para. 90 and 90/3 even include further regulations for infrastructure and specific public sectors such as telecommunication.

These legal frameworks allow two different procedures through which federal PPP projects can be installed in Austria (Gürtlich, 2000). First, a PPP firm is established by a certain law that has had to be passed by parliament. Second, such a firm, in this article called a legitimated state-owned firm, becomes legitimated to implement PPP projects itself, but has to be installed on a federal basis (Kühteubl, 1998). It acts as a legally separate vehicle similar to the international experience of the Dutch Arlandabanan or the Italian TAV (TEN, 1997). This second possibility is on its way to becoming the preferred procedure for installing PPP projects. An important example is SCHIG, which will be discussed in the following subsection.

11.3 Examples of PPP projects in Austria

The following sections analyse the framework for PPP projects in Austria over recent years, with a view towards exploring the potential for a standardised model.

11.3.1 SCHIG

To understand the two PPP projects presented in the following subsection as well as the organisational model in subsection 4, we first describe a legitimated state-owned company, the Schieneninfrastrukturfinanzierungs-Gesellschaft (SCHIG). SCHIG is an example of how PPP projects can be implemented successfully. In general, public enterprises are responsible for establishing PPP projects in Austria. They are owned by federal or local public corporations and funded by public means as well as by profit earnings. One of these enterprises is the SCHIG. In November 1996, SCHIG was established by law, Strukturanpassungsgesetz 1996, BGBl. Nr. 201/1996, and Artikel 94, Schieneninfrastrukturfinanzierungsgesetz as a private limited company. About 60% of its financial means are provided by the Austrian Government, the remaining 40% are earned by charges for the use of rail infrastructure. Its main task is to assure railway construction in the long run. Among others, SCHIG finances new railroad connections; their extensions and modernisation; terminals for cargo transportation and passenger traffic; and park and ride facilities. SCHIG also concentrates on the commercialisation of railroad relations by submitting tenders to private operators, on the promotion of a rail development free of discrimination, and on the co-financing of railroad infrastructure projects (Gürtlich, 1999; SCHIG, 2001). According to this last aim, it is responsible for PPP projects in the field of railroad transportation. SCHIG represents PPP in two ways. On the one hand, it is responsible for installing companies with mixed public and private ownership. On the other hand, it may become a PPP company itself when the government allows a privatisation of up to 49%. For a detailed discussion of SCHIG see Schaffhauser-Linzatti (2001).

11.3.2 Climate-Wind-Channel Vienna

The Climate-Wind-Channel Vienna is the only plant worldwide that can test rail vehicles in extreme climatic and aerodynamic conditions. These examinations of thermic convenience as well as the functioning of single devices and whole systems support research. Positive results of these tests grant international certifications for the companies producing such vehicles. The old Climate-Wind-Channel had been in use since 1960. Instead of renovating the old facilities, the best solution seemed to be to build a completely new plant that was opened in November 2001. In contrast to Cargo-Terminal Werndorf, the Climate-Wind-Channel Vienna is not a new project, but 'only' a replacement investment of a very successful company.

So far, the Climate-Wind-Channel has been a purely public entity. The public institutions concerned faced the following starting position when the new facilities had to be planned:

- The location of the company should remain in Vienna in order to keep the existing competence in the city
- The companies engaged in the railway industry were not ready to finance the whole project themselves. So they should at least participate as partners in order to tie them firmly to the Climate-Wind-Channel
- Due to the large investment volume, the company did not get credit facilities for more than 20 years. This period corresponds to the useful life of the plant
- The guidelines of the European Community did not allow public subsidies.

Figure 11.1 shows the PPP model for the Climate-Wind-Channel that met all these requirements. The different arrows indicate relations of ownership structure (➡) as well as contractual relations (▷) and fixed payments and

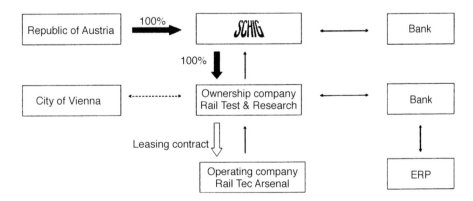

Figure 11.1 *PPP model Climate-Wind-Channel Vienna.*

repayments ($\rightarrow$, $\leftrightarrow$; broken lines symbolise variable payment agreements). Owned and built by a subsidiary company of SCHIG, Rail Test & Research GmbH, the Climate-Wind-Channel will be operated by Rail Tec Arsenal GmbH for at least 30 years. SCHIG as a public partner brings in about 10.9 million Austrian Shilling (AS) and takes a bank loan of 21.8 million AS. These figures are contributed to the Rail Test & Research GmbH as equity. The Rail Test & Research GmbH takes an ERP loan of 14.5 million AS with a pay-off period of five years and bank loans of 10.9 million AS with a pay-off period of ten years. SCHIG as a public partner collects market-oriented leasing fees of about 2.8 million AS each year with an underlying internal rate of return of about 7%. Hence, no hidden subsidies can be implied. Further, the Viennese Government had an intense interest in keeping this institution in Vienna. As the new Climate-Wind-Channel could have been build in any other place, the City of Vienna entered into the PPP project by an unusual credit contract. It granted a loan of about 13.2 million AS which has a payoff period of 40 years. Loan repayment and interest charges are arranged in a very flexible way.

The building and planning of the Climate-Wind-Channel was conducted by IGW, a consultant of Rail Tec Arsenal, and an international consortium of AIOLOS, VOEST, MCE, VATECH, and Elin EBG. Both contractors were authorised by the public Rail Test & Research in accordance with the private Rail Tec Arsenal.

It was a significant success to convince all large railway vehicle producers to become partners of the operating company Rail Tec Arsenal. Arsenal Research has a stake of 26% and ADtranz, Bombardier Transportation, GEC Alstrom Transportation, Siemens, and Ansaldo/Brede/Firema 14.8% each. Together, they contribute 2.9 million AS as working capital which corresponds to a one-year leasing fee. No regulations were made for the (unlikely) case that a large replacement of assets might be necessary during the useful life of 30 years (interview with Dir. Franz Hrachowitz, 2001; SCHIG, 2001).

11.3.3 Cargo-Terminal Werndorf

The Cargo-Terminal Werndorf is still under construction and will be opened in 2002. It is a terminal for cargo and combined transportation. Located 10 km south of Graz in the municipalities of Werndorf, Kalsdorf, and Wundschuh, it is optimally connected with railroad and motorway routes. The idea of the project came up at the beginning of 1990, when the Austrian Federal Railways (OBB) planned a cargo terminal in Southern Styria. Due to strategic considerations and also taking into account the Yugoslavian war near this region, the OBB withdrew from its engagement in 1996. Local government and industry have soon recognised the importance of this terminal as a main reloading point for the city of Graz, the capital of Styria. As the only large terminal in the south of Austria, federal government has a major interest in it, too. Hence, it was decided to change the already advanced OBB project into a PPP model. Figure 11.2 shows the organisation structure of Cargo-Terminal Werndorf (SCHIG, 2001).

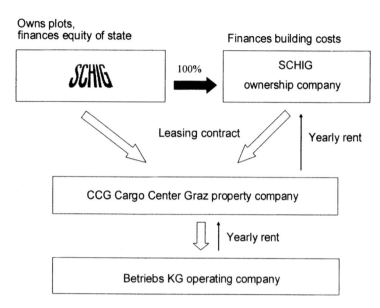

Figure 11.2 *PPP model Cargo-Terminal Werndorf.*

First, SCHIG as a subsidiary of the federal government founded its own project realisation company, the SCHIG ownership company, to finance the building costs. SCHIG and the SCHIG ownership company together then placed a leasing contract with Cargo Center Graz property company for 30 years. The Cargo Center Graz property company is owned by private companies, of which 51% is held by carriers, 25.1% by ESTAG, and 23.9% by the three main Styrian banks. When the project is finished, Cargo Center Graz property company will transfer the operation of the terminal to an own company, the Betriebs KG operating company, that has to pay rent. Another contract with Immorent GVG transfers the right to erect the main buildings on the premises and lease them to Cargo Center Graz property company. The SCHIG ownership company acts as the general developer. The construction itself is under the responsibility of the Hochleistungs-Eisenbahnen-AG that co-operates tightly with the SCHIG ownership company and Cargo Center Graz property company. Decisions are made by mutual agreement, a joint tracking system will be established. As the cargo terminal is still under construction, the financial structure is not yet fixed in detail. Probably being understated by 10%, the cost estimates for the whole project show a preliminary investment volume of about 109 million AS, whereof 34 million AS will be covered by equity contributed by SCHIG. Thirty-six million AS will be exclusively financed by the private partners who use it for buildings and equipment and may take recourse to ERP loans. Banks grant loans of about 39 million AS to SCHIG that in turn includes the corresponding interest and amortisation payments in the leasing fee of about 2.9 million AS per year. Hence, SCHIG invests about two thirds of the actual cash flow in advance, but, in fact, it is only participating with a contribution of about one third. If the economic success exceeds the estimated revenues, a clause in the contract will allow SCHIG to increase the leasing fee. At the end of the leasing period, an option enables the private partners to buy the terminal at favourable terms, i.e. approximately at the value of the valorised real estate (interview with Mag. Dr. Falschlehner, 2001; SCHIG, 2001).

11.4 An Austrian standardised model

So far, Austrian PPP projects have been established mainly as stand-alone solutions. Due to the increasing importance of this financial concept, their number will increase rapidly. From an academic and empirical point of view, it is now time to think about strategies that will help establish PPP projects efficiently and effectively. One possible strategy is the standardisation of certain elements such as the legal basis, special bank loans, or contractual frameworks with private partners. It accelerates the process of founding such co-operations and reduces the enormous costs in the early stages of these projects. We must be aware that such standardisations can only be a framework. It is obvious that each PPP project is not identical with undertakings in the same or even in a different sector. Deviations occur due to historical reasons, changing environments, the region, or simply different tasks. They force any prototypes to be adapted to the single project, but nevertheless, these adaptations will be more efficient than planning the project from the very beginning.

This section concentrates on standardising an organisational chart of a PPP project. It will be called the Austrian standardised model and aims at developing the basic structure on how public and private partners co-operate. Climate-Wind-Channel Vienna and Cargo-Terminal Werndorf served as role models for general considerations. Both projects have been successfully implemented by a legitimised state-owned company, SCHIG, that is responsible for the public part of financing and that authorises a public ownership company to rent the project to a private operating company. Structural differences between both projects result from their tasks and histories. The main reasons for recourse to Climate-Wind-Channel Vienna and Cargo-Terminal Werndorf are that both projects have simple structures with a clear task sharing clear responsibilities. They grant sufficient influence for the public institutions and at the same time do not restrict private partners. Advantages of both partners are exploited to the full. The Austrian standardised model merges their basic elements into an organisation chart and tries to keep out as many project-specific features as possible in order to be applicable universally.

Figure 11.3 presents the Austrian standardised model for PPP projects. It consists of three categories of involved parties, i.e. public corporations, private partners, and financial institutions:

(1) The Republic of Austria, the legitimated state-owned company, and the ownership company belong to the public corporations. Based on the

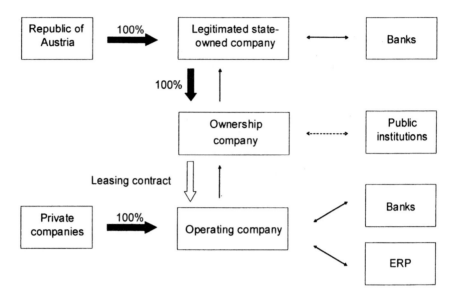

Figure 11.3 *Austrian standardised model.*

legal possibility discussed in section 11.2, the Republic of Austria establishes a legitimated state-owned company that will be responsible for a PPP project. The Republic of Austria transfers equity to the legitimated state-owned company to found an ownership company. This ownership company is responsible for planning and building the PPP project in co-ordination with the legitimated state-owned company and for finding private partners. The ownership company may also be supported by other public institutions, e.g. local governments or municipalities.

(2) The private partners are private companies that founded a joint venture, called the operating company. It concludes a leasing contract with the public ownership company that determines terms of fees, amortisation, responsibilities, or buyer's options.

(3) Banks and other financial institutions grant loans to public corporations and private companies by offering different credit terms. The Austrian standardised model takes into consideration a special form of credit that seems to be an Austrian peculiarity: public corporations like municipalities grant special loans to the public companies involved in the project. They offer flexible interest and repayment conditions that do not correspond to any regular credit terms. Hereby, they support the project in a way that is consistent with the EC directives of restricted subsidies.

When the Austrian standardised model is applied, the participating partners face the following advantages:

(1) The Republic itself is not involved in any operating business. This leads to two major consequences for the budget. First, the number of public employees is reduced, which is an important political argument in times

of budgetary cuts and reduction of civil servants. Second, as the Republic does not actively take part in the business as an operating partner, it does not face any unsure cash flows or obligations for loss compensation. On the contrary, it receives a contractually fixed income by means of leasing fees. These well-known, sure amounts can be used for other economically important projects in the budgetary forecast.

(2) Although publicly owned, the legitimated state-owned company is not under the direct authority of a ministry and political day-to-day business. It avoids the strict one-year budgets that are enforced by law. Further, it can follow economic criteria more easily, because it is not tied to public administration. The legitimated state-owned company will probably be entrusted with more than one PPP project. Therefore, it gains specific expertise that can be used for new projects.

(3) The ownership company is even more loosely bound to public bureaucracy than the legitimated state-owned company, but still has influential public support. It only concentrates on the establishment of one single project and knows the local conditions better than the headquarters. This expertise helps install the project in the region, for example by finding local partners or suitable locations.

(4) Usually, regional public institutions like municipalities or local governments are interested in the success of large infrastructure projects and will support them. The European Community prohibits subsidies or direct financial payments which have been in use before Austrian EC membership. Special loans granted by public institutions, however, enable the integration of both legal guidelines and economic interest in the project. These specific, flexible interest and repayment conditions do not correspond to any credit terms of banks. Hereby, the public institutions need not become partners, but engage actively in the economic policy. Moreover, they are not confronted with the popular reproach of still investing in nationalised industry or selling off national wealth by subsidies.

(5) Private partners support public activities with their efficient management know-how in the operating business. As private companies, they are not bound to the rules of public administration and hence can act more economically. At the same time, they benefit from the involvement of a strong public institution: many high-volume infrastructure projects can only be realised with the support of an influential legal and political partner. They could never realise such projects themselves. Furthermore, they can exploit the high creditworthiness of public institutions. Banks and other financial institutions usually grant better credit terms to public corporations because of their better credit ranking. The legitimated state-owned company raises high volume credits to favourable conditions and passes them on to the private operating company by means of higher leasing fees.

(6) Banks and other financial institutions do not face the high risk of PPP projects as they can take recourse on public corporations.

(7) The Austrian standardised model also faces certain disadvantages.

Among others, this model is at the edge of the EC directives concerning subsidies. Even small changes of these directives or an alternative interpretation would terminate its use. Then, political influence on the operation of the project still exists due to financial contracts or the involvement of the public ownership company in planning and building. Furthermore, the foundation of a PPP project depends on the interest of the legitimated state-owned company to install it, and on its financial resources. Lack of interest as well as lack of funds may prevent a quick establishment of the project or even the establishment itself, although the project may be favourable. And last, a complex instalment procedure is still necessary. The standardisation of the organisational chart does not prevent the public partners from detailed negotiations and contracts due to the individual character of each project. These critical points can hardly be avoided by any other model as they are induced by the PPP concept itself. The advantages of the Austrian standardised model, however, offer a great possibility to determine an organisational structure for PPP projects. It should be seen as a starting point for detailed and individualised models in order to save time and costs in the early stage of a project.

11.5 Aspects of risk

The Austrian standardised model concentrates on developing an organisational model for PPP projects. Its chart describes institutional processes, but implies multiple aspects such as economic success or social questions behind its ostensible structure that are not obvious at first glance. This chapter emphasises risk as one of these underlying aspects. It focuses its analysis on those risks that occur due to the specific co-operation between public and private partners. For a general classification and evaluation of risks in PPP projects, see Merna and Smith (1993), Bundesministerium für Verkehr (1996), UNIDO (1996), European Commission (1997b), TEN (1997), EIB (1999), Booz and Hamilton (2000), iC Consulenten (2000), Schaffhauser-Linzatti (2000a). Based on these analyses, risk can be roughly classified into four categories, i.e. political, legal, technical, and economic risks. These risks are disregarded here. For PPP projects, however, it is interesting to extract which uncertainties can emerge especially between the public and private partners due to their unequal co-operation.

In public opinion, political and legal risks are underestimated. In fact, the political and legal status within a country and its international partners is the basic condition on which a company is founded. It must trust the responsible leaders that these conditions are not changed at random. Such changes are initiated by political actors. Let us assume they know about the consequences of their agitation on their stakes in PPP projects *and* agree on them. Consequently, the political and legal risk in a PPP remains with the private partner that cannot influence these changes and has to take them as given.

Among others, the private partner faces the political risk of a cabinet reshuffle; politically induced referenda and public opinion against the project; as well as changes of policy makers, of relations to other counties, and of the political environment. For the private partner, the most important risk is public commitment to the project. A withdrawal of commitment or even a neutral position towards the project usually indicates its end. Another aspect that is never mentioned concerns elections. Politicians who are involved in a PPP project may be defeated in elections due to its imminent failure. Hence, they may change their position towards the project unexpectedly and in doing so cause damage to the private partner by acting against the projects that they have previously supported.

Legal risks for the private partner refer to changes in national and international legal positions or risks of legal finds without further appeal. These risks between public and private partners cannot be disregarded, but should not dominate risk management as the private partner may assume that public representatives do not harm their own projects. Nevertheless, the private partner loses more in the case of negative changes. It is concentrating on this single project with a high stake of its equity, while success or failure of smaller and politically less explosive projects do not noticeably influence public budgets.

Technical risks that affect the relationship between public and private partners in a PPP project depend on the character of each project. Most importantly, the private partner must trust the public promise to provide the missing infrastructure, for example feeder roads, while the public partner has to bear the risk of acquiring the necessary real estate.

Economic risks represent the major part of any risk management as most of all risks can be traced back to economic values. Important economic risks relate to early stage risks. They are defined in detail in TEN (1997) and comprise all risks during the first stage of the project, i.e. during the whole planning stage and the first years of operation. Although the European Union enforces PPP, its 'rules under the control of DG VII and XV could in fact increase such risks' (European Commission, 1997a) and hence go against the presumed intention. In this early stage, private partners depend very much on their public companions and face risks with regard to the public tendering process, instalment of competitive public projects, sunk costs if the project stops, changes in promised subsidies, or cost overruns due to political problems. When the project has left the early stage, the residual value and future reinvestments still depend on the decisions of the public partner. When the PPP company acts in a regulated market or a public institution is the dominating customer, uncertainties about price and volume of delivered performances should already be settled in the partnership contract to protect the private partner. Although most of these contracts include fixed cash flow payments to the participating public institutions, they still have to face certain economic risks against the private partner, for example its bankruptcy, reduced cash flows, exceeding the time limits for building the project, or lacking capacity to fulfil the project's requirements.

Organisational models can help reduce these risks between public and private partners by eliminating critical interfaces. As each project has different characteristics and structures, it is impossible to allocate all risks universally to a certain partner. All attempts are tied to specific models and define their own degree of accuracy. For example, an analysis of the Portuguese shadow toll experience shows the extent to which private and public partners can bear certain risks (MES-SEAOP, 1999). In general, risk allocation is the result of political agreements and not of theoretically derived ideal solutions. Partnership contracts are the major instrument in determining the tasks and consequently the risks of all partners. The European Union recommends contracts in which all partners 'commit themselves to respecting a predefined set of rules ... and which establish final responsibilities and risk-sharing' (TEN, 1997). However, they should be flexible enough to 'allow renegotiations of some terms and conditions if necessary' (European Commission, 1999).

For the two PPP projects in Austria that were discussed in section 11.3, the public and private institutions agreed on a strict separation of the business activities. They did not contract special agreements on how to deal with risk between both partners except one regulation for the Climate-Wind-Channel Vienna concerning market risk. The private partners are not allowed to cancel the partnership contract during the first ten years. Due to this regulation, the public partner only faces a reduced amortisation risk of its bank loans and does not have to fear intentional competition of another climate-wind-channel. Furthermore, the private partners must guarantee the utilisation of the facility by contract, otherwise they have to pay a fixed compensation. However, as all large railway vehicle companies participate as private partners and face a regular demand to test their engineering results, this risk is negligible. For the Cargo-Terminal Werndorf, the early risks were already covered as the project was launched a second time after the withdrawal of the ÖBB. SCHIG was made liable for the acquisition of real estate as it could gain releases from the land transfer tax and, if necessary, take advantage of a governmental expropriation law, the Eisenbahnenteignungsgesetz 1954. Consequently, it had to bear the risk of getting it in time and of paying excessive prices for it. For both projects, SCHIG bears the bankruptcy risk of the private partner. All other risks do not affect the direct co-operation between public and private partners, but are general risks that each single company and each joint venture has to face. These risks were estimated by well-known instruments like Value-Added-Risk models and are not relevant for this analysis.

The Austrian standardised model is derived from these two PPP projects. Hence, risk allocation is regulated in a similar way. In order to specify some peculiarities of the Austrian standardised model, we have to regard all political, legal, technical, and economic risks for each participating partner, i.e. the operating company, ownership company, legitimated state-owned company, banks and financial institutions, public corporations, private customers/contractors/suppliers, and the Republic of Austria. We will first list the most important risks that may occur for each kind of company, here called general risks. Then we will allocate these insecurities, that arise between the public and private partners and were discussed above, to the participating companies:

(1) Operating company: like any other company, the private operating company bears the full operational and financial risk, risks regarding permits and requirements, or enlargements and changes of the projects that are initiated by the operating company itself. It is liable for any buildings and utilisation of technology. Consequences of an economic failure are the loss of dividends, in the case of bankruptcy the loss of equity contributions. If strategic considerations have led to a participation in the project, targeted synergetic effects are lost too. Special risks with regard to the public partner, the ownership company, are all political risks as discussed above, among others, changes of the commitment and the opinion of policy makers or changeover of these policy makers. Furthermore, the private operating company faces legal risks such as changes in the law induced by the public partner; technical risks such as missing infrastructure or areas that have been promised; and economic risks such as profit margins, cash flows, capacity overload, reinvestment, or residual value that can be influenced significantly by the public partner.

(2) Ownership company: this public partner chooses the private partners and is responsible for their reliability. It has to perform and supervise the planning and construction of the project and bears the majority of the basic technical risks. As a special feature of the Austrian standardised model, the franchisee participates in the planning and construction decisions, and from the very beginning has to bear its share of risk. Consequences of economic hardship are a reduction or even loss of the leasing fee. In a worst case scenario, it loses all assets and goes bankrupt itself. The ownership company is an intermediary between the private operating company and its owner, the legitimated state-owned company. The specific risks between partners completely depend on the contracts on both sides, but in the end, the final responsibilities for these risks are with the legitimated state-owned company.

(3) Legitimated state-owned company: in the standardised Austrian model, the legitimated state-owned company initiates the PPP project. It has to care for its basic requirements and cover the large part of all early stage risks like a publicly induced stop or major changes to the project. It accounts for the ownership company by taking over all its risks in case of economic problems. Additionally, the partnership contract will probably include tasks such as permits and requirements in the start-up period of the project, availability of real estate, admission to basic infrastructure, and public support in case of unforeseen technical problems under construction or catastrophes during operation. The legitimated state-owned company is held responsible by the Republic for the whole portfolio of PPP projects. Hence, it bears the risk that one failed project may influence other subsidiaries in the holding, reduce financial

means for other projects or demand unfavourable reinvestment in similar projects to maintain infrastructure.

(4) Banks and financial institutions: in the Austrian standardised model, banks and financial institutions grant loans to public as well as private partners. In the case of bankruptcy, they will lose the outstanding loans and future interest payments less the contracted guarantees. With PPP projects, they do not face higher, but possibly even smaller risks than with purely private projects because of the higher creditworthiness of public institutions.

(5) Public corporations: those public corporations which grant preferred loans to the ownership company only bear the risk of delayed or reduced payment of interest and amortisation. In the worst case, they have to completely write-off their uncollected receivables, if the supported project does not succeed as required.

(6) Private customers/contractors/suppliers: public and private entities may be customers, contractors or suppliers of the PPP project. They face the same risks as doing business with any other company. Depending on the character of the project, these risks are reduced by the fact that public involvement leads to an increased public sense of responsibility against financial disturbances caused by its own economic activities.

(7) Republic of Austria: the Republic of Austria does not bear any risk from the operating business as it is not directly involved in any aspect of the project. It can be made directly liable in only a few cases, e.g. for the development of demand, if it is the dominating customer. However, the Republic has to bear all residual risks of the whole project as it is the last institution in the economic chain to be charged. It loses equity and cash flow in the case that the legitimated state-owned company breaks down or faces financial disturbances. When the private companies go bankrupt, the Republic suffers a reduction in income, for example for lower tax payments or unemployment. Furthermore, it will be necessary to substitute certain institutions immediately at higher costs, for example in the field of infrastructure. In the case of failure, it always risks political damage in a national or international context.

Summarising the risk allocation of the Austrian standardised model, the private partners predominantly bear the market risk, as they are engaged in the operational business. Banks and financial institutions only have to consider financial risks in regard to surety of interest payments and repayment of loans. The public entities and ultimately the Republic itself bear all residual risks as a failure of the private partners falls back on them completely. Therefore, customers and suppliers may face reduced risks as the involvement of public institutions gives a kind of back-up guarantee. As a result, the Austrian standardised model allocates those risks to the partner who can best bear them. Also from this point of view, it can be valued as a suitable organisational model for PPP models.

11.6 Conclusions

In Austria, the concept of PPP is gaining importance. It aims at reducing subsidies, raising new equity from private partners, circumventing guidelines about public budgets, and sharing risks. Most of the existing projects concern the transportation infrastructure. Investments in railroads especially have been conducted by PPP approaches. A specific enterprise, SCHIG, was founded by law to set-up and supervise such co-operations; others include the Climate-Wind-Channel Vienna and the Cargo-Terminal Werndorf. The organisational structure of both projects includes successful features of favourable credits, reducing possible conflicts between the public and private partners, or allocating risk to the partner who can best bear it. Hence, they serve as prototypes for a standardisation of an efficient and effective structure for PPP projects developed in this chapter. Under this so-called Austrian standardised model, a legitimated state-owned company is founded by law. It instructs a 100% subsidiary, the ownership company, to install and build a PPP project. The ownership company then rents this entity to the privately owned operating company. The co-operation between public and private partner results from the collective participation in planning and risk-sharing and hence emphasises aspects of risk as a dominating parameter in PPP.

There is much work left for further research. First, planned and existing projects in Austria should be compared and analysed in more detail. So far, only studies on federal projects have been published internationally, but local projects may also be of common interest. Several aspects of PPP have been described here. Theoretical approaches may address problems such as incentive structures, risk allocation, risk-bearing capacity, or principal-agent relations. Furthermore, deeper insights into the Austrian legal framework could lead to a clearer and easier application of PPP projects, especially for local or municipal institutions. As a whole, further research should consider standardisation of specific aspects such as organisational structure, legal framework, or risk evaluation instruments to a greater extent. This chapter is intended as a first step towards this discussion in Austria.

Acknowledgements

I am grateful to Mag. Wolfgang Anderl for supporting me with detailed information on the Austrian legal system. I would also like to acknowledge the assistance of the interview partners for discussing Austrian PPP models. In addition, I especially thank Professor Engelbert Dockner from the University of Vienna and an anonymous referee for helpful comments.

References

AK (Kammer für Arbeiter und Angestellte Wien) (1985) *Eigentumsverhältnisse in der österreichischen Wirtschaft*. AK, Wien.

Booz A. & Hamilton A. (2000) *PPPs in Practice Case Studies*. Presentation in SCHIG, December 2000.

Bundesministerium für Verkehr (1996) *Hochrangige Gruppe für öffentlich-private Partnerschaften 1996/97*, Präsentation durch Deutschland, Beispiele für private oder öffentlich-private Finanzierung von Verkehrsprojekten, Mdir. Dr. Sandhäger, Bonn.

EIB (European Investment Bank) (1999) *Framework for the Commercial Use of Outer Space. The Galileo Task Force on Public Private Partnership*. Brussels, December 1999, EIB.

European Commission (1997a) Directorate-General VII for Transport, Directorate A. *International Relations and Trans-European Network, Infrastructures and Financing*, Brüssel, VII/A3/AM/g, High-Level Group on Public-Private Partnership Financing of TEN Transport Projects. Brussels, February 1997, EC.

European Commission (1997b) Directorate-General VIII for Transportation, subgroup 4, *Final Report Sub-group 4*. March 1997, Brussels, EC.

European Commission (1999) Directorate-general transport. *Seminar on Public-private Partnerships*. Brussels, EC.

Gutknecht B. (1994) Die Privatisierung von Verwaltungsaufgaben, Bericht über die Tagung der Vereinigung der Deutschen Staatsrechtslehrer 1994 in Halle an der Saale. *Österreichische Zeitschrift für Wirtschaftsrecht*, 105.

Gürtlich G. (1999) *Modelle und Möglichkeiten einer langfristigen Finanzierung der Eisenbahninfrastruktur – gibt es Chancen für privates Kapital?* Federal Ministry of Transportation, Vienna, working paper.

Gürtlich G. (2000) *A Concept for Public Private Partnership in Austria*. Federal Ministry of Transportation, Vienna.

iC Consulenten, FCP, Fritsch, Chiari & Partner (2000) *Privatisierung der Infrastruktur, Projektstudien für den Schienenverkehr*. Presentation in SCHIG, 21.7.2000, Vienna.

Korinek St. (1998) Ausgliederung–Privatisierung–Beleihung, Bericht über das Symposium der Studiengesellschaft für Recht und Wirtschaft. *Zeitschrift für Verwaltung*, **296**.

Kühteubl St. (1998) Ausgliederung–Privatisierung–Beleihung, Bericht über das Symposium der Studiengesellschaft für Recht und Wirtschaft. *Österreichische Zeitschrift für Wirtschaftsrecht*, **56**.

Merna A. & Smith N.J. (1993) *Verband öffentlicher Banken, Leitfaden für die Erstellung und Beurteilung von Angeboten bei BOOT*. Internal project review, Vienna.

MES-SEAOP (1999) *Seminar on Public Private Partnerships, the Portuguese Motorway Concession Program*. Ministry of Public Works, Brussels.

Schaffhauser-Linzatti M. (2000a) *PPP-Algorythmus, Von der Projektidee zum Vertrag*. Presentation in SCHIG, 20.2.2001, Vienna.

Schaffhauser-Linzatti M. (2000b) *Ökonomische Konsequenzen der Privatisierung, Eine empirische Analyse der Entwicklung in Österreich.* Wiesbaden.

Schaffhauser-Linzatti M. (2001) Developing public private partnership models: the experience of Austria. In: *Public and Private Sector Partnerships: the Enterprise Governance* (ed. L. Montanheiro & M. Spiering). Sheffield Hallam University Press, Sheffield.

SCHIG (2001) Schieneninfrastrukturfinanzierungs-Gesellschaft mbH. *Business Report 2000.* SCHIG, Vienna.

Schüssel W. (1987) *Abmagerungskur für den Staat.* In: *Bundekammer der Gewerblichen Wirtschaft.* Wirtschaftspolitische Blätter, Nr 5/6, Vienna.

TEN (1997) High-Level Group on Public-Private Parternship Financing of TEN Transport Projects. *Final Report,* May.

UNIDO (1996) *BOT Guidelines.* UNIDO.

12 Risk assessment and management in BOT-type public-private partnership projects in China – with speed reference to Hong Kong

Mohan M. Kumaraswamy and Xue-Qing Zhang

12.1 Introduction

Other chapters in this book provide extensive coverage of risk management and public private partnerships (PPPs) in general. This chapter provides an overview of related trends and developments in China in general and in Hong Kong in particular. Although Hong Kong is now (again) part of China, the differentiation arises from the post-1997 'one country two systems' basis of the different legal and regulatory regimes and from the historical development of build-operate-transfer (BOT) procurement protocols in Hong Kong in the pre-1997 period. In this context, a brief background to the recent rethinking of construction project risk allocation in Hong Kong is provided at the outset.

McInnis (2000) cites a UK-based study which indicated that a sample of owners and contractors estimated that the way construction contracts are written can add about 5% to the cost of typical projects in general, while Shadbolt (2000) stresses that general principles of construction project risk allocation cannot be transplanted to private finance initiative (PFI)-type projects, identifying a need for more clearly identified and allocated risks in the latter scenario. On the other hand, Charlton (2001) draws attention to the development of two parallel (but apparently contrary) trends: with one moving towards 'extremely onerous conditions of contract where fixed prices and construction periods are demanded, with all risks placed on the contractor', while the other has moved towards risk sharing, for example through partnering and alliancing. These trends are examined in this chapter in the context of BOT-type procurement, where the risks are far greater in number, magnitude and complexity.

Whilst Renton (1997) says that 'the allocation of project risks in BOT projects is an art rather than a science', the present chapter examines how some 'science' may be profitably injected to complement 'state of the art' approaches to risk management in BOT-type PPP procurement of infrastructure. Experiences and 'best practices' from 'Mainland China' and Hong

Kong are drawn together with those distilled from other countries – enabling the formulation of strategies for improved risk management of BOT-type PPP projects, for example through more pro-active support frameworks, improved selection of concessionaires and the development of a distinct BOT body of knowledge.

12.2 Construction project risks and their management

Fellows (1996) provides a very useful summary of general approaches to construction risk identification, analysis (and quantification), allocation and response (by removing, reducing, avoiding, transferring or accepting). Smith and Bohn (1999) summarise previous literature classifying construction contract risks and mitigation measures. Risks are broadly classified as: (1) natural risks (e.g. from 'Acts of God'); (2) design risks (e.g. from scope changes or new technology); (3) logistics risks (e.g. from losses and delays due to late materials deliveries); (4) financial risks (e.g. from exchange rate fluctuations and inflation); (5) legal and regulatory risks (e.g. changes in regulations or problems with permits and licenses); (6) political risks (e.g. war or changes in trade laws); (7) construction risks (e.g. from quality, productivity, safety, weather or labour problems); and (8) environmental risks (e.g. from pollution). Corresponding methods of risk management have been proposed, e.g. through improved quality control procedures.

12.2.1 Hong Kong

The Hong Kong Government approach to construction risk management has been seen to have developed 'in a piecemeal fashion over the years' (Marriot, 2000), based on transparency and accountability for public money; with greater client price certainty being targeted by transferring most risks to contractors. The advisability of such large-scale transfers, e.g. of unforeseeable sub-surface (ground) conditions risks has been questioned from time to time (Kumaraswamy *et al.*, 1995). A government sponsored review of the General Conditions of Contract for Construction Works in 1998 (Grove, 2000) led to recommendations that included shifting back the unforeseen/unforeseeable sub-surface conditions risks to the Government. However, the Government has been slow to accept these recommendations and has apparently rejected some already. Some examples are listed in Table 12.1.

The recent 'Tang' Report of the Construction Industry Review Committee (CIRC, 2001), has, however, invited a revisitation of the Grove Report recommendations on risk re-allocation in government construction contracts. This would also help to align contractual arrangements with the more pro-active partnering-type co-operation advocated in the Tang Report. Risks and their allocation in private sector construction contracts could be similar, although a few differences are likely to be encountered. More differences will arise in different 'types' of contracts, for example in design and

Table 12.1 *Examples from a comparison of (1) present risk allocation and related practices, with (2) Grove's (2000) report recommendations and (3) initial Government responses.*

Risk	Existing provision	Recommendation (Grove, 2000)	Initial Govt. response
Changes in law	Contractor	Client	Accept
Ground conditions	Contractor	Client	Reject
Legal and physical impossibility	Client	Allow engineer to relax contractual requirement or issue variation	Reject
Third party interference	Contractor – cost, Client – time	Client should accept both	Reject
Breach of contract by employer	No specific provision	Should be introduced	Reject
Need to terminate	No provision to terminate without default	Should be introduced	Accept
Client's need to accelerate	No provision	Should be introduced with compensation to contractor	Reject
Care of the works	Contractor's risk except damage, loss 'excepted risks'	Require All Risk insurance coverage	Accept – on a needs basis
Notice and time bar provisions for claim	Notice – 28 days, particulars – 180 days after completion	Failure of notice should give rise to damages not forfeiture	Reject
Profit on claims for 'loss and expense'	No profit	Profit should be allowed	Reject
Global claim	No contractual prohibition	Should be contractually prohibited	Reject
EOT for the events not included in contract	Allowed for special circumstances	Should be avoided	Reject
Liquidated damages	Only for delay damages	Apply to performance deficiencies as well	Provided via special conditions
Dispute resolution	Engineer's decision, voluntary mediation and arbitration	Wider use of DRA and voluntary use of 'no-decision' mediation	Not yet decided
Contractor's post-contract alternative design	No provision to incorporate as a variation	Should be considered. Variations preferably issued on a daywork basis	Accepted

construct, BOT or management-type contracts (e.g. in management contracting or prime contracting). For example, Ahmed *et al.* (1998) compared the allocation of different risks in 'management' and 'traditional' type contracts, and the management of risks in the former, while BOT-type scenarios are examined in the rest of this chapter.

12.2.2 Other types of risks in BOT-type projects

The longer periods involved and wider responsibilities entailed in BOT-type contracts, together with the greater complexities and uncertainties, broaden the range of risk exposure of the project sponsor (e.g. the government), the concessionaire and the financiers. The 'non-recourse financing' principle of relying on project assets and cash flows alone, introduces further risks and imposes greater demands on risk management.

Analysis of such recently emerging risk portfolios has spawned a cluster of enlightening research exercises, for example: (1) on risks and risk management in project-related finance (Price & Shawa, 1997), which isolated seven factors that significantly influence bank finance decisions; (2) on identifying important risk factors in managing PFI projects (Akintoye *et al.*, 1998); (3) on risk identification frameworks for international BOOT projects (Salzmann & Mohamed, 1999) which compiled a comparative table of published risk factors, success factors and critical elements; (4) on risks encountered in major PPP infrastructure projects (Lam, 1999), which compared the risk mitigation measures, residual risks and risk consequences separately within different sub-sectors such as 'power' and 'expressway' projects; (5) on the risk management of specific projects (Zhang *et al.*, 1998), which examined the Yan'an Donglu Second Tunnel project in Shanghai; and (6) on the evaluation of specific risk such as foreign exchange and revenue risks in China's BOT projects (Wang *et al.*, 2000).

Focusing on South East Asian countries, Tam and Leung (1999) suggest that technical risks are the easiest to manage on BOT projects, with financial risks being a 'bit harder' and political risks being the most difficult to manage. This is borne out by their examination of BOT project breakdowns on the Bangkok Second Expressway System project, the Bangkok Don Muang Tollway System and on the Bangkok Elevated Transport System (Ogunlana, 1997; Tam, 1999). In another example, political risks arising from a change of state government severely disrupted the US$1.3 billion Dabhol power project in India (Kumaraswamy & Morris, 2002). Meanwhile, the realisation of other risks (other than political) has also precipitated project failures as outlined below.

12.3 Lessons from realised risks

Apart from the political changes that led to painful (if not fatal) upheavals in supposedly agreed BOT procurement protocols in the above scenarios (Tam, 1999), other obstacles to planned success have originated from inadequate identification and/or assessment of other risks. For example:

(1) While the Guangzhou–Shenzhen–Zhuhai highway in Southern China initially suffered a two year delay in financial closure due to the events in Tiananman Square in 1989, this was aggravated by delayed land acquisition, design changes, currency depreciation and incomplete control of post-completion traffic management (Lam, 1999).

(2) Over-estimated traffic flows on a bridge across the Nam Ngum River at Tha Ngone in the Lao PDR, led to the Government of that country buying out (at a loss) the Australian concessionaire's 50% share in the 50/50 PPP project (Kumaraswamy & Zhang, 2001), mainly because the expected traffic preferred a longer route to what were considered high tolls.

(3) A coal-fired power station in the south of Fujian Province in China is running at a loss since commissioning in 1999 because of an over-estimated power demand and a correspondingly over-estimated price which has been undermined by cheaper power now available nearby; and even the $3.2 billion invested in it by 'the Chairman of Taiwan's biggest private company' is now unlikely to be recovered within the agreed 20-year BOT period expiring in 2016 (O'Neill, 2001).

(4) Toll road operators of some BOT tunnel projects in Hong Kong (e.g. Tate's Cairn, Western Harbour Crossing and Route 3 Country Park Section) also complain of considerably lower revenues than estimated. Admittedly, elaborate toll adjustment mechanisms in the BOT agreements on the last two projects (these being more recent) permit toll increases if the actual net revenue drops below certain levels (Zhang & Kumaraswamy, 2001a). But toll increases could well lead to even lower traffic volumes, specially when alternative routes are available as with other cross-harbour tunnels (in the case of the Western Harbour Crossing), and the proposed Route 10 expressway may divert traffic from the Route 3 Country Park Section.

(5) Public opposition to some road projects on environmental or even personal/vested interest grounds could derail planned infrastructure development (Levy, 1996), who used the acronym NTFIMB to convey the 'no toll facility in my backyard' syndrome.

(6) The risk of 'running down' a facility closer to the handover, e.g. not spending enough time and money on maintenance towards the end of the operation period, needs to be guarded against (Zhang, 2001).

Some relevant questions that may be raised from a sponsor's (e.g. government) perspective are: (1) is a given infrastructure project really suitable for BOT-type procurement? (2) If so, which is the most appropriate vehicle from the many versions such as BOT, BOOT (build-own-operate-transfer) and BOO (build-own-operate) that are available? (3) How should the concessionaire be selected (against which criteria)? (4) What guidelines/conditions should be applied to ensure a satisfactory 'service' to the public, e.g. with regard to toll/tariff levels, quality of construction and operation? (5) What guarantees/assurances/'comfort letters' should be given to potential concessionaires to attract private investment?

Similarly, some relevant issues to be explored by concessionaires include:

(1) the critical success factors (CSFs) that would position them most appropriately to 'win' the concession from among competitors at bidding stage and to achieve the expected profits and other benefits; and (2) what guarantees/'comfort letters' to seek and which risk management strategies to adapt (Kumaraswamy *et al.*, 1999).

Thirdly, from the financier's perspective, a group of risks perceived by private lenders/financiers has been described by Woodward (1997), Fitzgerald (1998) and Wang *et al.* (1999), while potential mitigation measures for seven of these groups have been proposed by Fitzgerald (1998). These are consolidated in Tables 12.2 and 12.3.

12.4 Relevant BOT-type PPP trends in China

12.4.1 Enlisting private sector participation in rapid development

The economy of Mainland China has grown particularly rapidly since 1978. Urbanisation, industrialisation and population growth have also accelerated the demand for basic infrastructure; thereby leading the Chinese Government to emphasise infrastructure development even further in the eighth five-year plan (1991–1995), the ninth five-year plan (1996–2000) and the tenth five-year plan (2001–2005). However, national fiscal revenues and strengths have not been adequately boosted by this rapid economic development so as to independently finance the exponentially increasing demand for infrastructure facilities. The Chinese Government is therefore in active pursuit of supplementary private finance for its national infrastructure development. BOT-type procurement schemes have been tried, tested and are being improved. PPP in general is seen to be a convenient vehicle for this purpose and therefore also offers more business opportunities for foreign investors.

Pointers to infrastructure development programmes (that include, for example, 81 new power plants of at least 2000 MW capacity by the year 2010, 35 000 km of expressways and Class 1 highways over 30 years and 112 000 km of new provincial and country roads) are provided by Zhang and Kumaraswamy (2001b), who also describe the current laws and regulations governing foreign investments via three possible vehicles, all of which are usually mobilised as limited liability companies: (1) equity joint ventures; (2) co-operative joint ventures; or (3) wholly foreign owned enterprises.

12.4.2 Evolution of BOT procurement models in China

The BOT projects in China (in the 1980s) were initiated on the basis of sino-foreign (co-operative) joint ventures. A negotiated tendering system was used by provincial authorities to select a foreign partner for the co-operative JV, which was formed under the *Sino-Foreign Equity Joint Venture Law* and the *Sino-Foreign Cooperative Joint Venture Law*. The then State Planning Commission (SPC), Ministry of Power (MOP) and Ministry of Communications

Table 12.2 *General risk concerns of private financiers (based on Woodward, 1997; Fitzgerald, 1998; Wang et al., 1999).*

Type of risks	Descriptions
Currency risks	These include (1) inconvertibility risk – the host country does not have enough foreign exchange reserves; (2) transfer risk – the host country does not allow or restricts the transfer of foreign exchange out of the host country; and (3) local currency devaluation risk – whenever a lender lends in foreign exchange and relies for repayment on a borrower who generates revenues only in local currency that may depreciate in value, and may result in the borrower's inability to meet its foreign debt
Expropriation risk	The host country government may arbitrarily or discriminatorily nationalise a project without 'just compensation'. This type of risk is great in high profile projects that are often associated with public ownership. The expropriation can take the form of nationalisation through either 'wholesale' or 'creep' expropriation whereby the government changes laws to gradually control the project
Change in law risk	The host country government may introduce/change laws that consequently render a project unprofitable. These include changes and reinterpretation of laws and regulations, changes in the procedures to deal with inflation, currency conversion and transfer, taxation rates, tolls/tariffs, and imports/exports
Political violence risk	War, revolution, insurrection, civil strife, terrorism and sabotage can seriously affect the development process of a project, or can even destroy the project's ability to generate revenue streams for debt service
Delay in approval risk	The host government authority may not approve the project-related issues in time or even cancel those already approved. Obtaining approvals or permits for a project from various government departments can be extremely time-consuming and may even delay the entire project development process and impair the project's financial viability
Loan security risk	Most developing countries lack sufficient protection of creditor rights, because of (1) a primitive and rapidly-changing legal infrastructure; (2) a court system that may have no track record of enforcement of creditor rights due to the short history of underlying laws; and (3) the many countries that place significant restrictions on the ability of foreign entities to operate or purchase projects upon foreclosure
Law enforcement risk	The host country arbitration body may not be protective of foreign creditor's rights. Local law can make agreements with local entities problematic and enforcement virtually impossible
Host country entities' reliability risk	Many participants in a project are from the host country, such as contractors, suppliers, operators, guarantors, offtakers and the ultimate customers. The project success depends on the performance of all these entities
Corruption risk	The host country's government officials or representatives may use political, legal, or regulatory leverage to extract additional costs which no one will ever admit and the project developer can never recoup

Table 12.3 *Some possible risk mitigation measures from a lender's perspective (based on material from Fitzgerald, 1998).*

Type of risks	Possible mitigation measures
Currency risks	(1) Assess the host country's foreign exchange reserve position. (2) Obtain rights under local law to convert local currency into foreign currency and transfer the converted currency to the lender for payments of interest, fees and principal. (3) Establish an offshore account. (4) Obtain government support/guarantees on preferential access of the project to foreign exchange, conversion and transfer. (5) Index the purchase price of the output to inflation or to fluctuations in the exchange rate
Expropriation risk	(1) Internationalise the risk by co-financing the project with multilateral and bilateral agencies. (2) Establish an offshore account. (3) Provide safeguards against nationalisation of the project and guarantees of reasonable compensation in case of any nationalisation. (4) Lenders require the right to accelerate their loans upon any expropriation. (5) The borrower pledges all of its stock to the lender. (6) Lenders and shareholders insure their loans and equity investments with political risk insurers
Change in law risk	(1) Changes concerning import/export restrictions, price control and tax increase have significant effects on the project's profitable operation. Host government guarantees against these risks should be obtained. (2) Insure these risks with international political risk insurers. (3) Shift and share these risks with loan borrowers and output purchasers
Political violence risk	(1) Establish an offshore collateral account. (2) Obtain the agreement of the host government to provide security to the project. (3) Make political insurance with multilateral or bilateral political risk insurers
Government approval risk	(1) It should be a condition precedent to the 1st advance of the loan that all government approvals necessary for the development of the project shall have been obtained. (2) Host government support/guarantees on various permits should be obtained
Loan security risk	(1) Identify what type of security the local law provides and how the security is enforced. (2) Comply with all local formalities. (3) Determine how foreclosure and insolvency may work, and take appropriate measures
Law enforcement risk	Select an international arbitration body rather than one in the host country

(MOC) issued a joint circular in 1995 (the 'BOT Circular') entitled *Circular on Several Issues Concerning the Examination, Approval and Administration of Experimental Foreign Funded Concession Projects*. This BOT Circular identified the type of projects to be procured through wholly foreign-funded BOT schemes during the experimental period, and prescribed a general framework for the selection, approval, open tendering process and the setting up of wholly foreign-funded BOT project companies. A national pilot BOT programme was initiated based on this BOT Circular, demonstrating the commitment to its successful implementation.

However, the Asian Development Bank (1996, 1997) identified some critical issues to be addressed in order to attract foreign investments to Mainland China. For example, they identified: (1) the need for a comprehensive regulatory framework that is clear, transparent and predictable; and (2) the need for bidding procedures, documents and bid evaluation mechanisms that are efficient, effective and fair to all concerned parties. In dealing with these issues, the Chinese Government mobilised international consultants to develop standard documents for a set of five pilot BOT projects. These documents include a basic pre-qualification document, tender document and concession agreement. The United Nations Development Program also funded a review of China's regulatory environment and BOT initiatives, in order to assist in the formulation of new national guidelines, regulations and legislation needed for BOT, while the ADB provided technical assistance to two pilot projects: the Changsha power project and Chengdu water supply project. Many other power, water treatment and highway projects have been developed on a BOT JV basis.

12.4.3 Examples of BOT-type power projects in China

Evolving BOT-type project practices in China may be illustrated by the following comparison of examples of BOT-type power projects. The Shajiao B, Laibin B and Shandong Zhonghua power plants are typical power plants developed at different times and in compliance with different laws and regulations. Table 12.4 compares the main features of these three projects.

The Shajiao B power plant is the first BOT-type power project in China. It was developed in the mid 1980s by a JV [between the state-owned Shenzhen Special Economic Zone Power Development Co. and Hopewell Power (China) Ltd] under the *PRC Sino-foreign Joint Venture Law* and the *PRC Foreign Economic Contract Law*, and through a negotiated tendering system.

Although the planned operating period was only ten years, as in the original concession agreement, this period was reportedly extended due to various difficulties encountered in this pioneering BOT project. However, the construction and commissioning were carried out within the designated 33 months, also winning a UK construction industry award and leading to a follow-up contract for Hopewell to build the 3×660 MW cold-fired Shajiao C power plant (Electric Focus, 1992).

In 1997, the SDPC (State Development and Planning Commission, formerly SPC) and the SAFE (State Administration of Foreign Exchange) together formulated the *Administration of Project Financing Conducted Outside China Tentative Procedures* (Project Financing Procedures). The Shandong Zhonghua power project was the first project finance to be arranged on the basis of these Project Financing Procedures. The project feasibility was confirmed by the SDPC in March 1996, while the JV company, the Shangdong Zhonghua Power Company was formed in May 1997. This was before the promulgation of the Project Financing Procedures. To ensure compliance of the project with the 'Procedures', the lenders then required the JV company to make an adjustment report which was later approved by the SDPC, while the financing terms were approved by the SAFE.

Meanwhile, the Chinese Government launched a national BOT experimental programme based on the 1995 BOT Circular (issued by the former

Table 12.4 *Main features of typical BOT-type power projects in China.*

Features	Shajiao B	Laibin B	Shandong Zhonghua
Scope of project	2 × 350 MW coal-fired power plant	2 × 350 MW coal-fired power plant	Four coal-fired plants: Shiheng I (2 × 300 MW), Shiheng II (2 × 300 MW), Heze II (2 × 300 MW), & Liaocheng (2 × 600 MW)
Construction cost	US$512 million	US$650 million	US$2.15 billion
Contract award time	1984	1996	1994
Concession period	33-month construction & 10-year operation period	3-year construction period & 15-year operation period	BOO process (no transfer)
Contract type	Sino-foreign joint venture BOT	Wholly foreign-owned BOT	Sino-foreign joint venture BOO
Loan financing	HK$600 million syndicate loan; Japanese Yen 63 billion loan	12-year US$300 million French export credit; 10-year US$190 million commercial loan	17.5-year US$312 million export credit; 12-year US$350 million commercial loan; 15-year Rmb 6.8 billion yuan loan
Tender modality	Negotiated	Open competitive	Negotiated
Main governing laws or regulations	Sino-foreign joint venture law (1986); Foreign economic contract law (1985)	1995 BOT Circular; Wholly foreign-owned enterprise law (1986); Electric power law (1996); Foreign exchange regulations (1996); Company law (1994)	1997 Project financing procedures; Sino-foreign joint venture law (1986); Electric power law (1996); Foreign exchange regulations (1996); Company law (1994)

SPC, MOP and MOC) and the *Wholly Foreign-Owned Enterprise Law*. Five projects (two power plants, a water treatment plant, a bridge and a highway) were selected as pilot projects to be developed through BOT schemes. Standard project documents were developed by foreign consultants. These will be fine-tuned for future BOT projects. The Laibin B power project was the first wholly foreign-funded BOT project in which a public tendering process was used.

In terms of evaluation criteria, when evaluating Laibin B, the focus shifted from rate of return to tariffs and financial arrangements. The electricity tariff rate was given a 60% weighting. The initial tariff rate, annual adjustments, proportions of foreign exchange and the local RMB (Renminbi) currency in the tariff, and the tariff of additional net electricity output were all incorporated in this evaluation. The financing arrangement (the tenderer's financing schedule, financing cost, ability to finance, and extent of equity committed) was given a 24% weighting. The technical proposal and OMT

(Operation, Maintenance and Transfer) proposal were each given 8% weightings (Wang *et al.*, 1999). In terms of rights, obligations and risks, the concessionaire has the exclusive right to design, construct, operate and maintain Laibin B, and to sell electricity during the concession period. It is also permitted to mortgage or transfer the right to operate all assets, facilities and equipment of Laibin B for financing purposes. The concessionaire is required to take insurance to reduce and mitigate certain *force majeure* risks (beyond one's control), and risks related to construction and operation, but enjoys tax concessions. The Chinese Government guarantees the convertibility of RMB and assumes change-in-law risks, but stops well short of guaranteeing a specific rate of return (Zhang & Kumaraswamy, 2001b).

Instead, the concessionaire entered two important agreements with Chinese entities: (1) the Power Purchase Agreement (PPA) with the Guangxi Power Industry Bureau (GPIB); and (2) the Fuel Supply and Transportation Agreement (FSTA) with the Guangxi Construction and Fuel Corporation (GCFC). Under the PPA, the GPIB agrees to purchase a minimum output of 3500 million kWh of electricity in each operating year during the concession period. Under the FSTA, the GCFC agrees to supply fuel required by the concessionaire. The mechanisms for adjusting, as well as the initial setting up, of the electricity tariff and fuel price are incorporated in the PPA and FSTA respectively. The Guangxi Government accepts liability for contractual breaches by Chinese entities that are a party to project documents, including the PPA and FSTA.

Furthermore: (a) 'comfort letters' were provided to the tenderers (on foreign exchange risks) – to facilitate backing and export credit from their home country/Government and relevant bodies; and (b) the concession contract incorporates a clear and internationally recognised arbitration agreement. More information on these aspects are provided by Zhang and Kumaraswamy (2001b).

12.5 Relevant BOT trends in Hong Kong

12.5.1 Five BOT-based tunnel projects

The Hong Kong SAR has a population of a little less than 7 million within an area of about 1000 km^2, representing the highest population and traffic densities in the world. The territory is well known for having developed high calibre transportation infrastructure with 'state of the science' road networks, bridges and tunnels using 'state of the art' procurement systems, particularly with respect to the presently corporatised railway systems and five tunnels procured on a BOT basis. The first BOT-procured tunnel was in fact the first cross-harbour tunnel that was recently transferred at the end of the designated 30-year concession in 1999. This provides an excellent opportunity to review the full cycle and evolution of the Hong Kong BOT experience, which included two other cross-harbour tunnels: the Eastern

Harbour Crossing (EHC) and the Western Harbour Crossing (WHC) as well as the two land-based tunnels: the Tate's Cairn Tunnel (TCT) and the Route 3 Country Park Section [R3(CPS)].

The required legal framework for each of the five BOT tunnels was provided under a specific legislative enactment (an enabling ordinance). The body of BOT knowledge in the Hong Kong scenario has steadily grown in both public and private sectors. Commendable construction project performance levels have been achieved, for example in terms of high standards, early completion and few disputes. The fact that early completion would lead to earlier and longer revenue flows contributed to better teamwork, hence minimising some of the problems encountered in traditional construction project procurement.

'Smart' engineering solutions were developed to meet common objectives by more integrated teams. For example, considerably reduced construction periods on the Tate's Cairn tunnel project were achieved by replacing the originally planned single vertical shaft adit with two sloping adit tunnels that were constructed in advance and then used for construction traffic (Kumaraswamy & Morris, 2002). This enabled the opening up of more tunnel excavation faces, thereby facilitating more simultaneous tunnelling operations and reducing the construction period, thereby increasing the toll-earning period.

The completion of the BOT cycle on the Cross-Harbour Tunnel went smoothly with the transfer at midnight of 1 September 1999. The transfer plans were prepared from late 1997 onwards, and involved all relevant government departments. The Government paid for the reduced value of any machinery, equipment or plant forming part of the assets purchased by the franchisee with the agreement of the Financial Secretary within the five years immediately preceding the expiration of the franchise period. A new operator was selected through open tendering to take over the role of the former franchisee in running the CHT as a public tunnel under a MOM (management, operation and maintenance) contract. The MOM contract is for a period of two years (with a possible one-year extension).

The ownership of the CHT rests with the Government, and the operator is paid a fixed fee through deduction of tolls collected for its work in management, operation and maintenance. The Government sets up specific requirements for routine inspection and scheduled maintenance and repair work. Non-scheduled maintenance and repair work would be paid for separately. There was an initial concern that the operator may 'skimp' on its routine work in the hope that it may receive additional income by being instructed to carry out additional work when the situation deteriorates to a level that warrants treatment under the non-scheduled maintenance category. To address these concerns, a Government Monitoring Team (GMT) was set up to monitor its performance.

In fact, some post-transfer issues hit the headlines recently, when for example: (1) the response time to a minor vehicle fire was considered to be more than that expected of the operator; and (2) there were some delays caused by unexpectedly extended maintenance operations on the road surface and the tunnel lining in separate incidents. While these can be considered to be isolated incidents, they provide examples of issues to be better defined and enforced in the post transfer operation by a private operator.

12.5.2 BOT project procurement procedures

Whilst the Hong Kong Government was not short of capital for infra-structure for the major part of the 1980s and 1990s, it opted to divert funds to alternative uses, while mobilising private sector resources, including finance and expertise through the BOT vehicle, and sharing risks with the private sector. However, some major infrastructure projects may not merit BOT-type approaches, in that there may be too many long-term uncertainties, or they may not generate self-supporting revenue streams (that will cover construction costs as well). For example, it was realised that the 'Lantau Link' (the Tsing Ma bridge, Ma Wan viaduct and the Kap Shui Mun bridge) transport projects associated with the new HK International airport, would not be viable on a BOT basis.

The Government usually commissions a team of leading engineering, financial, legal and environmental consultants to conduct a multi-disciplinary feasibility study, in order to assess a project's suitability for BOT-type procurement. If found feasible, the Government prepares a Project Brief as part of the tender documents to: (1) explain the Government's general requirements with respect to the project and the concession, and to provide relevant information; (2) provide guidance in the preparation of tenders and explain the tender evaluation criteria; and (3) set out in detail the Government's requirements in design, construction, operation and main-tenance (Lloyd, 1996; Zhang & Kumaraswamy, 2001a).

The tendering process is monitored by the Independent Commission Against Corruption (ICAC), in order to guard against any possibility of corruption. The procedures used in the competitive tendering process for the five BOT tunnels include: (1) Gazette notification; (2) evaluation and shortlisting of tenders; (3) negotiations with the shortlisted tenderers and compilation of a draft project agreement; (4) Legislative Council approval and enactment of a special Ordinance; and (5) signing of project agreement and award of concession.

12.5.3 Concessionaire selection

The principal tender evaluation criteria used in selecting BOT con-cessionaires include: (1) the level and stability of the proposed toll regime; (2) the proposed methodology for toll adjustments; (3) the robustness of the proposed works programme; (4) the financial strength of the tenderer and its shareholders, their ability to arrange and support an appropriate financing package, and the resources they are able to devote to the project; (5) the structure of the proposed financing package including the levels of debt and equity, hedging arrangements for any interest rate and/or currency risks, and the level of shareholders' support; (6) the proposed corporate and financing structure of the franchisee; (7) the quality of the engineering design, environmental considerations, construction methods, including traffic control, surveillance, and tunnel electrical and mechanical, ventilation and lighting systems; (8) the ability to manage, maintain and operate effectively and efficiently; and (9) benefits to the Government and community (Hong Kong Government, 1992, 1993).

Tender evaluations examine the three main areas of (1) financial; (2) engineering; and (3) 'planning of operation and transport', assigning them relative weights of, for example, 65%, 20% and 15%, respectively. Tenders have been assessed in recent projects with the aid of the Kepner-Tregoe technique (Harris & McCaffer, 1995; Tiong & Alum, 1997). This technique first separates the 'musts' (essentials) from the 'wants' criteria, through which any tender that fails to meet any 'must' requirement is rejected at the outset. The degree of satisfaction of the 'wants' is next evaluated, with combined scores being derived for each tender, and tenders with the highest scores being then shortlisted for negotiation and possible award (Zhang & Kumaraswamy, 2001a).

12.5.4 Managing BOT project risks

Whilst the initial feasibility study and careful selection of a competent concessionaire are essential elements of initial risk management, details of other important tools that have been developed for this purpose in Hong Kong are described by Zhang and Kumaraswamy (2001a). These include descriptions of: (1) the purpose and principles of the enabling ordinance that is unique to each project; (2) the project agreement; (3) dispute resolution procedures; (4) management of contentious land issues that usually need special attention, design and construction processes where independent checking engineers are mobilised, and operation and maintenance issues, given that the design life of the tunnels is 120 years although the concession period only 30 years; (5) the toll adjustment mechanism which has been developed to enable a 'reasonable but not excessive return', but which is still a central issue; (6) transfer issues, which were recently tested in the transfer of the Cross-Harbour Tunnel in 1999; after a two-year preparation period involving many government departments and new legislation; and (7) post-transfer issues including formulating a MOM contract and selecting a new operator.

In terms of the basic risk allocation, the concessionaire takes on risks related to construction, engineering, geology, environment, climate, financing, inflation, and cost escalation. On the other hand, the Government takes on the risks of the timely provision of land for construction and operation, e.g. by agreeing to grant an extended concession period or compensation if land is not made available at the specified time, and the right to an extended time or compensation for any extra work required by the Government, or due to suspension of works, or provision of facilities in excess of the concessionaire's obligations. The Government also grants the concessionaire the right to terminate if special risks (such as war, riot or disorder) occur and prevent or impede progress for a long period. The Government or its agents work out a programme for implementing major infrastructure, road works, road improvements and traffic management in connection with a BOT tunnel, to facilitate smooth flow of traffic to and from the BOT tunnel. However, no guarantees or warranties are given with regard to the programme for commencement or completion. There is no tax exemption for the concessionaire, no governmental guarantees on minimum traffic flows or economic returns, and no guarantee against future competitive routes (Zhang & Kumaraswamy, 2001a).

The 'toll adjustment mechanisms' incorporated in the two most recent tunnel projects may provide some degree of relief for concessionaires, although of course market forces would drive away traffic if the toll is too high. This mechanism is a good example of evolutionary improvement of BOT practices in Hong Kong, in that while high operational revenue levels in the first cross-harbour tunnel were considered to attract investors for further BOT road tunnels, concerns soon arose on the adequacy of returns in the Tate's Cairn Tunnel and the Eastern Harbour Crossing. In the latter, a toll increase was agreed after arbitration (Tam & Leung, 1999).

Other than these toll-related matters, there is no clear evidence of any major dispute or arbitration between the Government and the concessionaire on any Hong Kong BOT tunnel project. Apart from their obligations being clearly defined in the contract documents, parties to the contract usually build up the good relationships that are needed for the long term and resolve most conflicts through negotiation. For example, a dispute on the EHC, where the Government asked for an additional traffic lane subsequent to the initial agreement was settled through negotiation. The only known arbitration encountered in the five BOT tunnel projects was in the R3(CPS), where the concessionaire was not even a party to the arbitration. The dispute related to the land for the new site of a relocated school, with third parties claiming ownership.

Still, it has been suggested by some well-positioned public and private sector participants in recent BOT-based projects in Hong Kong, that the Government should seriously consider providing more guarantees or 'comfort' clauses/letters, in order to increase the viability of potential BOT projects. There is of course a wide range of possible guarantees/warranties/ comfort cushions that may be provided to different degrees, e.g. foreign exchange and revenue repatriation guarantees, tax benefits, concession period extensions due to *force majeure*, as listed by Kumaraswamy and Zhang (2001).

12.6 Strategies for improved risk management of BOT-type PPP projects

This chapter focuses on China (including Hong Kong) in summarising examples of evolving and improving practices in risk management in BOT-type PPP projects, and their related strengths and weaknesses. However, the need for a broader overview (as provided by other chapters) is evidenced by dissatisfaction with, for example, the 'poor initial performance of toll road projects' in other countries, as described by Songer *et al.* (1997), who recommend enhanced risk analysis tools to provide improved information for pre-project decision making and performance improvements. Zhang and Kumaraswamy (2001c) describe details of critical improvements needed in PPP procurement frameworks under the 'heads' of: (1) suitable legal foundations (including laws, regulations and guidelines); (2) workable procurement processes (including timely mobilisation of suitable financiers); (3) a suitably supportive co-ordinating authority; (4) careful affordability assessments and proper marketing; (5) effective selection of the most suitable concessionaire; and (6) re-aligning mind-sets (including a shift of governmental mind-set from regulatory-judgemental to a more pro-active-liberal-dynamic mode).

In the context of the last item above (item 6), it is pertinent to draw parallels with significant trends towards partnering, alliancing and other forms of relational contracting – that are based on clearly identified common objectives, trust, co-operation, collaboration/coalescence and joint risk management (Rahman & Kumaraswamy, 2002). Mumford (1998), whilst suggesting 'complete' contracts for PFI projects that are most amenable to full specification *ex ante* with relatively predictable and measurable outcomes, goes on to recommend relational (and 'incomplete') contracts for PFI projects whose outcomes are less predictable and less easily monitored by third parties. Zhang (2001) has summarised Mumfords's comparisons of complete vs relational contracts as in Table 12.5. This draws attention to the usefulness of introducing 'relational' joint risk management approaches to deal with some of the less predictable and more complex risks in PPP that may be handled far more effectively if approached jointly by the relevant parties.

Table 12.5 *Comparison of complete and relational contracts (sources: Mumford, 1998; Zhang, 2001).*

Complete contracts	Relational contracts
Suit single projects	Suit series of projects
Require full pre-specification	Need less full specification
Ex ante agreement needed	Expect frequent renegotiations
Rules laid down in advance	Flexible rules agreed internally
Duties clearly defined	Duties may be indefinite and shared
Rewards pre-allocated	There may be bargaining over rewards
Risks assigned	Risks may be shared *ex post*
Use third party verification	Third party monitoring may be difficult
Enforcement by courts/ADR	Parties need to resolve their own disputes
External power invoked	Parties share power amongst themselves
NPV readily estimated	There are likely to be many 'real options'
Finite contract duration	Open-ended contracts

12.7 Conclusions

The greater complexities, as well as the longer term and broader risks encountered in PPP projects provide greater scope for 'relational' joint risk management approaches than in the traditional procurement scenarios, where they are already proving their worth. Lessons learned from the latter scenarios will provide guidance in adopting these approaches for PPP projects, in order to avoid some of the pitfalls (e.g. through abuse of trust) and to provide safeguards where necessary.

In providing a pro-active and suitable support framework for BOT-type PPP in general, lessons may be drawn from past (good and bad) experiences. The support framework itself may be structured as for example suggested and described by Zhang and Kumaraswamy (2001d) under specific areas of attention such as:

(1) The feasibility study (including suitability for BOT-type procurement; and type of BOT to be recommended).
(2) Establishing the project team (including legal, financial and technical experts and determining roles and relationships).
(3) Preparing project documents (including project brief, pre-qualification and tender documents).
(4) Pre-qualification and shortlisting (including consortium constitution, checks on BOT experience, strengths in finance, technology and other capacities).
(5) Tender evaluation (including multiple criteria for different packages; and evaluation methodology).
(6) Negotiations and award (including negotiations with shortlisted concessionaires to obtain the best value for money).
(7) Design and construction (including optimising inputs from various participants, design checks and works checks).
(8) Operations and services (including possible performance-based payments, penalties for low standards or delayed services).
(9) Dealing with long-term changes (including legislative, operational and financing changes, changing requirements of the concessionaire, technological and other changes arising during the project life cycle).
(10) Transfer and post-transfer management (including smooth transfer of assets and personnel where applicable/useful, and selection of new operator).

While Tiong and Alum (1997) deal at length with critical success factors and distinctive winning elements on BOT-type proposals from the perspective of prospective concessionaires, Kumaraswamy and Morris (2002) highlight some of the critical success factors in the win-win-win scenarios targeted for both public and private sector participants, as well as the ultimate public users, as: (1) careful and detailed feasibility assessments of suitability for BOT-type PPP procurement, e.g. in terms of stable and suitable political and legal regimes, and socio-economic conditions, with a project that is already in the public interest, and has good prospects of steady and adequate cash flows, and safeguards against all potential risks; (2) a reasonable but not excessive rate of return, again facilitated by sensible safeguards e.g. toll/tariff adjustment mechanisms to deal with certain risks; (3) a pro-active, stable and reasonable project sponsor and pro-active support framework; and (4) a financially strong and healthy, technically competent and managerially exceptional concessionaire.

A cross-section of potential risks, projects and managerial approaches have been scanned in this chapter. More examples and details are available in the references. The 'BOT body of knowledge' that is growing rapidly through consolidation and documentation exercises such as in this book, should also prove valuable when approaching PPP projects in other sectors and regions, and provide deeper and broader insights into potential risks and their management in similar scenarios.

Infrastructure development continues to surge in China, and Hong Kong is itself embarking on further PPP ventures, such as a planned 'logistics centre' and a new international exhibition centre at the new airport – to be developed and operated in a partnership where the Government will contribute up to 50% of the construction cost. Apart from the China- (and specifically Hong Kong) based BOT-type examples described in this chapter, it has even been remarked that Hong Kong itself may be considered as one of the largest ever DOT (develop-operate-transfer) projects given out by China on a 99-year lease that ended with a smooth transfer in 1997!

References

Ahmed S.M., Ahmad R. & de Saram D.D. (1998) Risk management in management contracts. *Asia Pacific Building & Construction Management Journal*, **4**, 23–31.

Akintoye A., Taylor C. & Fitzgerald E. (1998) Risk analysis and management of private finance initiative projects. *Engineering, Construction & Architectural Management*, **5**(1), 9–21.

Asian Development Bank (1996) *Technical Assistance to the People's Republic of China for the BOT Changsha Power Project*. TAR: PRC 30063. ADB.

Asian Development Bank (1997). *Technical Assistance for Legal Training in BOT/BOOT Infrastructure Development*. TAR: TRA 30150. ADB

Charlton, M.C. (2001) Risk allocation in construction contracts in East Asia. *CIOB (HK) Journal*, **3**, 13–17.

CIRC (2001) *Construct for Excellence*. Construction Industry Review Committee (CIRC), the Government of Hong Kong Special Administrative Region.

Electric Focus (1992) Electricity development in China: an exclusive interview with Gordon Wu. *Electric Focus*, p. 1. Electricité de France International and HKIE.

Fellows R.F. (1996) The management of risk, *Construction Papers*, No. 65, p. 8. The Chartered Institute of Building, UK.

Fitzgerald P.F. (1998) *Project Financing*. Practicing Law Institute, New York.

Grove J.B. (2000). Consultant's Report on Review of General Conditions of Contract, for Construction Works for the Govt. of the Hong Kong SAR, *Conference on 'Whose Risk?'*, Hong Kong, 20 and 21 November 2000, Hong Kong Institution of Engineers.

Harris F. & McCaffer R. (1995) *Modern Construction Management*, 4th edn. Blackwell Science, London.

Hong Kong Government (1992) *Western Harbour Crossing Project Brief*. Hong Kong, China.

Hong Kong Government (1993) *Project Brief of Route 3 Country Park Section – Tai Lam Tunnel and Yuen Long Approach Road*. Hong Kong, China.

Kumaraswamy M.M. (1998) Lessons learnt from BOT-type procurement systems, *Mainland and Hong Kong BOT Conference, Beijing, China, Oct. 1998*, pp. II, 238–247. China Highway & Transportation Society and Works Bureau of the Govt. of Hong Kong SAR.

Kumaraswamy M.M. & Morris D.A. (2002) BOT-type procurement in Asian megaprojects. *Construction Engineering and Management*, **128**(2), 93–102.

Kumaraswamy M.M. & Zhang X.Q. (2001) Governmental role in BOT-led infrastructure development. *International Journal of Project Management*, **19**(4), 195–205.

Kumaraswamy M.M., Yogeswaran K. & Miller D.R.A. (1995) A construction risks 'underview' of ground conditions risks in Hong Kong. *Construction Law Journal*, **11**(5), 334–342.

Kumaraswamy M.M., Zhang X.Q. & Dissanayaka S.M. (1999) The BOOT route to rebooting infrastructure development in Asia. *Constructor*, **4**(3), 59–64.

Lam P.T.I. (1999) A sectoral review of risks associated with major infrastructure projects. *International Journal of Project Management*, **17**(2), 77–87.

Levy S.M. (1996) *Build, Operate, Transfer*. John Wiley and Sons, USA.

Lloyd R.H. (1996) Privatisation of major road tunnels in Hong Kong through build-operate-transfer arrangements. *Conference on 'Highways into the Next Century'*, pp. 109–117. Hong Kong Institution of Engineers.

McInnis A. (2000) Form of building contract for Hong Kong. *Conference on 'Whose Risk?'*, p. 12. Hong Kong, 20 and 21 November 2000, Hong Kong Institution of Engineers.

Marriot A. (2000) Whose risk? Managing risk in construction – who pays? Introduction. *Conference on 'Whose Risk?'*, Hong Kong, 20 and 21 November 2000. Hong Kong Institution of Engineers.

Mumford M. (1998) *Public Projects Private Finance: Understanding the Principles of the Private Finance Initiative*. Griffin Multimedia, Welwyn Garden City, UK.

Ogunlana S.O. (1997) Build Operate Transfer procurement traps: examples from transportation projects in Thailand. *CIB W92 Symposium on Procurement*, pp. 585–594. Montreal, Canada, CIB, Rotterdam, Netherlands.

O'Neill M. (2001) Beijing Briefing. *South China Morning Post (Hong Kong)*, 5 November.

Price A.D.F. & Shawa H.H. (1997) Risk and risk management in project related finance. *Journal of Construction Procurement*, **3**(3), 27–46.

Rahman M.M. & Kumaraswamy M. (2001) Revamping risk management in Hong Kong construction industry. Vol. 1. *Cobra 2001 (Proceedings of the RICS Foundation, Construction and Building Research Conference)* (eds J. Kelly & K. Hunter), pp. 61–73. 3–5 September 2001, School of the Built and Natural Environment, Glasgow Caledonian University, Glasgow.

Rahman M.M. & Kumaraswamy M.M. (2002) Joint risk management through transactionally efficient relational contracting. *Journal of Construction Management and Economics*, **20**(1), 45–54.

Renton D. (1997) The identification, allocation and mitigation of risks in BOT projects. *Regional Seminar on Infrastructure Procurement – the BOO/BOT Approach*, Sri Lanka, Sept. 97, 6 page supplement to Proceedings. Institute of Engineers Sri Lanka, Colombo, Sri Lanka.

Salzmann A. & Mohamed S. (1999) Risk identification frameworks for international BOOT projects. *Joint CIB Symposium on 'Profitable Partnering in Construction Procurement'*, pp. 475–485. AIT, Thailand.

Shadbolt R. (2000) Hong Kong and the rest of the world – working towards best contract practice. *Conference on 'Whose Risk?'*, p. 12. Hong Kong, 20 and 21 November 2000, Hong Kong Institution of Engineers.

Smith G.R. & Bohn C.M. (1999) Small to medium contractor contingency and assumption of risk. *Construction Engineering and Management*, **125**(2), 101–107.

Songer A.D., Diekmann J. & Pecsok R.S. (1997) Risk analysis for revenue dependent infrastructure projects. *Construction Management and Economics*, **15**, 377–382.

Tam C.M. (1999) Build-Operate-Transfer model for infrastructure developments in Asia: reasons for successes and failures. *International Journal of Project Management*, **17**(6), 377–382.

Tam C.M. & Leung A.W.T. (1999) Risk management of BOT projects in Southeast Asian countries. *Joint CIB Symposium on 'Profitable Partnering in Construction Procurement'*, pp. 499–507. AIT, Thailand.

Tiong R.L.K. & Alum J. (1997) Distinctive winning elements in BOT tender. *Engineering, Construction and Architectural Management*, **4**(2), 83–94.

Wang S.Q., Tiong R.L.T., Ting S.K. & Ashley D. (1999) Political risks: analysis of key contract clauses in China's BOT project. *Construction Engineering and Management*, **125**(3), 190–197.

Wang S.Q., Tiong R.L.T., Ting S.K. & Ashley D. (2000) Evaluation and management of foreign exchange and revenue risks. *Journal of Construction Management & Economics*, **18**, 197–207.

Woodward D.G. (1997) Risk analysis and allocation in project financing. *Accounting and Business Review*, **4**(1), 117–141.

Zhang W.R., Wang S.Q., Tiong R.L.K., Ting S.K. & Ashley D. (1998) Risk management of Shanghai's privately financed Yan'an Donglu Tunnels. *Engineering, Construction and Architectural Management*, **5**(4), 399–409.

Zhang X.Q. (2001) *Procurement of Privately Financed Infrastructure Projects.* Ph.D. thesis, the University of Hong Kong.

Zhang X.Q. & Kumaraswamy M.M. (2001a) Hong Kong experience in managing BOT projects. *Construction Engineering and Management,* **127**(2), 154–162.

Zhang X.Q. & Kumaraswamy M.M. (2001b) BOT-based approaches to infrastructure development in China. *ASCE Journal of Infrastructure Systems,* **7**(1), 18–25.

Zhang X.Q. & Kumaraswamy M.M. (2001c) Procurement protocols for public private partnered construction projects. *Construction Engineering and Management,* **127**(5), 351–358.

Zhang X.Q. & Kumaraswamy M.M. (2001d) Proposed support framework for the procurement of public private partnered infrastructure projects. *1st International. Structural Engineering and Construction Conference, 'Creative Systems in Structural and Construction Engineering',* Hawaii (eds. A. Singh and A.A. Balkema), pp. 239–244. Rotterdam, Netherlands.

13 Public-private partnership projects in the USA: risks and opportunities

Arthur L. Smith

13.1 Introduction

The United States is an active participant in the worldwide trend toward use of public-private partnerships (PPP) to improve the quality and cost-effectiveness of government service provision. PPPs are now widely employed at all levels of government in the US, i.e. the federal, state, and local governments, although their level of acceptance varies across political, geographical, and functional boundaries.

A PPP may be defined as 'any arrangement between a government and the private sector in which partially or traditionally public activities are performed by the private sector' (Savas, 2000). In the United States, as in any nation, the spectrum of such potential arrangements is shaped by the existing socio-economic environment. For example, the US, even before the current trend toward PPPs began in the 1970s, maintained relatively few state-owned enterprises; thus the large-scale divestiture privatisations which have formed the backbone of many national programmes have played a lesser role in the US. [Nonetheless, certain high-visibility examples can be cited: on 5 February 1998, the US Department of Energy (DOE) concluded the largest divestiture of federal property in the history of the US Government when it sold the Elk Hills Naval Petroleum Reserve, an oil and gas field in the state of California, to Occidental Petroleum for $3.65 billion. Other examples included the sale of CONRAIL, a federally-operated freight railroad in 1987, and the United States Enrichment Corporation, which prepares uranium to be used as fuel in nuclear power plants, in 1988.]

The majority of the PPP experience in the US falls into two broad categories: contracting for government services, and infrastructure partnerships, in which private sector capital and expertise are used to develop, and in many cases operate and maintain, infrastructure needed by the public at large. In either case, a variety of contract/transaction types and approaches may be selected and the appropriateness of the choices made will have significant impact on a given partnership's probability of success.

Risk analysis is a critical part of this decision process. Any PPP involves some degree of risk transference to the private sector. This principle is demonstrated by Fig. 13.1, which depicts the degree of risk transference associated with different management approaches to provision of potable water in a US community.

In the lower left corner of Fig. 13.1, the government owns and operates the water plant and has full responsibility for the provision of water. In the event of negligent performance, catastrophic equipment failure, or unforeseen fluctuations in demand for water, the government bears the full financial and legal liability for the service. In short, the risks of service provision are borne by the government; there is risk, but no risk transfer.

A common form of PPP is the operation and maintenance contract. In such a contract, the private contractor would have overall responsibility for operating and maintaining the plant, and meeting defined standards for water quality, equipment maintenance, etc., and assumes the performance risk for the plant operations. The government retains ownership of the plant, and retains all capital risks, such as might be associated with plant obsolescence, fluctuating demand, and changing environmental regulations.

Each of the above examples assumes utilisation of an existing, government-owned facility. Suppose, however, that there is a requirement for a new water plant, to accommodate increased capacity requirements due to population growth. Many local governments find it difficult to fund major capital investments from conventional public funding mechanisms. And even if the funds can be raised, the short-term impact on taxpayers or ratepayers may be significant.

An alternative is a PPP in which a private partner assumes the risk of project capitalisation. (And, in fact, savings in capital cost is one of the top

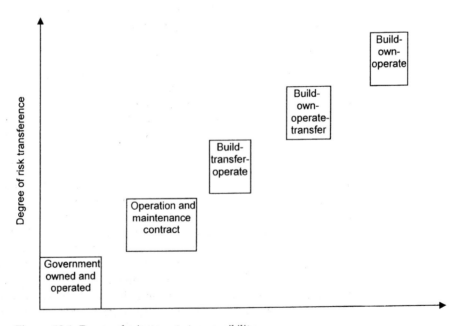

Figure 13.1 *Degree of private sector responsibility.*

two reasons cited by local government officials in the US for entering infrastructure partnerships; the other is access to private sector expertise.) One such form of partnership is build-transfer-operate (BTO). In a BTO transaction, a private partner would finance and build the water plant. Once construction is complete, the private partner transfers ownership of the plant to the government, which then leases the plant back to the partner, under a long-term lease during which the partner operates the plant. Thus the private partner assumes the risks of capital and construction, while the government assumes the risk of plant ownership and, typically, demand fluctuation.

Another common form of partnership is build-own-operate-transfer (BOOT). In this type of partnership, the private partner would finance, build, own and operate the water plant for a specified period of time, during which it would collect user fees based on water consumption. At the end of this period, ownership is transferred to the government. This transaction structure transfers significantly more risk to the private partner than BTO, since the private partner is exposed to various legal, environmental, and regulatory risks during the extended period of private ownership.

The final example from Fig. 13.1 is a build-own-operate (BOO) transaction. In this instance, the private partner would finance, build, own, and operate the water plant, with government regulatory control (e.g. review and approval authority over proposed rate increases). In this type of PPP, the risk transference to the private sector is maximised (unless, of course, one is willing to contemplate complete abrogation of government control, an approach rarely advocated in the case of water).

This depiction is, of course, an oversimplification. There are many more gradations of risk allocation than the five displayed, and many ways to mitigate the actual risk exposure of the parties. Also, in infrastructure partnerships, there are frequently more than two parties involved, not just a private partner and a government agency; for example, financial institutions which loan funds to the private water plant developer assume an element of risk as well.

The key point is that some degree of risk transfer is inherent to every PPP, with the private partner risk being largest in those projects with significant capital investment and the greatest degree of change from the pre-partnership status quo.

In every transaction, there is a premium for risk assumption. If the government asks the private partner to propose a price per gallon of potable water over a 20-year contract term, based on the government's consumption projections, but with a guaranteed minimum purchase each year, the partner will choose one pricing strategy. If the government asks for a price per gallon, but without a guaranteed minimum, so that the private partner absorbs the risk of unforeseen demand fluctuation, a different pricing strategy will likely be chosen. The extent to which a party can foresee, control, and address the risks which they are asked to assume influences the magnitude of the risk premium. Given the cumulative effect of such decisions on the project's financial structure, inappropriate or one-sided allocation of risk can render an otherwise viable PPP infeasible.

Risk analysis, therefore, is an important step in the structuring of a PPP. The analysis should include risk identification, risk assessment, risk allocation, and risk mitigation. The goals of this analysis are both to optimise risk allocation (i.e. to achieve an equitable and cost-effective mix of risk assumption), and to minimise the occurrence and severity of actual risk events.

To explore the state of risk analysis in PPPs in the United States, the following paragraphs provide some insights into current PPP activity at both the federal and state/local levels of government, followed by a discussion of associated risk analysis techniques. The discussion will focus on two of the primary areas of partnership activity in the US today, managed competition and infrastructure project partnerships. These two types of partnership encompass opposite ends of the risk transference curve shown in Fig. 13.1.

13.2 Managed competition

Managed competition is a process by which public agencies compete against private sector organisations to determine which sector can provide government services most cost-effectively. At the federal level, the US Government has formally recognised managed competition as a business management tool for 35 years. The government's primary mechanism for conducting such studies is the Office of Management and Budget (OMB) Circular A-76, 'Performance of Commercial Activities', originally issued in 1966, along with associated legislation and implementation guides. Circular A-76 establishes a managed competition process for determining in-house or contract performance, and provides general guidance on the types of functions appropriate for competition (US General Accounting Office, 2000a).

The US Department of Defense (DoD), the largest agency in the Federal Government, with a budget of over $300 billion, and well over 1 million military and civilian employees, has announced plans to study the functions performed by 203 000 employees for potential outsourcing through Circular A-76 over the period 1997–2005. The goal for this effort, based on DoD's historical savings through managed competition, is to generate savings of $9.2 billion over the eight-year period, with recurring savings of $2.8 billion per annum thereafter. This programme is now well under way, and appears on course to achieve the desired savings, albeit on a slightly longer timeframe.

Based upon this success, in February 2001, the federal OMB announced a new requirement for all federal agencies, civilian as well as military, to study 5% of their contractible positions for potential outsourcing by the end of the fiscal year 2002, and an additional 10% by the end of the fiscal year 2003. The long-term goal is to study half of the civilian positions currently designated as contractible, or approximately 400 000 positions total, by the end of 2005. The basic A-76 process is in Fig. 13.2

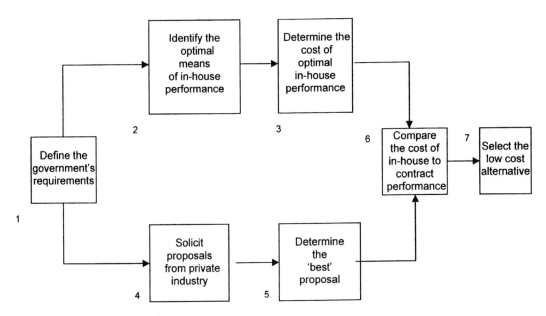

Figure 13.2 *The basic A-76 process.*

After a particular government function is identified for study under Circular A-76, the first step is to determine the government's true service requirements, to include the specific products or services, performance or quality standards, and projected workload. These requirements are compiled in a document called a Performance Work Statement (PWS), which will serve to govern future work execution. The PWS should be performance-based; i.e. focused on the performance required in terms of outcomes or products, not on how to perform the services. This approach is intended to allow maximum flexibility for the performing organisation, government, or private sector, to apply innovative or cost-cutting measures, while still meeting the government's performance objectives.

In Step 2, the government performs a management study to identify the optimal in-house organisation to meet the terms of the PWS. The government may re-engineer business processes, revise its future staffing and organisational structure, plan to implement new technologies, or identify other potential operational improvements. The culmination of the Management Study is the detailed description of the 'Most Efficient Organisation' (MEO), the government's specific plan for how it will organise, resource, and execute to meet the PWS requirements.

Step 3 is calculation of the In-house Cost Estimate (IHCE), the projected cost of the government's MEO performing the requirements of the PWS. In effect, this is the government's cost proposal. Since the IHCE is based on the MEO, not historical performance, the government's proposal fully reflects any efficiencies introduced by the MEO – the government must also be able to demonstrate the realism, i.e. the implementability, of these efficiencies.

In Step 4, contract clauses are added to the PWS to create a fully-fledged

Request for Proposal (RFP) which is advertised and distributed to interested vendors. Next, in Step 5, a responding vendor proposal is evaluated and selected for comparison to the cost of government performance. The proposal may be selected purely on a low-cost basis, or by a weighted consideration of both cost and technical merit. The total cost of performing the PWS via the selected vendor's proposal is then compared to the total cost of MEO performance, using the OMB costing guidelines (Step 6). 'Total cost' means that all relevant costs which would be incurred by the government under a given option are considered. For example, the cost of contract performance may include not only the price of the contract, but also contract administration costs, and one-time costs, e.g. severance pay to displaced government employees, associated with conversion to contract.

Finally, in Step 7, the government selects the low-cost alternative; on average, cost reductions of over 30% are achieved through this competitive process. The government then moves to implement the MEO, or to award a contract for future performance of the function. During this timeframe, the key decision documents, to include the Management Study and costing documentation, are made available for public review. Any affected party may file an administrative appeal if they determine, based upon their review, that the study documents violate the OMB A-76 guidelines in any way. A defined appeals process will then generate a final determination.

Interestingly, the A-76 process, despite having a detail-oriented, structured methodology, does not contain a formal risk analysis requirement. In part, this is because these managed competitions occur at the low-risk end of the risk curve shown at Fig. 13.1. Prior to initiation of a study, the function to be competed is reviewed to determine its contractibility. Vendor proposals may be subjected to a risk analysis, based on past performance, financial capacity, employee turnover rate, etc., as part of the process of selecting the final private sector proposal to be compared to the government. However, the final selection between contract and government performance is based purely on cost, without a direct comparison of risk between the two alternatives; each has been determined, formally or informally, to represent an acceptable level of risk, so that anticipated cost is given precedence as a decision criterion.

Post-decision audits, conducted by organisations such as the US General Accounting Office (2000b), show that contractor failures resulting from the managed competition process are relatively infrequent, despite the lack of formal risk analysis. When the private partner does fail to perform as desired or to meet cost targets, it is most frequently due to inadequacies in the PWS, with actual contractor non-performance the second most frequent cause. A more common problem is non-performance by a winning government organisation. The degree of business process re-engineering required to render the in-house organisation competitive can be extensive. Without adequate transition planning and change management, a government organisation which plans to reduce costs by, in many cases, 30% or more through introduction of new technology and work processes, can have a difficult time achieving successful self-transformation.

In addition to the federal A-76 programme, many state and local governments have employed formal managed competition programmes. Examples include the State of Michigan PERM Program (for Privatise, Eliminate, Retain, or Modify) and the managed competition methodologies of the cities of Charlotte, North Carolina, and Phoenix, Arizona, and the County of San Diego, California. Most of these programmes, like Circular A-76, contain no formal risk assessment component; those that do, such as Michigan's PERM, do so only on a limited basis. Yet these programmes, like A-76, have achieved a number of notable successes.

It appears, therefore, that at the low risk end of the risk curve, it is possible to consider, or even implement, PPPs without an overt risk assessment process. Other procedural devices, such as source selection procedures designed to limit contract awards to private sector partners who are financially stable and offer a fully responsive technical solution, serve as risk minimisation techniques. This does not mean, however, that risk assessment should not play a role. As noted above, the most common failure in the A-76 process is the public sector's inability to achieve full compliance with technical and cost performance goals. One possible solution would be to perform a risk analysis, in addition to a cost comparison between the government MEO and the selected private sector proposal, so that likelihood of acceptable performance within the proposed cost, not just projected cost of performance, could be considered in the decision process.

13.3 Infrastructure partnerships

Like managed competition, infrastructure-related PPPs are being crafted at all levels of government in the US. In contrast to managed competitions, however, only a minority of the infrastructure-related partnerships occur at the federal level.

At the federal level, the most ambitious current programme is the military housing privatisation programme. The DoD owns, operates, and maintains over 500 000 units of family housing for military personnel with dependents. DoD estimates that approximately 200 000 of these units are aged, lack modern amenities, or are in need of renovation or replacement. Faced with a projected US$16 billion construction requirement to remedy these problems, DoD sought and obtained permission from the US Congress to involve the private sector. The new programme, termed the Military Housing Privatisation Initiative, allows private sector financing and ownership of military housing. Through the possible use of direct loans, loan guarantees, and other incentives, DoD hopes to encourage private developers to add 31 000 new units of available housing by 2006.

Other federal efforts include an investigation of full privatisation of utility systems on US military installations through sale/transfer agreements with long-term service provision requirements. Over 1400 individual utility systems have been identified for privatisation, subject to case-by-case demonstration of economic viability. Aside from these two large DoD programmes, federal infrastructure partnerships have been initiated on an individual project basis.

At other levels of government, the primary focus has been on various types of design-build transactions, particularly for water and wastewater treatment facilities. The capital requirements of these high-cost facilities frequently exceed the fund-raising capacity of local governments, and the complexities of project design and implementation may require skills not resident in the government organisation. Such projects have been successfully completed for both large and small local government entities. At the larger end of the scale, the City of Seattle, Washington, utilised a design-build-operate (DBO) approach to develop the Tolt River Water Treatment Plant, a 120 million gallon-per-day filtration facility. The final DBO contract of $101 million, including 25 years of facility operations, was 40% below the originally estimated costs, due to both utilisation of cutting-edge technologies and mutual-advantage contract provisions to optimise risk allocation between the public and private partners.

Not all of these projects involve water/wastewater, however; some innovative public-private transactions occur in other areas of infrastructure, such as toll road construction and schools. In September 2001, the District of Columbia opened a new school, built without expenditure of public funds, through a unique PPP (Richard, 2001).

The James F. Oyster Bilingual School, a public school constructed in 1926, was in poor repair, and not designed in accordance with modern educational concepts, but the District lacked funds for construction of a replacement educational facility. However, the school was sited on a 1.7-acre tract of land, more land than it required, in a desirable part of town. Through a PPP, a private developer, LCOR, financed, designed, and built for the District a new $11 million school on the site of the old Oyster School. The District, in turn, deeded part of the campus to LCOR, which built a $40 million, nine-story, 211-unit apartment building on the deeded grounds. The developer's proceeds from operation of the apartment building will be used, in part, to pay off the bonds used to finance the school.

As these examples demonstrate, infrastructure-related partnerships offer the potential of significant economic benefit to government agencies, as well as the provision of services and capacity which might not otherwise be financially attainable. For private partners, these transactions offer the potential for profitable, long-term agreements. This win-win scenario, i.e. maximisation of economic benefit to all parties, is most likely to be achieved through an equitable allocation of risk. By assigning specific risks to those partners best able to bear, control, and/or mitigate those risks, a balance can be struck which optimises project cost.

Infrastructure projects, with their requirements for capitalisation and construction, typically have an inherently higher level of risk than service-oriented partnerships, such as result from most managed competitions. As a consequence, formal risk analysis is a standard component of such projects. The initial step of this analysis is to identify the risks associated with the specific project under consideration. While many risks are, of course, project-specific, a typical set of risk categories is displayed in Table 13.1. Each category of risk, and others as applicable, must be considered. Multiple risk factors may be identified within a single category.

Environmental risk, for example, may include risks associated with site conditions. If the government provides the land for an infrastructure project (e.g. a water plant), who will be liable, the government or the project developer, for the cost of unforeseen environmental issues with the site? Who will assume the liability for a change in the quality of the water inflow to the plant, a factor which may well be beyond the control of any party to the transaction?

Another example is regulatory risk, a significant consideration in the US. A project may be designed to meet or exceed current regulatory requirements for emissions, effluent quality, or other factors, only to have the regulatory requirements for environmental quality increased. Who will bear the cost of expensive plant modifications to meet the new standards, and/or the costs of fines or penalties which may be incurred until the standards can be met?

Demand for a project output or service can be difficult to assess, and this can impose a significant risk if the project financing is based, in whole or in part, upon usage fees. In 1995, the Dulles Greenway opened, the first private toll road development in the State of Virginia in 170 years. This 14-mile, limited-access highway was designed to give Washington, DC area commuters a high-speed alternative to the existing congested, toll-free roadways. The Greenway struggled initially, as traffic volume failed to meet projections. It took several years, a refinancing, and innovative approaches (such as a frequent rider programme, inspired by airline frequent flyer models) for the Greenway to achieve financial stability.

Following identification of the risk categories applicable to a specific project, the degree of risk associated with each must be assessed and allocation alternatives considered. For example, in the wastewater treatment example discussed above, unforeseen site conditions were a risk consideration. The parties will need to consider the quality and extent of current information about the site, whether it is a greenfield or brownfield site, and any known site issues requiring mitigation. Which party will assume the risks associated with cost growth related to site conditions? Can the developer propose an alternative site?

Table 13.1 *Typical risk categories.*

Political	Demand
Force majeure	Credit
Environmental	Interest rate
Regulatory	Inflation
Legal	Foreign exchange
Design	Cost
Construction	Tax
Permitting	Labour

Assessment of specific elements of risk can be extraordinarily complex and assessment methods vary depending upon the type of risk involved. Environmental risk analysis, for example, may require sophisticated analyses of multiple variables, performing probability distribution functions to characterise the uncertainty in each risk element and risk driver, and performing simulations, to define adequately the risk spectrum. Capitalisation risks can be clarified through exploratory discussions with the project finance community to determine capital availability and interest in the proposed project, and to identify potential investor concerns.

Equally important is the consideration of risk mitigation methodologies. Demand risk, for example, was identified above as a significant factor, as in the Dulles Greenway example. Demand risk mitigation measures can include: a moratorium or ceiling on competing projects; a guaranteed minimum purchase level, backed by a government entity; indexing of unit prices, to reflect fluctuations in demand level; or insurance. The effectiveness or desirability of each is situation-specific. A limit on competing projects may serve to slow economic growth (or may be unfeasible, due to the potential for competition from neighbouring jurisdictions or existing local facilities). While in some cases it may be able to purchase and resell 'excess' output (e.g. electricity), this is difficult in the case of a toll highway. For this reason, guaranteed minimums are frequently referred to as falling into the 'take and pay' or 'take or pay' categories. The assessment of mitigation measures for other risk factors can be equally complex.

Such issues, and the manner in which they are addressed across the full spectrum of risks, play a major role in determining the viability of PPPs. Below, two recent PPP transactions in the US are explored, and the treatment of risk is discussed as a contributor to their ultimate success or failure.

13.4 US Department of Energy, Hanford tank waste remediation project

The Hanford site, near Richland, Washington, was originally constructed in 1942 as a plutonium-manufacturing facility by the US Department of Defense. Following World War II, the facility was transferred to the Department of Energy (DOE), but continued its manufacturing role. Today, due to the long-term accumulation of the manufacturing byproducts, it is one of the world's largest repositories of hazardous waste.

Radioactive waste has been stored in large underground storage tanks at the Hanford site since 1944. Approximately 56 million gallons of waste containing approximately 240 000 metric tons of processed chemicals and 177 mega-curies of radionuclides are currently being stored in 177 tanks. These caustic wastes are in the form of liquids, slurries, saltcakes, and sludge. Since the 1960s, DOE has been aware that some of these tanks were leaking radioactive waste into the soil, with the potential for ground water contamination. In 1991, the Tank Waste Remediation System (TWRS) programme was established to manage, retrieve, treat, immobilise, and dispose of these wastes in a safe, environmentally sound, and cost-effective manner. The execution of this programme was by a private firm using government-owned, contractor-operated facilities, under a cost-reimbursable contract monitored by DOE (Di Prinzio, 2000).

As the TWRS effort progressed, the DOE grew increasingly concerned about cost growth in the programme. The cost-reimbursable contract structure gave DOE flexibility and control in directing the contractor's efforts, but provided the contractor little overt incentive for cost control. While acknowledging the technological challenges and unique liability considerations associated with private sector participation, DOE decided to pursue a PPP to obtain private sector capital and expertise, limit DOE's liability, and establish better cost control. The project concept, as announced in 1995, was to select a private partner to design, engineer, construct, and operate a state-of-the-art facility to remove the waste from the tanks and immobilise it into a less volatile form (via a process known as vitrification, the transformation of materials into a glass form). Through this partnership, DOE would transfer many of the risks then borne by the Government to the private partner.

The capital risk of project construction would be borne by the private partner, but mitigated by a long-term operations contract from DOE. Construction risk would be borne by the private partner, with both penalties for construction and incentives for success in meeting project milestones. The private partner would bear responsibility for all permits, but DOE would assume responsibility for changes in law, or denial of permit due to pre-existing conditions not caused by the private contractor. DOE would commit to providing steady streams of low and high-level waste which met the contract's technical specifications; the private partner would commit to process this waste, again to the contract's technical parameters. The private partner would perform waste processing at a fixed price per unit of waste, assuming that DOE met its contractual, technical, and volume obligations. The private partner would assume the liability for developing and implementing effective radiological, nuclear, and process safety procedures, subject to DOE's approval. However, DOE would indemnify the private partner for nuclear incidents, up to $9.7 billion (a statutory ceiling), and provide significant coverage for *force majeure*.

To facilitate this complex arrangement, and after extensive discussions with potential partners, DOE structured the procurement to select the private partner(s) into two phases.

Phase I consisted of two parts. Part A was a 20-month period to establish the technical, operational, regulatory, business, and financial elements required by privatised facilities that would provide tank waste treatment services on a fixed unit-price basis. Phase I, Part B was then envisioned as a period of 10–14 years during which the authorised contractor(s) would finance, design, construct, and operate new facilities, and deactivate the old waste-treatment facilities. During Part B, fixed unit prices would be paid only for completion and acceptance of waste-treatment services meeting contract specifications. Part B had an estimated value exceeding $5 billion.

Assuming that Phase I demonstrated the technical and cost adequacy of the proposed solution, DOE planned a second competitive procurement for Phase II activities. Phase II would be the full-scale production phase, and was scheduled to begin in 2004. The objectives of Phase II included implementing the lessons learned from Phase I, processing all tank waste into forms suitable for final disposal, and meeting or exceeding regulatory performance milestones.

To maximise competition, two competing teams were both awarded contracts for Part A, with their final product being detailed project plans. Following extensive review of the two sets of plans, DOE selected a team led by BNFL, Inc. (the US branch of the British government-owned nuclear fuel and waste firm) as its partner for Part B. Negotiations between the DOE and the BNFL team then began in earnest, with the Part A plan serving as the starting point for discussions.

At this critical point, a number of factors converged to impact on the originally projected cost of the project, as calculated by DOE. First, DOE's original plan was for the waste-processing facility built in Phase I to be a short-term pilot or demonstration facility. However, it became clear as the designs matured that, given the strict requirements for nuclear processing facilities, it made little economic sense to construct a short-term facility which would have to be designed with the durability (and at the cost) of a permanent structure.

A more complex set of issues arose around various contractual matters. The partners jointly agreed to a pricing structure which, while maintaining a fixed unit-price as the basis for payment, established a multi-tier system of fixed prices. In other words, the private sector partner would receive a higher fixed price per unit for quantities processed above the contractual minimum, to provide an incentive for more rapid processing. More intractable problems arose where standard federal contracting clauses came into direct conflict with standard private sector financing approaches. For example, the US Government typically insists on the right of 'Termination for Convenience', i.e. the right to terminate a contract at any time, for any reason. In such an event, the Government will pay the contractor termination costs, but the Government has a degree of latitude in determining what costs it will pay, and to delay payment pending its determination of these costs. Thus, the Government could not provide assurance that in the event of termination, all outstanding principal and accrued interest would be covered, significantly increasing the perceived risk to lenders. With an estimated $3.7 billion in financing required (in addition to BNFL's equity funding) for a project involving relatively untested technology, the project was already a challenging one for the project finance market. Ultimately, DOE agreed to support a majority of the debt.

In this project, DOE's goals of maximum risk transference to the private sector, private financing, state-of-the-art technology, and cost control proved to be in inherent conflict. Although a technically qualified team agreed with DOE on an apparently acceptable risk allocation plan, and the project finance community appeared willing to commit the requisite funds, the projected cost of the project grew by 120% from DOE's original projection. In May 2000, DOE announced the cancellation of the proposed PPP.

This experience demonstrates that in a high-risk transaction, the good-faith efforts of both public and private sector parties may not be enough to ensure the negotiation of an economically viable project plan. In addition, the government contract procedures' lack of flexibility contributed to the problems with the Hanford tank waste remediation system; these procedures were not designed to support private sector financing requirements, a non-traditional contracting environment for the federal sector. Establishing a contracting regulatory framework with the appropriate authorities would help to support the creation of such PPPs in the future. It is important to note, however, that both the magnitude and technological risk of the Hanford project contributed to make this an issue; despite their lack of flexibility, federal procurement regulations have successfully supported other PPPs, such as privately financed, designed, and built housing at US military installations.

13.5 Tampa Bay Water seawater desalination project

Tampa Bay Water (TBW) is a regional utility which supplies drinking water to more than 2 million customers in six jurisdictions (three counties and three municipalities) in southwest Florida. Current (2001) demand is 182 million gallons of water daily, but the region's population is growing rapidly, with demand expected to increase 25% by 2005. The current supply is from wellfields which are approaching maximum capacity. Additional pressure was placed on TBW by the Southwest Florida Water Management District (SFWMD), the body which plans and regulates water use in southwest Florida; SFWMD wanted TBW to reduce its wellfield pumping to relieve the environmental impact on the already stressed aquifer. Lacking other sources of drinking quality water, TBW began researching competing water purification technologies to meet the projected demand. Somewhat reluctantly, given the projected price differential versus the existing wellfields, TBW settled upon desalination of water from nearby Tampa Bay (on the Gulf of Mexico) as the preferred alternative. Based on metrics from existing desalination plants around the world, TBW anticipated that the cost for desalinated water would be roughly four times the $1.20 per 1000 gallons cost of its current wellfields (Public Finance Works, 1999).

In response to this situation, TBW decided to explore a PPP to provide desalinated water through a privately financed design-build-own-operate-transfer (DBOOT) facility. TBW hoped to utilise the partnership process to obtain state-of-the-art private sector design expertise, as well as financing.

Under the TBW contract, the private partner would provide 25 million gallons per day (mgd) of desalinated seawater, with no more than 100 mg/l of chlorides (the US standard for drinking water is a maximum of 250 mg/l), and an alternative price for an even tighter standard of 30 mg/l of chlorides.

Water would be furnished at a fixed price per 1000 gallons, with a higher fixed price if additional output were needed. The fixed price per 1000 gallons includes all private sector capital costs, operation and maintenance, and debt service on the project construction bonds. The private partner would bear all construction, cost, development, and performance risks. TBW committed to purchase the maximum of 25 mgd for 30 years, so long as the output met the quality standards, but has an option to purchase the plant before the expiration of the 30-year contract, if it so desires.

Following 14 months of negotiations with four competing teams of private partners, TBW announced a result which exceeded all expectations; a decision to award a contract to a team led by Poseidon Resources, for a cost of $1.71 per 1000 gallons, or roughly 40% of the anticipated cost. In fact, all of the final offers were significantly below the projected unit cost. How was TBW able to achieve its risk transference goals, and at the same time exceed its cost expectations? A number of factors contributed, some related to TBW's risk allocation decisions, others to unrelated factors, such as limitations in the comparability of the cost metrics used.

First, TBW did not dictate a technical solution; its solicitation identified the performance requirements, quality, and volume and let the potential private partners (who would bear the performance risk) propose the most appropriate technical approach. In the event, all four offerers proposed reverse osmosis membrane systems, a technology whose capital cost has greatly decreased in recent years. This decrease was not fully reflected by cost metrics based on existing plants used by TBW. In addition, operating costs for this technology have decreased as energy recovery techniques have improved.

Second, energy costs remain a significant component of desalination plant operating costs, even with the technology improvements noted above. Electricity costs in Tampa are approximately $0.04 per kilowatt hour, a fraction of the cost in many countries.

Third, since there were relatively few desalination plants in operation in the US, the cost metrics used largely reflected Middle Eastern and Caribbean experience. Tampa Bay is a relatively low-salinity marine environment, with an average of 26 000 mg/l total dissolved solids. Desalination plants in the Mediterranean must process 50% more total dissolved solids per litre and in the Middle East, salinity is even higher. The unit cost of desalination is naturally lower when there are fewer dissolved solids to be removed.

Fourthly, TBW allowed the desalination plant to be collocated with an existing power plant, allowing the private partner to share existing intake and discharge lines and to utilise the power plant's effluent to dilute the brine concentrate which constitutes desalinisation's primary waste stream.

Finally, the project benefited from low interest rates due to TBW's use of tax-exempt private activity bonds, and the relatively low technological and political risk associated with the project. Applied to the readily financible capital cost of just over $100 million, these factors helped contribute to favourable rates. TBW's contracting policies also allowed it to adopt, assisted by its financial advisory team of RBC Dain Rauscher and Scully

Capital Services, a contract structure fully compatible with the expectations of the private capital market.

Comparing the Tampa Bay Seawater desalination plant to the Department of Energy's Hanford tank waste remediation project, why was the former so successful, when the latter could not be brought to fruition? Both projects utilised formal risk analysis and allocation processes, competition among potential private partners, and good-faith negotiations between the public and private sector parties. The TBW project benefited from unforeseen cost advantages (e.g. the cost impact of Tampa Bay's low salinity), and the ability to select an optimal plant site from among a variety of alternatives. The TBW project had inherently lower technological, environmental, and construction risks than the DOE project, although the TBW effort certainly entailed risk.

The cost of risk transference to the private sector in the high-risk DOE project increased the project price. In addition, TBW's ability to minimise perceived risk to lenders through its contracting procedures was helpful in controlling project costs. Ultimately, the DOE project, while possible, could not achieve DOE's risk transference goals while simultaneously achieving its cost objectives.

13.6 Conclusions

PPPs in the US have demonstrated the opportunity to achieve billions of dollars in savings and enable the construction of new infrastructure. The involvement of the private sector in the provision of government services requires, of necessity, some degree of risk transference to the private partners. A structured process of risk identification, assessment, allocation and remediation can help to optimise the partnership development process. As the level of risk to be transferred increases, the necessity for formal risk analysis grows commensurately. The optimisation of risk allocation may require some restructuring of risk transference objectives to ensure that other project objectives may be met. In any event, a clear understanding of project risks, as well as the opportunities, should be a part of the partnership planning process.

References

DiPrinzio R. (2000) The US Department of Energy and the privatization of the Hanford tank waste remediation system. *Journal of Project Finance*, **6**(3), 54–60.

Public Works Financing (1999) *Private Desalting Prices Stun Tampa*. **126**, 1–5.

Public Works Financing (2001) *Poseidon-Ogden Confirm Tampa Desal Pricing*. **147**, 1–6.

Richard A. (2001) Developer, D.C. district team up to build new elementary school. *Education Week*, 5 September.

Savas E. (2000) *Privatization and Public-Private Partnerships*. Seven Bridges Press, LLC, New York.

US Department of Energy (1998) *Hanford Tank Waste Remediation System Privatization Contract*. DoE, Washington, DC.

US General Accounting Office (2000a) Report to Congressional Committees, *DoD Competitive Sourcing: Results of A-76 Studies Over the Past 5 Years*. GAO-01-20. GAO, Washington, DC.

US General Accounting Office (2000b) Report to Congressional Committees, *DoD Competitive Sourcing: Savings are Occurring, but Continuing Challenges Remain in Meeting Program Goals*. GAO/NSIAD-00-107. GAO, Washington, DC.

14 Public-private partnership in South African local authorities: risks and opportunities

Pantaleo D. Rwelamila, Lucy Chege and Tjiamogale E. Manchidi

14.1 Introduction

South Africa's 284 municipal authorities face tremendous service delivery challenges that are in many ways unique among emerging economies. With the end of the apartheid era, the South African Government began investigating a variety of innovative approaches to municipal (local government) service delivery that many other countries had refined and tested during the long period of South Africa's international isolation. Often grouped together as 'public-private partnerships (PPPs)', also referred to as municipal service partnerships (MSPs) to include possibilities for PPPs, these approaches include long-term concession and lease contracts, management and service contracts, as well as the outright sales of government assets.

In 1998, after two years of preparation, the South African National Government paved the way for PPPs, by creating the Municipal Infrastructure Investment Unit (MIIU), a non-profit company tasked with providing technical assistance and grant funding to municipalities investigating innovative service delivery partnerships. The long-term aim of MIIU is to develop a market place in which informed local authority officials and professionals can obtain the services of private sector advisers, investors and service providers as well as other public sector service providers and experts, to find more cost-effective ways of providing urban services to citizens.

Discussions with stakeholders in the government and the private sector, as well as analysis of policy and guidelines documents, procedures and international literature, have highlighted a number of constraints, opportunities and necessary conditions for PPPs in South Africa (Jackson & Hlahla, 1999; Rwelamila & Nangolo, 2002). For a greater number of PPPs to be implemented successfully, it is essential for these issues to be dealt with decisively. Specifically, it has been argued that there is:

- A need for policy reform (the need for cross-departmental policy coherence and consistency)
- A requirement for legal and procedural reform (the need for reform that would make the legal environment more PPP-friendly)

- A need to address public finance issues (the need for an appropriate framework to balance priorities between sectors and to ensure prudent control over the Government's financial commitments)
- Capacity and training issues (functional capacity to engage in partnership-type transactions)
- Institutional arrangements (a regulatory and support framework, which is driven and monitored by effective institutions).

This chapter focuses on two aspects, which are directly related to the above issues: risks and opportunities. These aspects are discussed based on understanding and dealing with the continuing negative perceptions of the role of the private sector and addressing those issues necessary to produce effective and efficient PPPs in South Africa.

14.2 PPP risk areas

Every PPP project like any other project is risky; meaning that there is a chance things will not turn out exactly as planned. PPP project outcomes are determined by many factors, some of which are unpredictable and over which project stakeholders have little control. Risk levels are associated with certainty about technical, schedule, and cost outcomes. High-certainty outcomes have low risk; low-certainty outcomes have high risk. Certainty derives from knowledge and experience gained in prior projects, as well as from a manager's ability to control project outcomes and respond to emerging problems.

According to Charoenpornpattana and Minato (1999), PPP-induced risks could be divided into five categories as indicated in Table 14.1.

In order to have a clear focus on PPP-induced risks faced by local authorities in South Africa, Charoenpornpattana and Minato's (1999) classification is divided into two parts:

- Typical risks
- Risks related to negative perceptions.

14.2.1 Typical risks

For this chapter, typical risks are defined using Charoenpornpattana and Minato's (1999) classification and they primarily include economic risks, legal risks, transaction risk and operation risk (Table 14.1).

Participants' risks relate to the reputation of the project sponsor. This includes the strength of the sponsor's balance sheet and their commitment to the development. This is important from the implementing agent side and financial institutions (lending organisation) responsible for funding the project.

Risks should also be considered by assessing the position of the implementing agent, or public partner – by asking a number of fundamental

Table 14.1 *PPP-induced risks.*

(1) Political risks	Internal resistance
	Labour resistance
	Nationalisation
	Political influence
	Uncertainty of government policy
	Instability of government
(2) Economic risk	Devaluation risk
	Foreign exchange risk
	Inconvertibility of local currency
	Inflation risk
	Interest rate risk
	Small capital market
(3) Legal risk	Changes in law and regulation
	Inefficient legal process
	Legal barrier
(4) Transaction risk	Delay of privatisation programme
	Improper privatisation programme
	Incapable administration body
	Reluctance to proceed
	Too small number of interested investors
	Unfair selection of state-owned enterprises (SOEs) to privatise
	Unfavourable investment environment
	Valuation of asset
(5) Operation risk	Associated infrastructure risk
	Demand and supply risk
	Incapable investor
	Improper regulation
	Liability risk
	Management risk
	Price escalation risk
	Technical risk

questions: Is the implementing agent well managed? Is the agent committed to the successful implementation of the project?

Design, construction and operating risk require the sponsor to carry out detailed research and statistical analysis as input into the design process. Guarantees should be given by the sponsor that the performance of the structures or equipment will be in accordance with design specifications, so as to meet the obligations of the contractor, as well as to ensure that the structures and equipments will remain in good condition until the end of the project defect liability period. Construction completion risk is mitigated by a fixed price, turnkey contract for the engineering, procurement and construction works associated with the project. Sponsor support must be available for project completion.

Operating risk should be mitigated by a contract with the operator for operation, management and maintenance. It is important that the operator of the facility should be responsible for providing the concessionaire with an operating performance bond. Provisions for adequate insurance, as well as liquidated damages, should be included in the contract.

The technology used by the concessionaire should be guaranteed by the sponsor to deliver the specific quality of service or product as required by the concession contract. It is important to establish the status of the operator, by looking at the operator's track record on the organisation's ability to operate the facility; the bidding and selection processes used to select the sponsor should be competitive and fair.

In order to reduce market risk, take-or-pay service or product supply agreements should be negotiated to the satisfaction of lenders. Exchange and interest rate risks fall into this category.

Environmental risk arises where a project has a potential impact on the environment. In such a case all environmental review procedures should be followed. Public sector officials charged with reviewing these findings should confirm the status of environmental risks associated with the project.

Termination risk can be assessed by looking at different scenarios of termination. For example, in the event of termination, what should happen to the party in default? Will the party in default suffer material financial penalties? If premature termination occurs, what will be the arrangements to compensate the lenders?

Risks related to negative perceptions in Charoenpornpattana and Minato's (1999) classification fall under 1, 3 and 4 (Table 14.1). According to Jackson and Hlahla (1999), there have been many debates in developing countries, which have focused on the negative aspects of private sector participation in the delivery of municipal services. Issues have been raised regarding possibilities of services price rise, falling quality of services offered, increase in unemployment levels, excessive profits from monopolies, which will go unchecked because of corruption (for example, Bishop & Kay, 1989; Miller, 1995; Sundaram, 1995; Samia, 1996; Gupta, 1997). Similar debates are taking place in South Africa.

Taking a closer look at PPP projects in a number of developing countries, each of the above points of concern has some validity, and the negative results described have happened elsewhere in the world where some form of private sector participation has taken place. Since they also have been known to happen where the public sector has been the service provider, care is being taken in South Africa not to condemn a system just because abuses have occurred. Instead, efforts are being made to ensure that checks and balances are in place to reduce the likelihood of abuses of the system.

There are strong indications to suggest that the primary mechanism for securing the best practice for a service delivered by a PPP is market competition. According to Jackson and Hlahla (1999), this could be in the form of competition within the market (two or more companies offering the same service) or competition for the market by competitive bids for a service contract or concession to render the service. As most municipal services in South Africa do not lend themselves readily to competition by duplicating infrastructure, the following discussion will be confined to monopoly situations.

14.2.2 Views on price escalation

The kind of efficiencies possible under private sector participation has meant that for the majority of PPPs for municipal services in South Africa, the price bid by the private sector has been similar to, or less than, that of the public sector provider for the same or better level of service (Jackson & Hlahla, 1999). This has been possible even with a respectable return for investors and improved compensation packages for workers, and may even include substantial investments in system expansion. There have been well-publicised examples of price rises in developing countries, but a careful analysis almost invariably shows good reasons for this.

The South African situation provides some examples of this. In this respect, adequate cost recovery is one of the pivotal issues. One typical situation is where the previous service provider simply did not base its price on actual costs and the service was benefiting from large hidden subsidies (e.g. salary and pension costs, or debt was serviced by another budget). Another scenario might be that prices were kept low for political reasons, which meant that the service was grossly underfunded and frequently underprovided because of lack of funds for operations and maintenance.

Within a PPP project, however, poor service could be a breach of contract and the costs of proper operations and maintenance would need to be raised by the service provider, typically through sales.

The other major reason why prices may well rise under a new arrangement (PPP) would be to cover the often significant investments needed to expand the system. PPPs are often used as a way of redressing years of underinvestment and therefore it should not be surprising if prices rise to fund such investment. A municipality contemplating a PPP must apply its corporate mind to the planning of service expansion. Planners must establish what improved levels of service and what rate of improvement and/or expansion consumers will be able to afford when such investments are translated into the price of the service. Some form of financial modelling of future scenarios is essential to check whether the council's aspirations will generate price increases that are simply not affordable, or will produce a serious political backlash.

As indicated above, the quality of service rendered by a PPP would normally be a contractual matter. Contracts should spell out the quality of service required in terms of performance criteria and set out sanctions for non-compliance. The municipality must then ensure that services are properly monitored and sanctions are applied. Every PPP should have a designated contract compliance officer who is specifically accountable for the smooth running of each contract. The cost of monitoring should be adequately provided for, perhaps by some form of levy on the cost of the service being rendered. Likewise, the contract must ensure that there is sufficient income to maintain service quality. If prices are kept artificially low, something will suffer – either quality of service, continued service provision or expansion of the service.

14.2.3 Views on potential job losses

In South Africa, primarily amongst trade unions, there is much concern over the potential for job losses in a service taken over by the private sector. Critics point to experiences overseas where large numbers of public sector workers were retrenched following some form of privatisation. In South Africa, however, such a situation is far less likely and experience in some PPP projects has revealed the contrary. According to Jackson (2000), in two water services concession contracts, Nelspruit and Dolphin Coast, the numbers of employees increased (Nelspruit, from 150 to 180, and in Dolphin Coast, from 22 to 38). Looking beyond this and revisiting the issue in terms of overall human resource development, Jackson (2000) quotes the commercial director of Saur (a French company that is part of the Dolphin Coast concessionaire) as stating that local persons, including those from disadvantaged backgrounds, have been employed at senior levels and are included in succession planning and further training.

Many PPPs under consideration are being driven by the need to expand a service. This may well result in an increase in employment in order to build, operate and maintain the enlarged system. There may also be a need for some turnover of staff in order to match skills with needs, but no one who wants to stay on and acquire new skills need lose their job. It is important that contractual arrangements for a PPP address labour issues and it should be standard practice that workers are not materially worse off under new arrangements, and that training and reskilling opportunities are made available. Under these conditions, labour turnover would then occur through normal processes such as retirement and voluntary retrenchments.

14.2.4 Views on unaffordable user charges

In South Africa, many PPPs are set up so that finance can be raised for extending services. In such cases, the bidding documents and subsequent contract must be very clear as to what system or service expansion is required, and by what dates. Careful planning is required to check that the envisaged extensions will not generate unaffordable user charges, but once agreement is reached on the price implications of the proposed investment, the specified extensions become contractual obligations to be rigorously monitored and enforced. The agreed levels of service may not be as high as those demanded by the population or promised by politicians, but they should represent best value for money within the affordability of the community as a whole, in addition to any (reliable) running cost subsidies that may be available to the municipality.

If services are improved and extended in a way that was not possible before, if the price is right and users are happy, and the politicians can take credit, is there cause to complain if the service provider makes more profit that originally envisaged? If, on the other hand, service quality is questionable, prices rise without due cause and the service provider still makes a handsome profit, then questions may well need to be asked. If prices need to be renegotiated for any reason, a municipality is advised to ensure that its negotiators are supported by experienced advisers who will be able to analyse and, if necessary, counter the service provider's case. There is enough evidence to suggest that evenly matched negotiating teams contribute to good contracts, which benefit all.

14.2.5 Views on potential for corruption

The threat of corruption is always present in most projects. It affects many aspects of public life and administrations must be rigorous in creating systems which reduce its incidence. In a 1998 World Bank publication on the subject, Klitgaard (1998) advances an argument:

> 'Consider two analytical points. First, corruption may be represented as following a formula: $C = M + D - A$. Corruption equals monopoly plus discretion minus accountability. Whether the activity is public, private, or non-profit, and whether it is carried on in Ouagadougou or Washington, one will tend to find corruption when an organisation or person has monopoly power over a good or service, has the discretion to decide who will receive it and how much that person will get, and is not accountable.
>   Second, corruption is a crime of calculation, not passion. True, there are both saints who resist all temptations and honest officials who resist most. But when bribes are large, the chances of being caught small and the penalties if caught meagre, many officials will succumb.'

Combating corruption, therefore, argue Jackson and Hlahla (1999), begins with designing better systems. Monopolies must be reduced or carefully regulated. Official discretion must be clarified. Transparency must be enhanced. The probability of being caught, as well as the penalties for corruption (for both givers and takers) must increase.

Transparent project delivery systems are essential. Well-organised competitive processes are highly desirable in almost all circumstances. Careful monitoring is indispensable, and all officials and councillors must be accountable to the consumer and the electorate. One way of making such accountability more widespread is to publicise the rights and obligations of consumers and to set up channels for them to register their complaints, plus an appeals process if they feel they have not been taken seriously.

14.2.6 Views on excessive tariff increases

According to section 94(1)(c) of the Local Government: Municipal Systems Act (MSA) (Department of Local Government, 2000), the minister responsible for local government is empowered to place limits on tariff increases by making regulations or issuing guidelines. The concern, according to Ashira

Consulting (Pty) Ltd (2000), is that if the Minister places limits on tariff increases, this may undermine the security of lenders in respect of municipal revenue streams.

From the foregoing, it can be argued that the identification and analysis of factors affecting the revenue stream over the PPP project period, and the accurate allocation of risk, is fundamental to the successful servicing of loans by lenders. According to Ashira Consulting (Pty) Ltd (2000), the level of tariffs has a direct impact on the revenue stream (the asset on which the PPP project is banked) and uncertainty in respect of tariff levels is a political risk, which the parties to the contract allocate between themselves. The concern prompting this argument is that section 94(1)(c) of the MSA increases this risk by making future tariffs levels unpredictable.

14.2.7 Summary

A PPP-based project delivery system is a very complicated matter for which successful implementation requires the application of risk management. Details of risk management principles are described elsewhere (e.g. Kangari & Boyer, 1989; Schuler, 1994; Hullet, 1995; Akintoye & Taylor, 1997). While most forms of traditional privatisation include 'transfer of risks' to one party, risk sharing between the public and the private sectors may improve the efficiency and effectiveness of privatisation. In this section, an attempt has been made to identify the major PPP-induced risks in South Africa.

14.3 PPP opportunities within the local authorities sector

There have been several estimates of South Africa's municipal infrastructure investment needs. The first version of the MIIF (Ministry in the Office of President (MOP), 1995) estimated an amount of US$5–8 billion (depending on the level of service provided) for urban municipal infrastructure in 1995. This included an allowance for some rehabilitation, plus addressing backlogs in domestic infrastructure and catering for new household formation over a ten-year period. It did not allow for economic infrastructure, i.e. for commerce and industry, apart from a share of bulk and connector infrastructure that a municipality would create for both domestic and economic purposes.

The National Infrastructure Investment Framework (NIIF) of 1996 (Department of Constitutional Development (DCD), 1996) estimated an amount of US$3.5–6 billion for municipal infrastructure (inclusive of bulk and connector, and local economic infrastructure) over a five-year period (1997–2001). This included the urban or urbanised areas.

A second version of the MIIF was prepared in 1996 (MOP, 1996) and included estimates of infrastructure needs in rural areas. It estimated capital expenditure at US$6.7–11.4 billion. Later that year, the Department of Water Affairs (DWAF) (Biwater & Murray-Roberts, 1996) produced their own estimate of costs for rural water supplies and sanitation of US$1.4 billion over ten years – higher than that envisaged by the new MIIF.

Although estimates vary and are notoriously difficult to improve on, the investment needs of South African municipalities are very large indeed (Manchidi & Merrifield 2001). To this, according to Jackson and Hlahla (1999), might be added the problem of the state of existing infrastructure. There is growing evidence that its condition is worse than was assessed by the MIIF (Table 14.2). Some infrastructure in many former townships, (for example in Soweto), has suffered years of neglect owing to underfunding of maintenance and is prematurely approaching the end of its useful life. There is evidence of impassable roads, overflowing sewers and massive amounts of unaccounted-for water are becoming increasingly common.

Terms in the table are defined as follows:

- *Water supply:*
  Full LOS: piped water in the dwelling
  Intermediate LOS: piped water on site or in the yard
  Basic LOS: public tap, water carrier, tanker, borehole, rainwater tank, well
  Inadequate: dam, river, stream, spring, other.
- *Sanitation:*
  Full LOS: flush toilet
  Basic LOS: ventilated improved pit latrines, chemical toilet
  Inadequate: pit latrine, bucket latrine, other.
- *Electricity:*
  Full LOS: electricity direct from an authority or electricity from other sources used for heating
  Intermediate LOS: difference between electricity direct from an authority or electricity from other sources used for heating and cooking
  Basic LOS: electricity direct from an authority or electricity from other sources used for lighting
  Inadequate: remainder of energy source.
- *Refuse removal:*
  Full LOS: removed by local authorities at least once a week
  Intermediate LOS: removed by local authorities less than once a week

Table 14.2 *Household levels of service (LOS) in South Africa (adjusted for 2000).*

Services	Full households	%	Intermediate households	%	Basic households	%	Inadequate households	%
Water supply	3976 900	44	1491 328	16	2319 131	26	1272 555	14
Sanitation (estimate)	4502 706	50	–		860 759	10	3696 405	41
Electricity	4030 943	45	234 561	3	955 623	11	3838 843	42
Refuse removal	4641 215	51	200 577	2	287 299	3	3930 879	43

Statistic South Africa 1996 census information, interpreted by DBSA and MIIU.
Note: the number of households with access to services is based on Statistics South Africa's 1996 census household figure of 9059 (approx.) million. Estimates for access to sanitation are based on the 1994 and 1995 October Household Survey and SSA 1996 census information. See text for definitions of other terms.
Source: Jackson and Hlahla (1999).

Basic LOS: communal refuse dump
Inadequate: own refuse dump, no rubbish disposal.

The foregoing description of services backlog in South Africa has made it imperative for the Government of the Republic of South Africa through MIIU to pursue a long-term goal of developing a marketplace in which informed local authority officials can obtain the services of private sector advisers, investors and service providers – as well as other public sector service providers and experts – to find more cost-effective ways of providing urban services to citizens. According to Hlahla (1999), a key element of this process since 1996 is to find ways of providing appropriate service provision to disadvantaged areas. The MIIU long-term objectives are as follows:

- To encourage and optimise private sector investment in core local authority services on a basis that is sustainable for both local authorities and at a national level
- To assist the development of an established market containing informed local authority clients, private sector advisers, private sector investors and service providers, so that the MIIU can be wound up no later than five years after the date of its original establishment.

14.3.1 Sectors and types of PPP in South Africa

According to Hlahla (1999), the service sectors within which the MIIU operates are primarily the essential urban services such as water, sanitation and solid waste. The MIIU also focuses on urban transport (e.g. municipal airports and bus and fleet management), power generation and distribution, information technology and a handful of other functions. The types of PPP addressed are quite comprehensive and include the following:

(1) Corporatisation: under this PPP, a municipality forms a separate legal corporate entity with the private sector to manage municipal service provision. The municipality continues to own the enterprise, but it operates with the freedom and flexibility generally associated with a private sector business.
(2) Management contract: here a municipality pays a private firm a fee for assuming overall responsibility for operating and maintaining a service delivery system, allowing it the freedom to make day-to-day management decisions. Typical duration: five years.
(3) Service contract: a municipality pays a private firm a fee for providing specific operational services such as meter reading, billing and collection, and operating facilities. Typical duration: one to three years.
(4) Concession: under this PPP arrangement a private firm handles operations and maintenance, finances investments (fixed assets) and provides working capital. Assets are usually transferred to the private company for the duration of the contract, but continue to be owned by the municipality and must be returned to them in the condition specified

at the end of the contract. Concession projects are designed to generate sufficient revenues to cover the private firm's investment and operating costs, plus an acceptable rate of return. The municipality exercises a regulatory and oversight role, and receives a concession fee for this arrangement, which usually focuses on operating and financing the expansion of existing system components. Typical duration: 15+ years.

(5) Municipal debt issuance: under this arrangement the municipality issues bonds or borrows from lenders to raise capital directly from private investors for financing the capital cost of the structure or expanding an infrastructure system. The municipality maintains total control of the project and bears all associated risks. The issuance process is usually facilitated by underwriting firms (public or private banks) and may involve financial advisory service providers. Typical maturity of debt: 5–20 years.

(6) Lease contract: here a private company rents facilities from a municipality and assumes responsibility for operation and maintenance. The lessee finances working capital and the replacement of capital components with limited economic life but not fixed assets, which remain the responsibility of the municipality. Typical duration: ten years.

(7) Build-operate-transfer: this form of PPP concession emphasises the construction of new, stand-alone systems. The municipality may or may not receive a fee or share of profits. Typical duration: 15+ years.

14.3.2 PPP project development process through MIIU

The MIIU is tasked with providing technical assistance and grant funding to municipalities investigating innovative service delivery partnerships. It produced a flow chart for MIIU PPP project development process which indicates the actors and actions needed to initiate each project (see Hlahla, 1999). The process covers project conceptual phase, formal application for MIIU assistance, feasibility studies, request for proposals, preparation of PPP bid documents, signing of PPP contract and the associated actors and action required at every stage of the process.

14.3.3 Project conceptualisation and funding application

After a municipality has expressed an interest, the MIIU assists with a preliminary conceptualisation of the nature of the problem for which MIIU assistance is required. The outcome of this process is a project concept for which the municipality may formally apply for MIIU development assistance, and which the MIIU can begin to evaluate using the decision-making criteria outlined below (Hlahla, 1999). The project concept is expected to evolve considerably throughout the development process as described in Table 14.1 above, but at this early stage it is a starting point for further consideration by and assistance from the MIIU.

After project conceptualisation, the next step is for interested municipal officials to fill out and return the MIIU application questionnaire, using whatever data or estimates are available. The application is required to state that the municipality has designated a project manager responsible for interacting on a day-to-day basis with the MIIU with regard to its provision of grant funding. The MIIU recommends that a project committee of in-house experts be formed to support the project manager.

Applications are processed on a first-come, first-served basis; the MIIU reserves the right, however, to apportion grant assistance so that as many eligible applicants as possible may receive at least some assistance. All applications are subject to the availability of funds by the MIIU for grant assistance purposes.

The application questionnaire is designed to help the municipality identify the assistance needed as clearly and concisely as possible in terms of the following information:

- The assistance needed and, as far as possible, the ultimate PPP project should be defined and justified, using the step-by-step forms provided
- The ultimate PPP project should be included in the municipality's prioritised list of planned projects, or its urgency assessed with the forms provided
- The municipal service operation targeted for a possible PPP project should be subjected to a 'notional ring-fencing' exercise that attempts, using available information and best estimates, to account for all current revenues and expenditures associated specifically with that service. The application questionnaire includes forms to aid in the collection of data necessary for this exercise
- There is a requirement for the application to include a formal letter of transmittal from the municipality; the letter should be signed by the highest municipal official and should demonstrate the organisation agreement with the decision to apply for MIIU assistance.

14.3.4 Addressing sustainability issues

Assessing the feasibility of various PPP project options is the first thing a municipality does with MIIU funding. If a grant is made for this purpose, the MIIU and the municipality negotiate the costs to be shared. The MIIU assesses the eligibility of the potential PPP project for grant funding using its internal resources, as well as the information provided by the municipality in the application. The focus of the assessment is on the potential for concluding a successful, sustainable project. The assessment includes the following:

- Financial sustainability: examination of potential apparent revenues from all sources, including cost recovery mechanisms. These are measured against all known costs over the likely duration of the potential project
- Technical sustainability: assessment of the apparent technological alternatives as to their suitability for providing the municipal services sought over the likely duration of the potential project

● Level of the amount of assistance: the amount requested should bear a reasonable ratio to the total potential project cost. Applying a general guideline that the amount of grant assistance requested should be no more than 3–8% of the estimated total cost of the project contemplated.

14.3.5 Summary

Opportunities for PPP-based delivery systems in South Africa are wide ranging. The infrastructure backlog within the municipal sector, which includes upgrading, maintenance and rehabilitation will not be funded from the fiscus. PPPs are therefore considered as a solution to the problem. Such partnerships do not only provide much-needed finance, but also introduce new technology, expertise and the efficiencies that usually accompany private sector operations.

14.4 Conclusions

As noted above, municipal infrastructure finance is one area where mis-allocation of resources has had a particularly painful impact. South Africa requires huge amounts of capital for infrastructure improvements that are critical for sustaining national productivity and competitiveness. Water supply and wastewater management are of course other elements of this overall infrastructure challenge. As a result of the size of such projects and the long-term nature of the financing required, however, neither bank borrowing nor current revenues can compete in terms of cost-effectiveness with the long-term financing available via 30-year PPP project delivery systems or long-term debt financing. Commercial banks can be highly averse to risk. They are not immediately attracted to, nor do they readily understand, long-term projects that depend for their viability on the goodwill of local-level politicians and the unspectacular revenue streams generally associated with urban and rural services such as water and sanitation. On the other hand, long-term financing arrangements, like PPPs, increase the affordability of such projects by allowing the high cost of construction to be amortised over long periods that more closely match the lifespan of assets. PPP project delivery systems also provide a more equitable method of financing such facilities than pay-as-you-build financing that uses current revenues. Most of the citizens, who pay off long-term debts incurred in financing construction, be it via taxes or charges, are paying during the useful lives of the facilities and are thus more likely to be direct beneficiaries.

From the foregoing, it is clear that opportunities for PPP project delivery systems within the South African municipal sector are great. However, PPP project delivery systems, as indicated in this chapter, are often difficult to implement and their successful implementation requires careful risk management. While most forms of traditional privatisation include 'transfer of risks' to one party, risk sharing between the public and the private sectors may improve the efficiency and effectiveness of PPP. In this chapter an attempt has been made to identify the PPP-induced risks and to classify them broadly into typical risks and risks related to negative perceptions.

The major conclusion obtained from this chapter is that government or any other public institution entrusted with the responsibility of co-ordinating PPP-based delivery systems should control some of these risks. It can also be concluded that other risks (transaction risks) might be shared between the major parties to a project.

PPP project delivery systems have been well-accepted tools for financing municipal infrastructure projects in South Africa. PPP project delivery systems reduce investment expenditure by the government. When PPPs are successful, privatisation enhances efficiency and effectiveness of the investment.

The future of sustainable municipal service provision will depend on the ability of local authorities to have enough capital and capacity for effective delivery of essential services. Under the current circumstances, it is highly unlikely that the municipal authorities will meet this daunting challenge on their own. Hence, the need for partnership with the private sector but before this can be achieved there is a dire need to address the negative perceptions on the role of the private sector.

References

Akintoye A. & Taylor C. (1997) *Risk Prioritisation of Private Sector Finance of Public Sector Projects*. Proceedings of CIB W92: Procurement systems symposium – the key to innovation. University of Montreal, Montreal, 18–22 May, Canada, 1–10.

Ashira Consulting (Pty) Ltd (2000) *The Relationship between Legislation and Municipal Undertakings Regarding Tariffs*. Unpublished paper, 6 November.

Bishop M.R. & Kay J.A. (1989) Privatisation in the United Kingdom: lessons from experience. *World Development*, **17**(5), 643–657.

Biwater & Murray-Roberts (1996) *Community Water Supply and Sanitation Strategic Study (Planning): National Assessment*. Biwater & Murray-Roberts, Heywood, Lancashire and Johannesburg, November.

Charoenpornpattana S. & Minato T. (1999) *Privatisation-induced Risks: State-owned Transportation Enterprises in Thailand*. Proceedings of the CIB 92 and CIB TG 23: Profitable Partnering in Construction Procurement, Chiang Mai, Thailand (ed. S.O. Ogunlana), pp. 429–439. E & FN Spon, London.

DCD (1996) *Municipal Infrastructure Investment Framework*, 2nd edn. DCD, Pretoria.

Department of Local Government (2000) Local government: municipal systems. Act No 32. *Government Gazette*, Pretoria, South Africa.

Gupta J.P. (1997) The privatisation process: a survey of state owned enterprises in Vietnam. *Sasin Journal of Management*, **3**, 1–11.

Hlahla M. (1999) The municipal infrastructure investment unit: the governments PPP-enabling strategy. *Development Southern Africa*, **16**(4), 565–583.

Hullet D.T. (1995) Project schedule risks assessment. *Project Management Journal*, **26**(1), 21–31.

Jackson B.M. & Hlahla M. (1999) South Africa's infrastructure service delivery needs: the role and challenges for public-private partnerships. *Development Southern Africa*, **16**(4), 551–563.

Jackson B. (2000) Workers are benefiting from MSPs in South Africa. *IMIESA*, **25**(11), Nov/Dec.

Kangari R. & Boyer L.T. (1989) Risk management by expert systems. *Project Management Journal*, **20**(1), 40–47.

Klitgaard R. (1998) International co-operation against corruption. In: *Fighting Corruption Worldwide. Finance and Development*, Vol. 35(1). World Bank, Washington, DC.

Manchidi T.E. & Merrifield A. (2001) Public-private partnerships, public infrastructure investment and prospects for economic growth in South Africa. In: *Empowerment through Economic Transformation* (ed. M.M. Khosa), pp. 409–421. African Millennium Press, Durban, South Africa.

Miller A.N. (1995) British privatization: evaluating the results. *The Columbia Journal of World Business*, Winter, 82–98.

MOP (1995) *Municipal Infrastructure Investment Framework*. With Department of Housing, October. MOP, South Africa.

MOP (1996) *Municipal Infrastructure Investment Framework*. March. MOP, South Africa.

Rwelamila P.D. & Nangolo I. (2002) *Innovative Delivery Systems for Municipal Infrastructure Projects in South Africa – How Prepared are Municipal Authorities?* Proceedings of the CIB-W92 – Procurement systems and technology transfer (ed. T.M. Lewis), pp. 403–416. Department of Civil Engineering, the University of the West Indies, St Augustine, Trinidad & Tobago.

Samia A.S.J. (1996) *Privatising State-owned Enterprises: Experiences of Asia-Pacific Economies* (ed. K. Yanagi). Asian Productivity Organisation, Tokyo.

Schuler J.R. (1994) Decision analysis in projects: Monte Carlo simulation. *PM Networks*, **8**(1), 30–36.

Sundaram K.S. (1995) *Privatising Malaysia: Rents, Rhetoric, and Realities*. Westview Press, Malaysia.

15 Private sector participation road projects in India: assessment and allocation of critical risks

Satyanarayana N. Kalidindi and A.V. Thomas

15.1 Introduction

The provision of quality infrastructure services at a reasonable cost is a necessary condition for achieving sustained economic growth. The growing mismatch between the available facilities and the actual need of various infrastructure services is hindering the economic and industrial growth of India (Ministry of Finance, 1996). The estimated investment required over the next five to seven years to keep infrastructure on track and to achieve the projected economic growth of 7%–8% is in the range of Rs. 20 000 billion (one UK £ = Rs. 69.7 in March, 2002). The share of budgetary support for infrastructure development as a proportion of total plan outlay has been decreasing in India over the last ten years. The scale and structure of huge infrastructure requirements and the shortage of public sector financing have compelled many developing countries/emerging economies like India to consider other avenues for investments including private sector participation (PSP) in infrastructure. Recognising the need to attract more investment in infrastructure, India opened itself to private investment as part of the country's 1991 reform programme. Though public expenditure is still the prime mover of infrastructure development in India, alternative resource mobilisation such as private investment and international aid channelled through government are being explored (Venkataraman, 2000).

15.1.1 Indian roads – overview

India has moved from being a rail dominant economy in the 1950s to a road dominant economy in the 1990s. The share of road transport in total freight traffic has increased from 11% to 70%, while passenger traffic has reached 82% from 28% during the same period. This increase is primarily due to the high flexibility of the road transport system in reaching any remote corner of the country, which is spread over an area of 3.29 million km^2. Roads in India for the purpose of their management and administration are divided into the following five categories: National Highways (NH), State Highways (SH), Major District Roads (MDR), Other District Roads (ODR), and Village Roads

(VR). The responsibility for the development and maintenance of NHs rests with central government, while all other roads are the responsibility of the respective state governments. Though primary roads (NH, SH, MDR and ODR) contribute only 15% of the total road length, they cater for 90% of the total road transport demand, in which the share of NH alone is 45%. A brief summary of Indian road transport development during 1951 and 2001 is given in Table 15.1.

The public sector financing and management model was the only way for road infrastructure development in India till recently and was often influenced by social and political considerations rather than economic requirements. The continuous reduction in public investment through plan allocation for the last 50 years coupled with poor implementation of road development plans led to the Indian road infrastructure being highly inadequate and inefficient. Logistical problems associated with the geographical size of the country also contributed to the complexity of the problems. Negligence in maintenance activity due to poor funding, overloading of vehicles beyond their rated capacity and sanction, and problems of encroachments and ribbon developments are some of the main reasons for the poor condition of roads in India (Ministry of Finance, 1996; Jain, 2001). In a recent project report, the World Bank has highlighted that serious road capacity constraints on the NH network, worsening road safety amid growing motorisation, ineffective delivery of road infrastructure services by the public road agencies, and a weak enabling environment for private sector participation in road financing are the important issues affecting the Indian road sector (World Bank, 2000).

15.1.2 Need for privatisation of roads in India

As economic liberalisation gains momentum, the crucial operation of moving goods from origin of production to customers is being stymied by the rapidly increasing road transport bottlenecks. Recognising the inade-

Table 15.1 *Growth of various components on Indian roads (Jain, 2001).*

	1951	1995	2001 (expected)
Total vehicles	0.3 million	27.5 million	53.1 million
Total road length	0.4 million km	2.18 million km	3.3 million km
Freight traffic	6 BTK	350 BTK	3000 BTK
Passenger traffic	23 BPK	1500 BPK	3000 BPK
Freight traffic road: rail share	11:89	70:30	NA
Passenger traffic road: rail share	28:72	82:18	NA
Fund allocation as % of the plan outlay for roads	6.5%	3%	Less than 2%

BTK: billion ton km; BPK: billion passenger km.

quacies of the road transport system as a serious constraint on economic and social development, governments at central and state levels have drawn up massive road development plans. Total investments needed over the next six years for the development of the National and State Highways are estimated to total Rs. 1600 billion (Ministry of Industry, 2001). An ambitious National Highway Development Program (NHDP) for the development of 13 000 km in three phases with an estimated cost of Rs. 580 billion is in progress. The Government has realised that it cannot afford to finance all required road infrastructures from budget allocations or by borrowing. Funding from private investors has become necessary to overcome the inadequacy and for rapid development of road infrastructure. The private sector's role in infrastructure development is vital in providing wider access to capital markets, better management skills, access to the latest technology, and implementing projects faster than possible in the public sector. About one quarter of the total fund requirement in the road sector is expected to come from private investments.

15.2 Evolution of the BOT framework in the Indian road sector

Build-operate-transfer (BOT) and build-own-operate-transfer (BOOT) types of concessions (with and without public-private partnership) are the two PSP models that have evolved in the Indian road sector during the last decade. In the BOT framework, the right of way land acquired by government is leased to the concessionaire for a fixed period without transferring the ownership. In BOOT, the concessionaire is given absolute ownership right to the land with permission to mortgage it to obtain the necessary debt finance. In some cases, a conditional developmental right along the corridor is also given. Road infrastructure development through private participation started in 1995 when the Ministry of Surface Transport (MOST; now the Ministry of Road Transport and Highways or MORTH) of the Government of India invited tenders from the private sector for the construction of selected bypasses and bridges on NHs. Many states like Maharashtra, Gujarat and Rajasthan have also taken initiatives for private participation in road infrastructure during this period.

Most of the projects executed in the initial period of privatisation were in BOT format with minimal public sector participation. Government support was limited to land acquisition and a state assurance for necessary support. Until 1999, about 17 road projects under the BOT setup were awarded by MOST. The majority of these projects were small to medium, with the investment size ranging from Rs. 150 million to Rs. 1200 million which were mostly either bypasses, bridges or railways over bridges (Ministry of Finance, 2000). In 1995, the National Highway Authority of India (NHAI) was established under the National Highway Authority of India Act (1988) and entrusted with all major works programmes on the NH network including outsourcing of services, works, and financing to the private sector. At present the NHDP under both public and private participation is being co-ordinated by NHAI. During the above period two major projects under the BOOT concept were also developed and implemented under a higher level of public-private partnership. Infrastructure Leasing & Financial Services Ltd (IL&FS), an all India financial institution, has developed the 32-km Vadodara-Halol toll road at a cost of Rs. 1600 million and the 550-m long Delhi–Noida toll bridge at a cost of Rs. 4080 million with equity participation from respective state governments. In both the projects, the concessionaire was permitted to mortgage the project land to obtain the necessary debt finance. Since the investment requirement of both of these projects was high and there were few of the early projects in the sector, it took almost six to seven years for its development, including obtaining necessary approval, development of the concession framework, identification of equity partners and debt providers. Both projects have successfully completed the construction phase and are in the initial phase of their 30-year operation period. However, they are facing heavy traffic revenue loss due to reduction in traffic demand and revenue.

In India, so far seven road projects have been implemented under public-private participation with varying ranges of government support (five projects in operation phase and two under construction). The details of BOT/BOOT road projects implemented by MOST/NHAI and various states are given in Table 15.2 and Table 15.3. About 17 road projects, with a total cost of about Rs. 25 billion, are also at various stages of development. Some of the major BOT road projects are the Tada–Nellore road (Rs. 7.6 billion), the Second Vivekananda bridge (Rs. 6 billion), and the Jaipur–Kishangarh road (Rs. 6 billion). The major difficulty in identifying more road projects for privatisation is that, except for a few high density corridors/stretches, city bypasses and bridges, the majority of the highway projects are economically not viable due to large capital investment for long periods, toll subsidy, and low traffic revenue realisation (Policy Impediments to Financing, 1999).

15.2.1 BOT model

The initial BOT model practised in Indian roads was very simple, with the Government giving concession to a concessionaire (individual company/ joint venture/special purpose vehicle). The concessionaires may either give separate contracts for engineering, procurement, and construction (EPC) and operation or execute it by themselves. The framework adopted by MOST is given in Fig. 15.1. During the initial phase of BOT road project development in India, the participation of the Government was limited to preliminary feasibility studies, land acquisition and finalising the concession. The procurement process followed in Indian BOT road projects initiated by MOST/NHAI is given in Fig. 15.2.

Although project data was provided with the bid document, the Government did not give any guarantee for the data related to traffic or any other unforeseen ground conditions. In MOST/NHAI projects, the basis for the concession award was the estimated lowest cost (ultimate) to facility users. Since the tolls and traffic levels used in the bid were pre-

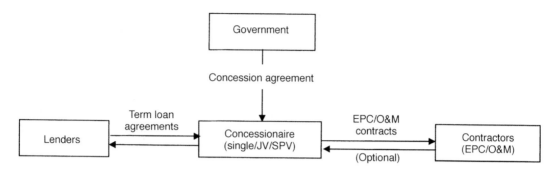

Figure 15.1 *General BOT framework adopted in Indian roads.*

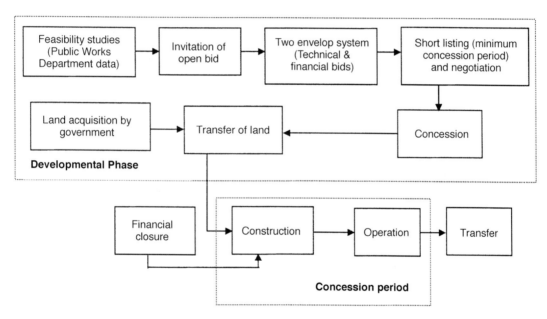

Figure 15.2 *Procurement process in Indian BOT road projects.*

specified, concessions were awarded based on the shortest concession period. Though transfer of land free from encumbrances was an essential precondition for the start of construction work, it was not so for financial closure. Since the total concession period was counted from the date of signing the agreement, the construction work was started before financial closure in several BOT road projects. The major provisions incorporated in the concession agreements were land lease, Material Adverse Effect (MAE) provisions, toll rate revision mechanism, step in rights for government and lenders, change in law, termination clauses, exclusivity provisions and dispute resolution mechanisms. A brief review of the above provisions with respect to five BOT/BOOT projects executed in India is shown in Table 15.4.

Table 15.2 *List of BOT projects executed by central government in India.*

Name of the project	Public agency	Project structure	Length (km)	Concession period	Cost (Rs. million)	Status of the project and details of government participation
Thane–Biwandi bypass two-laning (NH3/4)	MOST	BOT	24	7 years, 8 months	170	In operation
Udaipur bypass (NH8)	MOST	BOT	11	10 years, 2 months	240	In operation
Coimbatore bypass (NH17)	MOST	BOT	28	32 years	1050	In operation. Revenue sharing with existing bridge
Durg bypass (NH6)	NHAI	BOT-PPP	18.4	30 years	700	In operation. Subordinated loan and debt guarantee from NHAI
Nellore bypass (NH5)	NHAI	BOT	18	29 years	730	Concession signed. Work never started
Hubli–Dharward bypass (NH4)	MOST	BOT	30.35	26 years	750	In operation
Six bridges Andhra Pradesh (NH5)	NHAI	BOT	–	35 years	500	Majority of the bridges under operation. Final completion target April 2002
2nd Narmada bridge (NH8) + approach	MOST	BOT	1.37 + 4.63	12 years	1130	In operation. Revenue sharing with existing bridge
Chalthan ROB (NH8) + approach	MOST	BOT	ROB + 4.0	23 months	100	Concession completed
Nardhana ROB (NH8) + road	MOST	BOT	1.42 + 11.64	12 years, 10 months	340	In operation
Patalgana bridge & ROB (NH6)	MOST	BOT	–	15 years, 10 months	330	In operation

Koratalaiya bridge (NH5)	MOST	BOT	–	10 years	300	In operation
Nasirabad ROB (NH6)	MOST	BOT	30	10 years, 11 months	105	In operation
Wainganga bridge (NH6)	MOST	BOT	530	18 years, 9 months	326	In progress
Mahi bridge (NH8)	MOST	BOT	0.56	7 years, 8 months	420	In operation
Kishangarh bypass ROB (NH4)	NHAI	BOT	–	4 years, 3 months	167	In operation
Khambadki tunnel + road	MOST	BOT	8	–	400	In operation
Kaman-paigon (NH8.3)	MOST	BOT	–	–	240	In operation
Thane-Biwandi four-laning (NH3/4)	MOST	BOT	24	–	900	Work completed
Watrak bridge (NH8) + approach	MOST	BOT	0.33 + 10	10 years, 8 months	482	In operation
Derabassi ROB	MOST	BOT	18		360	In operation
Moradabad bypass	NHAI	BOT-PPP	18.4	–	1000	Phase 1 in operation. Equity participation from NHAI
Kharpadian bridge + ROB (NH17)	NHAI	BOT	–	–	330	In operation

NH, National Highways; ROB, Railway Over Bridge; MOST, Ministry of Surface Transport; NHAI, National Highway Authority of India; PPP, Public-Private Partnership.

Table 15.3 *List of BOT projects executed by state governments in India.*

Name of the project	Public agency	Project structure	Length (km)	Concession period	Cost (Rs. million)	Status of the project and details of government participation
Pali bypass, Rajasthan	Government of Rajasthan	BOT-PPP	7	–	102.5	In operation, traffic guarantee
Karaunti bridge, Rajasthan	Government of Rajasthan	BOT	–	–	22.5	In operation
Rao-Pitanpur bypass, Madhya Pradesh	MIDC	BOT	11.5	–	150	In operation
Delhi Noida toll bridge + approach road	New Okhala Industrial Development Authority (NOIDA)	BOOT-PPP	0.55 + 5.5	32.5	4080	In operation, equity participation from NOIDA, rate of return guarantee, debt guarantee for all events, conditional developmental rights
Bangalore–Mysore expressway	Government of Karnataka	BOOT	120	–	7870	Corridor development project. Land acquisition delayed for long period
Mahakali flyover (Mumbai)	Corporation of Mumbai	BLT	–	–	1400	
Vadodara–Halol bypass (Gujarat)	Government of Gujarat	BOOT-PPP	35	30	1600	In operation, equity participation, rate of return guarantee, debt guarantee for all events, conditional developmental rights
Ahmedabad–Mahesana (Gujarat)	Government of Gujarat	BOOT-PPP	63	30	3150	Under construction, equity participation, rate of return guarantee, debt guarantee for all events, conditional developmental rights

Project	Authority	Type		Duration		Status
Mattancherry bridge, Kerala	Cochin Development Authority	BOT	0.50	–	300	In operation
East Coast road	Government of Tamil Nadu	BOT-PPP	113	31 years	610	Construction completed. Waiting for Government orders to start operation. Equity support from Government
Bharatpur bypass, Rajasthan	Government of Rajasthan	BOT	10.85	9 years, 3 months	130	In operation
Sikkar bypass, Rajasthan	Government of Rajasthan	BOT	19.5	–	110	In operation
Kekri–Nasirabad road, Rajasthan	Government of Rajasthan	BOT	25	5 years, 6 months	40	In operation
Raila ROB, Rajasthan	Government of Rajasthan	BOT	–	–	80	In operation
Bata Chowk ROB, Hariyana	Government of Hariyana	BOT	–	–	240	In operation
Eluru bypass	Government of Andra Pradesh	BOT	–	–	80.9	In operation
Yanam–Godavari bridge	Government of Andra Pradesh	BOT	–	–	1100	In operation
Chitoor bypass	Government of Andra Pradesh	BOT	–	–	23	In operation
Karur bridge, Tamil Nadu	Karur Municipality	BOT	0.7	14 years	160	In operation

MIDC: Madya Predesh Industrial Development Corporation.

Table 15.4 *Details of provisions in selected BOT concessions in Indian road projects.*

Project Name	Coimbatore bypass	Vadodara–Halol toll road	Bharatpur bypass	Second Narmada bridge	Delhi–Noida bridge	Nardana ROB
Land lease	Simple lease from Government to concessionaire	Lease with power to mortgage the land to lenders	Simple lease from Government to concessionaire	Simple lease from Government to concessionaire	Lease with power to mortgage the land to lenders	Simple lease from Government to concessionaire
Material adverse effect (MAE) provisions	Toll charge adjustments, varying the concession period and direct financial contribution	Toll charge adjustments and varying the concession period	Varying the concession period through negotiation	Toll charge adjustments, varying the concession period, direct financial contribution	Toll charge adjustments and varying the concession period	Varying the concession period, compensation
Toll rate setting mechanism	Annual revision linked to WPI (all commodities)	Linked to CPI or WPI for inflation (whichever is higher)	Pre-fixed enhancement once in 4 years	Annual revision linked to WPI (all commodities)	Annual revision linked to WPI (all commodities)	Same rate for the entire concession period
Step in right provisions for Lender	In case of termination due to concessionaire default	In case of termination due to concessionaire default	In case of termination due to concessionaire default	In case of termination due to concessionaire default	In case of termination due to concessionaire default	In case of termination due to concessionaire default
Step in right provisions for Government	Yes	In case of national/state emergency	Yes	Yes	In case of national/state emergency	Yes
Provisions related to change in law	Yes	Treated as a GOG default and in case of MAE provision for toll revision	Compensation through negotiation	Yes	Treated as a NOIDA default and in case of MAE provision for toll revision	Compensation

Termination mechanism	In case of default by either party	For GOG default, concessionaire default and FM events (political and non-political)	In case of default by either party – not clearly specified in the concession	In case of default by either party	For NOIDA default, concessionaire default and FM events (direct political, indirect political and non-political)	In case of default by either party
Exclusivity provisions	Yes – there will not be any parallel facility within 20 km radius	Not to levy toll from any new road within 50 km of the project	Nil	Yes – there will not be any parallel facility within 20 km radius	Exclusivity till the bridge achieves 60% of the rated capacity or 10 years, whichever is later	Nil
Dispute resolution mechanism	2 stage on mechanism Steering group & Arbitration panel	3 stage on mechanism Consultation panel, Expert panel, & Arbitration panel	Arbitration panel	2 stage on mechanism Steering group & Arbitration panel	2 stage on mechanism Consultation panel & Arbitration panel	Sole arbitrator appointed through mutual consent

CPI, Consumer Price Index; WPI, Wholesale Price Index.

With increasing experience in the implementation of BOT road projects, many modifications to the initial BOT framework have evolved. The annuity system of BOT procurement is also being tried by NHAI in selected road projects, where Government returns the capital investment of the private sector (with an agreed rate of return) through deferred payments during the operation period. In this system, the entire traffic demand and revenue risk is borne by the Government. NHAI may, at its sole discretion, levy, demand, collect, retain and appropriate the fee either by itself or authorise any person by contract. So far six road projects costing Rs. 17.6 billion have been awarded under the annuity scheme. In all these projects, the construction and operation were combined as a single contract and the concessionaire was selected based on minimum annuity payments quoted.

15.2.2 Major issues in road procurement under the BOT setup

Key issues related to private infrastructure development in India are project-related risks, financial management, legal issues and organisational problems (Thomas *et al.*, 2001). In a survey of 1800 investors from all over the world, risk (bureaucratic, legal, regulatory, social/cultural) has been identified as the most deterring factor for poor FDI flow to the Indian infrastructure sector (FDI Confidence Audit: India, 2001). Against the requirement and the opportunities existing in the Indian road sector, the level of participation from private investors is far below the expectations of the Government. At present there is no uniform lending policy of the financial institutions. This is done on a case-by-case basis with the onus entirely on the entrepreneur to establish the economic and financial viability of the project and their financial standing in the market to qualify for loans. Most of the BOT road entrepreneurs in India are medium-sized contractors and have limited access to project-based financing. In many projects, entrepreneurs have mortgaged their fixed assets (other than project assets) for availing debt because the concessions did not permit for mortgaging the project land as collateral for finance. Due to a tendency for heavy subsidisation of tolls in the Indian BOT system, in most cases user charges alone are not sufficient to recover the capital investments. Apart from funding and profitability problems, reasons for poor investor response in Indian BOT road projects include inordinate delay in project finalisation, inadequate regulatory policy framework, lack of long-term finance products, limited availability of risk cover products, inadequate dispute resolution mechanisms, high traffic revenue and demand risk, and poor risk allocation and sharing among major stakeholders (Policy Impediments to Financing, 1999). The foreign equity investment in the Indian road sector has also been minimal. Except for two BOOT projects promoted by IL&FS and Banglore-Mysore Expressway, none of the projects executed till 2001 under the PSP framework could attract foreign equity. By the year 2001, the performance data of the initial phases of BOT road projects under operation had started coming in and many of them had serious revenue realisation problems, which added to the apprehension of investors.

15.2.3 Government initiatives

The issues discussed above slowed the privatisation of the road sector and governments both at the centre and state level have initiated many legal, regulatory and policy changes including increases in their level of participation and sharing of risks. The Indian Tolls Act (1851), the National Highway Act (1956) and the Land Acquisition Act (1894) have been suitably amended to facilitate private sector participation. The amended Toll Act authorises states to levy tolls on users for using any bridge or road (excluding national highways) which has been made or repaired under the expenses of state or central government. Tolls may be levied directly by the State or through its lessee. The amended National Highway Act permits levy of fees on sections of the national highway and also enables the Government to involve the private sector directly in development and maintenance of national highways for a specified period. The Land Acquisition Act now empowers the Government to quickly take possession of land for National Highway development. Once the Government decides to acquire the land, the owners have to surrender it. They can only raise disputes about the quantity of compensation. The land for highway construction and *en route* facilities will be licensed to the enterprise during the concession period. License documentation expenses such as stamp duty will be borne by the developer. The concessionaire is not allowed to sub-lease the land to anyone, but can license *en route* facilities. The Arbitration and Conciliation Act (1940) was amended in 1996 to incorporate the United Nations Commission of International Trade and Law (UNCITRAL) provisions. Other incentives/ initiatives taken by central government for improving private participation in Indian BOT road projects are as follows:

- Establishment of a specialised financial intermediary known as the Infrastructure Development Finance Company (IDFC) with equity participation from government, domestic financial institutions and foreign investors
- Government to provide capital grants up to 40% of project cost on a case-by-case basis
- Government to carry out all preparatory work including land acquisition and utility removal at its cost
- Right of way (ROW) to be made available to concessionaires free from all encumbrances
- 100% tax exemption allowed for five years and 30% relief for next five years. May be availed for 20 years
- Foreign direct investment up to 100% is permitted for construction of roads and bridges
- External Commercial Borrowings (ECB) exposure up to 50% of the highway and other infrastructure subject to fulfilment of other ECB guidelines
- Promoters of BOT road projects allowed to dilute their stake after the construction phase, however, original bidders required to maintain majority stake throughout the life of the project

- NHAI to participate in equity in BOT road projects up to 30% of total cost
- Revision of fee linked to Wholesale Price Index (WPI)
- Permission for private sector to retain the toll money
- Concession period is allowed up to 30 years
- The NHAI provided with a capital of Rs. 7000 million to leverage funds for road development from the capital market
- Automatic approval of foreign equity investment up to 74%
- FIDIC conditions of contracting for EPC contracts
- Duty free import of modern high capacity equipment for highway construction
- The housing and real estate development which is an integral part of the highway project will be treated as infrastructure and will be entitled to the same tax benefits
- Creation of a non-diversionary fund (Central Road Fund) levied on sale of petrol and diesel
- Internal rate of return up to 20% is permitted for privatised road projects
- Modified BOT frameworks like shadow tolling and annuity is proposed for stretches with relatively low traffic or which are not commercially viable
- Model Concession Agreements for BOT roads in India to improve the project procurement environment/attract more private investments.

Many states like Maharastra, Gujarat, Andhra Pradesh, Tamil Nadu, Kerala, and Karnataka have incorporated state road development corporations to work as nodal agencies for undertaking/co-ordinating the development of bridges and roads through alternate financing mechanisms. Gujarat state has enacted a law governing BOT (Gujarat Infrastructure Development Act 1999) which is the first of its kind in India. The law provides a framework for participation by persons other than the State Government and government agencies in financing, construction, maintenance and operation of infrastructure projects. The road policies formulated by these states for BOT road projects are in line with the policy initiatives taken by the Government of India.

15.3 Risk assessment in Indian BOT roads

15.3.1 Risk management in projects

Risk is defined as exposure to the possibility of economic or financial loss or gains, physical damages or injury or delay as a consequence of the uncertainty associated with pursuing a course of action (Cooper & Chapman, 1987). Risk can be characterised by three components (Al-Bahar, 1989): (1) the risk event – what might happen; (2) the uncertainty of the event – the chances of the event occurring, and (3) potential gain/loss consequence of the event happening. Risk is often used in several distinct senses such as opportunity, hazard/

threat or uncertainty. A brief summary of project risks and their management in different contexts is given in Table 15.5.

Risk management is the management of assets and liabilities to minimise the adverse changes in future cash flows from external shocks. When there are several risks, they are often interrelated and an integrated approach should be adopted (Claessens, 1993). The main task of risk management in a project can be approached systematically by breaking it down into the following three stages: risk identification, assessment, and risk response and control (Al-Bahar, 1989; PMBOK, 1996).

A thorough analysis, comprehensive risk identification and assessment is required at the preliminary stage in order to be able to correctly manage project risks (Salzmann & Mohamed, 1999). The maximum benefits of risk management can be derived only if the process is applied continuously throughout the project life cycle (Raftery, 1994). Risks in BOT road projects could be viewed (from the concessionaire's view point) as an opportunity in the bidding phase for getting the best deal from the Government and as a hazard or threat in the implementation phase. The main objective of risk management in BOT road projects in the implementation phase should be aimed towards reducing the probability of negative events which are likely to affect the economic performance of the project and goal of the organisation. There is not much scope for upside gain in a regulated environment other than what had been agreed upon in the concession.

The success of a project is very much dependent on the extent to which the risks involved can be measured, understood, reported, communicated and allocated to the appropriate parties (Tah & Carr, 1999). There is a worldwide shortage of equity investment in transportation projects in comparison to other infrastructure sectors because of uncertainty in the revenue stream. Investors are nervous about projects that involve interaction with the public (Raghuram *et al.*, 1999). A study carried out by the World Bank shows that the road sector in Asia is subjected to the highest market risk (Alexander *et al.*, 2000). Equity beta was used to measure the relative riskiness of the

Table 15.5 *Risk and risk management in different contexts.*

	Description	Aim of risk management in given context
Risk as opportunity	The greater the risk, the greater the potential for return as well as loss	Maximise the upside and minimise the downside within the constraints
Risk as hazard or threat	Potential negative events which affect the goal and the economic performance of an organisation	Reduce the probability of negative events at minimum cost
Risk of uncertainty	Refers to the volatility of the outcome	Reduce the variance between anticipated outcomes and the actual

company's equity compared to the market as a whole. Private infrastructure projects (BOT) in general have a complex risk profile due to several factors like high investment and long pay back period, length of the term of the loan, susceptibility to political and economic risk, low market value of the security package, non-recourse/off-balance financing, complex contract mechanism involving many participants with diverging interests, and limitations on enforcing security. The additional risk dimensions of road projects under the BOT setup are high developmental efforts and upfront cost, long stretches of land acquisition and associated resettlement issues, absence of underwriting from the customer, and cyclic variation in volume/demand.

15.3.2 Risk management studies in BOT/BOOT projects

Infrastructure projects in general

UNIDO (1996) developed a risk checklist under two major categories (general/country risk and specific project risks) with three sub-categories in each. Political risks, commercial risks and legal risks make up the first category, whereas developmental risks, construction/completion risks and operating risks make up the second. Akintoye *et al.* (1998) have done risk assessment/prioritisation in private finance initiative (PFI) projects in the UK. The ten most important risk factors identified (based on surveys among clients, contractors and financial institutions) are design risk, construction cost risk, performance risk, risk of delay, cost overrun, commissioning risk, volume risk, operating/maintenance risk, payment risks and tendering cost risks. The land acquisition, debt risk, bankers risk and political risks are found to be least important. Salzmann and Mohamed (1999) have presented a risk identification framework for international BOOT projects based on four super factor groupings, i.e. host country, investors, projects and project organisation and also reviewed the published risk and success factors.

The Asian Development Bank (ADB Report, 2000) has reported a list of PSP risks in expressways. But developmental risk issues such as project identification, bidding cost, land acquisition, arrangement of finance in time etc. are not covered. Wang *et al.* (2000) have studied in detail the criticality of political risks and effectiveness of the various management strategies in China's BOT projects. The critical risks identified are credit worthiness and reliability of Chinese parties, change in law, *force majeure*, delay in approval, expropriation, and corruption. Ye and Tiong (2000) proposed a project appraisal/risk assessment method for privately financed infrastructure projects (NPV-at-risk) combining the weighted average cost of capital and dual risk-return method which can effectively incorporate financial, political and market risks.

Transportation/road projects

Malini (1997) proposed a simulation-based financial risk assessment model for the evaluation of BOT transport infrastructure wherein the assessment of risks was demonstrated with the use of risk profiles. The input parameters used were policy parameters (construction and concession period, and toll rates), stochastic variables (probability distributions of cost components, traffic volume) and macro economic indicators (interest rate, discount rate, inflation rate, debt-equity norms and traffic growth norms). Lam and Tam (1998) used Monte Carlo simulation for risk analysis of traffic and revenue in privately financed road projects. Charoenpornpattana and Minato (1999) have identified privatisation-induced risks in transportation projects in Thailand and proposed criteria for risk allocation between government and private sectors. They also reviewed privatisation-induced risks of various countries. Arndt and Maguire (1999) have carried out research in risk identification and allocation of Australian BOOT projects. The risks in private sector toll roads operating under public sector road networks were identified and classified as design and construction risks, operating risks, market risk, sponsor risk, legislation risk, network risk, technology risk in operation and external risks.

15.3.3 Identification and classification of risks in Indian BOT road projects

The criticality of risks in a BOT project are influenced by the legal, political, economic and social environment of the country. They are also sensitive to the sector/nature of service for which the facility is provided (Akintoye *et al.*, 2002). The majority of the risk assessment-related literature is generic or country-specific and may not be applicable directly to other counties. Kalidindi and Thomas (2002) and Thomas *et al.* (2002a) have carried out detailed investigations into the risk identification and assessment of BOT road projects in India. About 200 risk factors falling under 22 risk categories have been identified based on extensive literature reviews, case studies and discussions with Indian BOT road project participants. It is assumed that the realisation of respective risk factors will contribute with a varying degree of probability/impact to the risks under which it is classified. These risks have been classified based on the project phase during which they occur (Table 15.6). Project life cycle risks are those which can occur at any time during conception and completion.

15.3.4 Risk criticality assessment study

Survey approach

A two-stage survey among major participants (Government representatives, promoters/developers, lenders and consultants) of Indian BOT road projects was carried out to evaluate the criticality of risks and risk factors. In the first stage eight 'very critical' risks were identified. In the second stage 'very critical' risk factors falling under each critical risk were short-listed. The survey respondents (very senior level officials who have experience in BOT road projects in India) were asked to rate the criticality of risks on a five point Likert scale based on their experience with Indian BOT road projects.

Table 15.6 *Classification of risks in BOT road projects (Thomas et al., 2002a).*

Project phase	Risk category
Developmental phase	Pre-investment risk, resettlement and rehabilitation risk, delay in land acquisition, permit/approval risk, delay in financial closure
Construction phase	Technology risk, design and latent defect risk, completion risk, and cost overrun risk
Operation phase	Traffic revenue risk, operation risk, demand risk, debt servicing risk
Project life cycle	Legal risk, political risk (direct & indirect), partnering risk, regulatory risk, financial risk, environmental risk and physical risk

Criticality is assumed to be the combined effect of probability of occurrence and the impact of occurrence of the risk. Twenty-seven 'very critical risk factors' falling under eight 'very critical risks' were identified in Indian BOT road projects (Table 15.7). The risks in the order of criticality are traffic revenue risk, land acquisition delay, demand risk, delay in financial closure, completion risk, debt servicing risk, cost overrun risk and direct political risk.

Risk assessment – case study approach

In the light of the survey-based criticality assessment, six BOT road projects executed in India, currently under the operation phase, were studied for actual risk occurrence and their impacts (Table 15.7). The projects examined were the Coimbatore bypass (28-km bypass road on NH-7 + 33-m Attupalam bridge), Vadodara–Halol highway (widening and strengthening of 32-km state highway between Vadodara and Halol), Bharatpur bypass (10.85-km bypass road from Bharatpur–Mathura road to NH-11), second Narmada bridge in Gujarat (1.37-km bridge + 4.63-km approach roads), Delhi–Nodia toll bridge (552-m long, 8-lane bridge across Yamuna River + 6-km approach) and Nardana railway over bridge (ROB) (1.42-km railway over bridge + widening and strengthening of 11.64-km road). There is a fair degree of agreement in the survey-based risk assessment and the actual risk impact in the projects examined. The majority of risk factors fall under traffic revenue and demand risks, land acquisition risk, completion risk, and overrun risk. Though respondents of the survey evaluated delay in financial closure as very critical, none of the projects examined faced serious problems over this issue. Even though there was delay in financial closure for all these projects, the promoter's equity was utilised till the debt component was made available.

15.3.5 Critical risks in Indian BOT road projects

Traffic revenue risk

Major factors contributing to this risk are public resistance to paying tolls, toll enforceability problems due to inadequate state support, modification in the toll structure/system by Government (under public pressure) and inadequacy in the toll tariff and its subsequent increase. Some of the BOT road projects faced problems due to resistance from local users of the facility to pay tolls. In India, traditionally infrastructure services were made available to the public free of cost. One of the main reasons for public resistance to user charges for infrastructure is the introduction of the privatisation concept prior to public acceptance of the commercialisation concept. In two cases, the concessionaires were compelled to reduce the toll level due to public agitation, which in turn affected the financial viability of the projects. In one BOT road project, the Government declared 100% free access for local road users on the day of commissioning and in one case reduced the toll tariff under public pressure during the initial period of operation.

Since law and order is a state subject, active support from state governments is required for toll enforceability. Some of the road projects had serious revenue realisation problems due to the poor state support in toll enforceability. Though Section 8B of the National Highway Act 1995 has a provision for punishment for mischief by injury to National Highways, State Governments have not prescribed a penalty for this. Similarly, section 8A(3) of the the National Highway Act has provision for control and regulation of traffic as per chapter VIII of the Indian Motor Vehicle Act, 1988. This power is entrusted to Regional Transport Officers of the State Government. No separate notification/system for implementing the Act by the concessionaire exists.

Once the toll road is opened, there is no proper tool to establish whether the appropriate tariff levels have or have not been set (ADB Report, 2000). In Indian BOT roads, the concession period is fixed by the bid and cannot be adjusted for fluctuations in demand. Some projects do have provisions for compensating the MAE through the extension of the concession period and increase in toll rate subject to the approval of the toll review committee and separate notification by the Government. This could take a long time because of the absence of any automatic/established system for such revision.

Land acquisition delay

Road projects under the BOT setup require large stretches of land free from all encumbrances prior to the beginning of construction. Delay in acquisition of a small stretch of land may affect the entire schedule and viability of the BOT project due to delay in the start of the commercial operation date (COD). Though the amended NHAI Act and Land Acquisition Act empowers the Government to quickly take possession of land for highway

Table 15.7 Risk assessment of Indian BOT roads (Kalidindi & Thomas, 2002).

Risks	Risk factors	Criticality index		Impact of risks in BOT road projects					
		Risk	Risk factor	P-1	P-2	P-3	P-4	P-5	P-6
Traffic revenue risk	Non-acceptability of toll rate to user (agitation/ public interest litigation etc.)	0.88	0.84	***	**		*		*
	Lack of toll enforceability due to inadequate State Government support		0.80	***		***			*
	Revenue loss due to scrapping/alteration in toll system by Government		0.78	***		***			
	Toll enforceability problems due to inadequate Toll Act provisions		0.73	***					**
	Agreeing to inadequate toll rate/toll increase		0.70						
Land acquisition delay/risk	Public/political interference for changing the scope/alignment	0.83	0.79					***	*
	Public interest litigation/litigation by project-affected prices		0.78	*	*			**	*
	Land acquisition delay due to encroachment problems		0.77	*	*	*		***	*
	Interference of environmental activists		0.71						
Demand risk	Development/upgradation of alternate toll free roads	0.81	0.74	*			*		**
	Over estimation of socio-economic development of the influencing zone		0.72					***	
	Short fall in the traffic due to recession		0.72	***	***	***	***	***	***
Delay in financial close	Project not bankable due to inadequate concession provisions	0.77	0.87						
	Lender(s) not comfortable with robustness of project financial viability		0.79						
	Failure to arrange equity in time		0.79						
	High change in the cost of debt		0.72						

Risk	Value	P-1	P-2	P-3	P-4	P-5	P-6
Completion risk							
Suspension of sanctioned loan by lenders	0.77						
Delay due to public interference (agitation/litigation)	0.79						
Delay in obtaining environmental clearances	0.74			***		***	
Liquidation of joint venture during construction period	0.74						
Delay in approvals from government agencies	0.73				*	**	**
Debt servicing risk							
Cash flow inadequacy to meet debt servicing due to traffic revenue decline/leakage	0.81		***	***	**	***	*
Cost overrun risk							
–	0.74			**	*		**
Direct political risk							
Payment failure by Government	0.78		***	***		*	**
Sanction for competing facility by Government under political pressure	0.77						
Government backtracking from provisions of concession agreement	0.77			***	**	*	**
Change in government policies/law	0.75					**	
Reduced support from future governments	0.71		**	**	*		

P-1, Coimbatore bypass; P-2, Vadodara–Halol toll road; P-3, Bharatpur bypass; P-4, Second Narmada bridge; P-5, Delhi–Noida toll bridge; P-6, Nardana ROB.
Impact of risk on project: *** High impact   ** Medium impact   * Low impact

development, it has not been effective because the land is still a state subject and the acquisition process is very time consuming (Fig. 15.3).

The processes involve notification of intention of land acquisition for the project, survey, file public objections if any, and hearing, decision by competent revenue authority, declaration, compensation and processing of the land. The land details, such as the extent of the land required, classification, cost, ownership etc. furnished by the divisional engineer, will be cross-verified by Thahasildar prior to the notification in the *Government Gazette* by the collector. Under normal circumstances, the process could take one to three years. It could be further delayed due to various reasons like public interest litigation, encroachments, legal suits for compensation, rehabilitation and resettlement issues, political patronage, environmental and forest clearance etc. Even after successful completion of land acquisition, vested parties are taking cases to the courts on one pretext or another prolonging the whole process for long periods.

Four risk factors have been rated as 'very critical' in delaying the land acquisition process: public/political interference for changing the scope/alignment; public interest litigation/litigation by project-affected parties; encroachments; and interference of environmental activists. Though transfer of land prior to commencement of work is a condition precedent, often the concessionaire starts construction if a major portion of the project land is made available. In the Delhi–Noida toll bridge, part of the project got delayed for almost one year due to encroachment problems and delay in

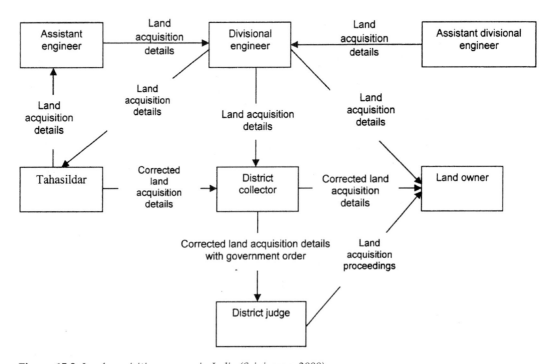

Figure 15.3 *Land acquisition process in India (Srinivasan, 2000).*

permission for cutting trees. The Rs. 7870 million Bangalore–Mysore express highway BOOT project has serious problems due to many litigations and a long delay in environmental clearance as it involved the acquisition of 168 acres of forestland.

In the Hubli–Dharward bypass project, the Government could not provide the land free from all encumbrances/utility clearances even after the commissioning of phase 1 of the project. Inordinate delay in various utility clearances, error in acquisition and additional land requirement due to scope changes were the reasons reported for the delay (Ramesh, 1999). Even if the land for the road is acquired, physical possession of land free from existing utility services and other encumbrances involves co-ordination of many government departments. The problem of ribbon development has existed for a long time on Indian roads and land acquisition along the road for widening/strengthening often faces encroachment problems. Proper rehabilitation of encroachers is insisted on in World Bank-funded projects, and identification and procurement of sites for rehabilitation of displaced persons near urban areas could lead to delay as well as a substantial increase in the land acquisition cost.

Demand risk

The demand for the use of a toll road to a great extent depends on the clear economic value/return attached to the usage of the toll road. In India, small isolated stretches of highways, bypasses and bridges have been developed under BOT concession and explicit economic gain is not visible to vehicle users because of the bottlenecks remaining in the undeveloped stretches of the highway. The unaffordability of tolls, and absence of understanding/ accepting commercialisation concepts among the public can also cause reduction in demand. Elasticity of demand due to the introduction of tolls has not been adequately studied in the Indian road project context (Malini, 1997). Very few BOT road projects in India have realised the projected traffic demand. The total traffic demand loss varies between 20% and 50% in various projects. The survey investigation revealed that the severity of demand risk in Indian BOT projects is very high. Development/upgradation of alternate toll free roads, over estimation of socio-economic development of the influencing zone and demand reduction due to recession are rated as very critical. For all the six BOT road projects studied, demand was affected most by the recession. There could also be forecasting errors in the traffic projections. The error could be due to data, model specification, model inputs or network assumptions. The continuous fluctuation in factors like government funding, growth of industrialisation and agriculture, land use pattern and population growth rate will make long-term demand forecasting a difficult task in the Indian context.

Delay in financial closure

Arranging for both debt and equity in Indian BOT projects is difficult and many BOT road projects planned in India became non-starters mainly due to this problem. Infrastructure projects require huge capital investment upfront and the major portion of the resources is arranged through non-recourse type debt financing. Promoters of Indian BOT roads are frequently medium-sized contracting companies, which themselves are not well capitalised. Availability of long-term financing for infrastructure projects is very limited in India. The Indian capital market is also not adequately developed for this type of financing. Only one private road project in India so far has used publicly placed discount bonds for financing the capital investment. Maharahtra State Road Transport Corporation (MSRTC) has used government guaranteed public issues for part financing its road projects. In the earlier BOT road projects, even though financial closure was not a necessary precondition for start of the work, many entrepreneurs/promoters started the process of arranging debt and equity prior to the signing of the concession. Financial institutions took a lot of time sanctioning the necessary loan for BOT road projects because of lack of precedence in debt syndication, documentation and creation of securities. Among all the factors evaluated, the factor 'poor bankability due to inadequate concession provisions' was rated highest. In the majority of Indian BOT road projects, lenders are not involved in contract negotiations prior to concession and they often raise objections to the security package offered in support of the transaction leading to re-negotiation, arrangement of additional collateral etc. Over the years, the Government has taken many initiatives to improve the legal, regulatory and institutional environment in BOT road projects aiming to improve the investment climate. In some BOOT projects promoted by ILFS in partnership with the Government, even the debt due to lenders is guaranteed for all events including the concessionaire default.

Completion and cost overrun risks

Though these risks were identified as very critical in the survey, the case study investigation of BOT road projects revealed that most of them were completed well ahead of schedule with marginal cost overrun. This is in contrast with the general construction environment in India where time and cost overruns are a feature of most projects. Out of the 421 major projects implemented during the Eighth Plan (1992–97), 169 projects had reported cost overruns (total cost overrun of Rs. 234 billion) and 202 projects reported time overrun (Ministry of Statistics and Program Implementation Report 1999, Government of India). Promoters of BOT road projects were very keen to complete the construction well before the stipulated completion time even at an additional cost, as this will increase the operation period within the total concession period. The increase in cash inflow due to early start of operation is much more than the cost of additional resources required for planned enhancements. The Mattachery toll bridge near Cochin was completed within half of its scheduled completion period and the Delhi–Noida toll bridge was completed four months ahead of schedule.

The major risk factors contributing to completion risk are suspension of a sanctioned loan by lenders, delay due to public interference (agitation/litigation) against the project as well as land acquisition, delay in obtaining environmental clearances, liquidation of joint venture during the construction period and delay in approvals from government agencies. Except for the possible joint venture failure, all the critical factors attributed to completion risk are external in nature and the promoter/contractor have no direct control over them. None of the risk factors attributable to cost overrun is rated as 'very critical'. Unlike other risks in BOT roads, construction risks are easy to manage through insurance and other time and cost control techniques. Focusing on South East Asian countries, Tam and Leung (1999) also observed that technical risks were the easiest to manage in BOT projects in comparison to financial and political risks.

Debt servicing risk

The major reasons for debt servicing risk are traffic revenue shortfall, sudden increase in operation and maintenance costs, inadequate cash flow due to faulty debt servicing structure, due diligence, increase in taxes or increase in interest rates. The cash flow inadequacy due to traffic revenue shortfall is the most critical risk factor which affects the debt servicing. The case study investigation shows that the present revenue declines in BOT road projects varied between 25% and 65% and all the projects have cash flow inadequacy to meet the debt service obligations. The second Narmada bridge that is located in one of the busiest traffic corridors in India has 25% reduction in the projected traffic revenue whereas in the Delhi–Noida toll bridge, which is located in the national capital of India, it is almost 65%. Many projects have floating interest rates linked to the prime lending rates (PLR) of respective banks. Though interest rates in India have been continuously coming down for the past few years, the banks have not reduced their PLR in tune with Reserve Bank of India (RBI) interest rates.

Political risks

Political risk may be defined as changes in the operating conditions of private enterprises arising out of political process, either directly through war, political violence, insurrection or through change in government policies that affect the ownership, profitability and behaviour of the firm (Bubnova, 1999). Since investments in road infrastructure are large, long-term, irreversible and domestic market dependent, change in government policies adversely affects profitability.

Payment failure by the Government, sanction for competing facility by the Government under political pressure, Government backtracking from provision of concession agreement, changes in government policies/law and reduced support from a future government are rated as 'very critical risk factors' contributing to political risks. In the Coimbatore bypass project, a cost overrun of Rs. 63.1 million (Government-induced scope change) claimed by the concessionaire is yet to be compensated by the Government. The decision on the nature of compensation is not finalised even two years after completion of the construction. In the Bharatpur bypass project, the compensation for MAE due to Government concessions to local users is also not finalised and payment has been pending for a long time. In the case of the Coimbatore bypass and Mattanchery toll bridge, under political pressure, the Government renegotiated the toll tariff finalised in the concession. In the East Cost Road (ECR) project near Chennai, the Government did not issue the necessary toll notification order which has delayed the start of toll collection.

15.4 Risk allocation

15.4.1 Introduction

Success in a BOT project, to a great extent, is influenced by the degree to which a win–win situation for all participants is attained through an equitable risk allocation/sharing framework (Walker & Smith, 1996). The basis of risk transfer/allocation in a BOT project should be stakeholder support, financial feasibility, regulatory and monitoring framework, and knowledge of the state of operating assets and their current performance. Improper allocation/sharing of risks among stakeholders may lead to sub-optimality, and result in higher than necessary prices for risk transfer (ADB Report, 2000). Though the commercial risk should have been assigned to the private sector in PSP projects, the Government is often compelled to assume it (at the expense of tax payers), which it cannot control or can control no better than the private sector, thus creating large contingent liabilities (Arndt & Maguire, 1999). On the contrary, if the promoters are burdened with risk, over which they have no control or capability to manage, it will result in either the failure of the project or increase in the project cost. Both these situations will not help the Government, promoter or users of the road. The dogma that privatisation involves the transfer of all risk to the private sector (prevalent in many countries until very recently) is no longer respectable. Most governments now recognise that privatisation is a partnership in which the State must retain some risk, whether in the form of financial subsidies or the assumption of contractual responsibilities or contingent liabilities (Projects Group Publications, 1996).

15.4.2 Risk allocation preference of stakeholders

The rationality of risk sharing in projects is based on the principle that the party best capable of managing the risk should assume it. Often this principle is not followed/accepted in many of the projects. The summary of the studies conducted by authors to evaluate the perception of major Indian BOT road project participants with respect to risk management capability of various project participants, their preference of risk sharing/allocation and current risk allocation practice is given in Table 15.8.

Table 15.8 *Perception of project participants on capability of risk management and risk allocation (Thomas et al., 2002b).*

Risks	Best party capable of managing the risk[1,2]	Allocation preference[1,2]	Allocation practice in Indian BOT roads
Traffic revenue risk	P, G	P, G, RU, L	P
Delay in land acquisition	G	G	G
Demand risk	P	P, G, L	P
Delay in financial closure	P	P, G	P
Completion risk	CO	CO, P	P, CO
Cost overrun	P, CO	P, CO	P, CO
Debt servicing	P	P, L	P, G[3]
Political risks	G	G, P	G, P

[1] P, promoter; G, government; CO, contractor; L, lender; I, insurer; RU, road user.
[2] Combined opinion of all the categories of respondents. Included only if the combined opinion is more than 30%.
[3] In a few BOOT-PPP projects, debt is guaranteed by the government.

A particular group is included for allocation, only if more than 30% of the respondents in the corresponding category prefer to allocate/agree to share the risk. The opinion clearly indicates that except for construction risks and delay in land acquisition, all other critical risks should be shared among a wider group (Government, lenders and road users) even if they are not the best capable parties. However, except for traffic revenue risk and political risk, the current practice of allocation is rational. This is probably because of the criticality of these risks in the Indian context.

15.4.3 Risk allocation in a model concession agreement

The NHAI developed a concession specifically for the Jaipur–Kisangarh road section of NH-8 and published it as the Model Concession Agreement (MCA) for BOT road projects in India (one for project cost greater than Rs. 1000 million and one for project cost less than Rs. 1000 million). The MCA lays down the principles governing risk sharing between various shareholders. The NHAI envisions that the document will serve as a template for awarding contracts for BOT roads. The key features of risk allocation include revenue short fall, loan from government, indexing of toll revision to inflation, debt guarantee for all *force majeure* and termination events including concessionaire default. The summary of risk allocation provisions of the model concession provision is given in Table 15.9. The NHAI has also published a separate MCA for annuity-based BOT road projects with provision for a bonus in annuity for early completion

Table 15.9 *Important risk allocation provisions in Model Concession Agreement of BOT road projects (Model Concession Agreement, 2000).*

Risk	Risk allocation provisions
Delay in financial closure due to concessionaire default (other than *force majeure* incidents)	Penalty for delay beyond 180 days. Beyond 270 days NHAI can terminate the concession and forfeit the performance security/bid security
Construction delays and cost overruns	Provisions for equity grant support with a cap of 25% of the project cost/maximum paid in cash by shareholders. Delay in completion date will attract payments to NHAI if the reasons for delay are not attributable to NHAI
Land acquisition	NHAI will make payment if delays arise due to failure to provide right of way
Change in scope	Any number of changes by NHAI with a cap of cost increase of 5% of the total cost if the same does not affect the commercial operation date
Additional toll way	Guarantee from Government not to open a parallel facility for a specified period from start of concession. In case of government default, provision for extension of concession period as well as guarantee for 33% higher toll in the new facility
Toll tariff	Annual revision is linked to the extent of variation in WPI. Toll concessions up to 75% for local personal traffic and 50% for local commercial traffic. No indexation of toll to the exchange rate movements
Quality deficiencies in operation and maintenance	Provision for penalty/termination by NHAI. Concessionaire is not required to maintain performance bond (after 25% of the total project cost has been spent) or maintenance reserve fund
Increase in operation and maintenance cost	Operation and maintenance support grant will be provided from the balance of the equity support grant (if any)
Equity holding	Minimum 51% during construction. Concessionaire can sell down equity stake from 51% to a minimum of 26% after 3 years of COD
Issue of completion certificate	Commercial operations can be started with provision for provisional completion certificate subject to the approval of the independent consultant even if work is not fully completed
Traffic revenue decline	NHAI will provide revenue shortfall loan (only to cover senior debt repayment and operation and maintenance cost) if the realisable toll revenue falls below subsistence level

State support

Force majeure (after financial closure)

Provision for state support agreement

- Non-political events: the parties shall bear their respective costs
- Indirect political risks: Concessionaire and NHAI shall share any loss excess of the insurance claim equally
- Direct political risk: NHAI shall bear the cost

The *force majeure* (FM) cost shall not include any loss of revenue/debt repayment obligations. Shall cover interest on debt, O&M expenses and any other cost directly attributable to *force majeure* event. If FM takes place before financial closure, the parties shall bear their respective costs

Termination payments

- For indirect political event: NHAI will pay total debt due less due insurance claim, 100% subordinated debt and opportunity cost of 10% above equity (total 110%) from the date of operation with depreciation of 7.5% per annum after adjusting for the changes in the WPI
- For direct political event: NHAI will pay total debt due less due insurance claim, 120% subordinated debt, opportunity cost of 50% above equity (total 150%) from the date of operation with depreciation of 7.5% per annum after adjusting for the changes in the WPI
- Non-political events: NHAI will pay 90% of senior debt
- Concessionaire default: NHAI will pay 90% of senior debt

and reduction in annuity for delay in completion and for not meeting the prescribed service quality and other obligations set forth in the agreement.

15.4.4 Evaluation of the model concession agreement

Though the Government claims that the model concession document for BOT road projects was prepared after extensive consultations with major BOT stakeholders, not many BOT projects have so far been finalised based on this agreement. Investigation was carried out among BOT road project participants to identify the inequitable provisions in the MCA. A maximum number of project participants perceive provisions in the termination payments, toll tariff fixation/regulation and compensation for political *force majeure* are inequitable. The greatest number of comments were from promoters because 70% to 90% of the debt component is safeguarded in the MCA whereas the concerns of equity holders are not adequately addressed. Some promoters indicated that they were not consulted or their opinion was ignored while preparing the document.

As per MCA, the termination payment to concessionaire for NHAI default is:

'150% (one hundred fifty per cent) of the Equity (subscribed in cash and actually spent on the Project but excluding the amount of Equity Support from Government) if such Termination occurs at any time during three years commencing from the Appointed Date and for each successive year thereafter, such amount shall be adjusted every year to fully reflect the changes in WPI during such year and the adjusted amount so arrived at shall be reduced every year by 7.5% (seven and a half per cent) per annum.'

The depreciation of 7.5% used in equity calculation is not appropriate for BOT road projects because the promoters are not likely to earn any dividend during the initial years of operation and the majority of the revenue is utilised for debt repayment. Moreover, the equity calculation specified in the agreement does not cover for the cost overrun. The 'total project cost' is defined as the minimum of estimated cost, actual construction cost certified by the statuary authority and the total project cost specified in the financing documents. When the government grants and debt components are deducted from the above project cost (minimum of three), the promoter is left with much less than what he had actually invested. Government estimates are conservative and based on conventional methodology and in most cases they are likely to be less than the actual project cost. This provision does not take care of the cost overrun/actual cost of the project while payment due is evaluated.

The definition of 'debt due' covers only the principal amount of the debt provided by senior lenders under the financing document which is outstanding as on the termination date but excluding any part of the principal and its interest that had fallen due for repayment one year prior to termination date. Many project promoters and lenders feel that the definition of debt due should include all the outstanding dues up to the termination date. There is no provision in the termination clause to make *ad hoc* payments to lenders/concessionaire without waiting for the claim settlement. In normal circumstances, the settlement of insurance claims in India will take a long time. In case of delay in termination payment beyond 30 days NHAI will give only the State Bank of India prime lending rate (SBIPLR) plus 2% whereas the cost of borrowing for infrastructure financing in India is much above the SBIPLR.

Though the toll tariffs and subsequent escalation criteria are fixed at the time of concession, the agreement restricts the concessionaire from collecting the toll/review until the Government issues the toll notification. In Indian BOT road projects, upfront toll notification is not issued. Many times the toll notification is delayed due to political and administrative reasons. Timely notification of toll/toll review is not included in the MCA as an obligation of the NHAI. The compensation payable to the concessionaire on account of political *force majeure* will include only interest on debt, O&M expenses and any other cost directly attributable to the *force majeure* event. The lost revenue during the period of *force majeure* is not considered. Issues like compensation for an alternative toll free road and change in the law, absence of traffic guarantee, inadequacy of State support agreement, time limit for financial closure, inability to assign project assets, handing over of land, design and quality control are also highlighted as inequitable/unsatisfactory by some participants of Indian BOT road projects.

15.5 Conclusions

The inadequacy of infrastructure as a whole and the road sector in particular is one of the major constraints for meeting the desired growth rate in the Indian economy. The major reason for inadequate infrastructure procurement is lack of funding. The Government of India has accepted the commercialisation of infrastructure as one of the solutions. Massive developmental programmes are being planned/implemented in the Indian highway sector with public as well as private finance.

So far, most of the private participation in the road sector is in small- to medium-sized projects including bypasses, bridges or small stretches of busy corridors. Build-operate-transfer (BOT) and build-own-operate-transfer (BOOT) types of concessions (with and without public-private partnership) are the two PSP models that have evolved in the Indian road sector during the last decade. The initial enthusiasm observed among private participants for investing in road projects is coming down after the experience of high developmental and operational risks in many of the BOT road projects. The level of participation from private investors is far below the expectations of the Government.

Based on the survey conducted, the critical risks identified in Indian BOT roads in order of criticality are traffic revenue risk, land acquisition delay, demand risk, delay in financial closure, completion risk, debt servicing risk, cost overrun risk and direct political risk.

Governments at both central and state levels have taken many initiatives to improve the legal, regulatory and institutional environment of the Indian road sector. With increasing experience in implementation of BOT roads, many modifications have been evolved with increase in Government participation and sharing of risks with private investors. Annuity systems of BOT procurement are also being tried by the NHAI in selected road projects, where Government returns the capital investment of the private sector (with agreed rate of return) through deferred payments during the operation period.

The participants of Indian BOT road projects feel that the allocation of critical risks should be shared among a wider group (Government, lenders and road users) even if they are not the best capable parties to manage the risks. Except traffic revenue risk and political risk, the current practice of risk allocation is rational in Indian BOT roads. An evaluation of the MCA published for Indian BOT roads shows that concerns of equity holders are not adequately addressed whereas a substantial portion of the debt component has been guaranteed by the Government.

References

Akintoye A., Taylor C. & Fitzgerald E. (1998) Risk analysis and management of private finance initiative projects. *Engineering, Construction and Architectural Management,* **5**(1), 9–21.

Akintoye A., Fitzgerald E. & Hardcastle C. (2002) *Public-Private Partnership Projects in the UK – Treatment of Associated Risks by Local Authorities.* Proceedings of CIB W92 Symposium on Procurement Systems, Trinidad & Tobago pp. 297–316. The Engineering Institute, University of West Indies, Trinidad and Tobago.

Al-Bahar J.F. (1989) *Risk Management in Construction Projects: a Systematic Analytical Approach for Contractors.* Ph.D Thesis, University of California, Berkely, pp. 6–27.

Alexander I., Estache A. & Adele O. (2000) *A Few Things Transport Regulators Should Know about Risk and Cost of Capital.* World Bank Publication, Washington, DC.

Arndt R. & Maguire G. (1999) *Risk Allocation and Identification Project-Survey Report.* The University of Melbourne, the Department of Treasury and Finance, Melbourne.

ADB Report (2000) *Development of Best Practices for Promoting Private Investment in Infrastructure: Roads.* Asian Development Bank, pp. 101–106.

Bubnova N. (1999) *Guarantees and Insurance for Re-Allocating and Mitigating Political and Regulatory Risks in Infrastructure Investment: Market Analysis.* International Conference on Private Infrastructure for Development: Confronting Political and Regulatory Risks, Rome.

Charoenpornpattana S. & Minato T. (1999) Privatisation-induced risks: state-owned transportation enterprises in Thailand. In: *Profitable Partnering in Construction Procurement* (ed. S.O. Ogunlana), pp. 429–440. E&FN Spon, London.

Claessens S. (1993) *Risk Management in Developing Countries*. World Bank Technical Paper Number 235. World Bank Publication, Washington, DC.

Cooper D. & Chapman C. (1987) *Risk Analysis for Large Projects*. John Wiley, Chichester.

FDI Confidence Audit: India (2001) A.T. Kearney, Inc. http://www.atkearney.com/main.taf?site=1&a=5&b=3&c=1&d=31 visited on 16.10.2001.

Jain J.P. (2001) Adversity of Indian Roads. *Indian Highways*, **29**(12), 37–47.

Kalidindi S.N. & Thomas A.V. (2002) *Identification of Critical Risks in Indian Road Projects Through Build, Operate and Transfer (BOT) Procurement Approach*. Proceedings of CIB W92 Symposium on Procurement Systems, Trinidad and Tobago, pp. 339–354. The Engineering Institute, University of West Indies, St Augustine, Trinidad and Tobago.

Lam W.H.K. & Tam M.L. (1998) Risk analysis of traffic and revenue forecasts for road investment projects. *Journal of Infrastructure Systems*, **4**(1), 19–27.

Malini E. (1997) Evaluation of financial viability of BOT transport infrastructure projects. *Journal of Indian Road Congress*, **58**(1), 87–123.

Ministry of Finance (1996) *The India Infrastructure Report*. Ministry of Finance, Government of India, Vol. 2.

Ministry of Finance (2000) Economic Survey. Ministry of Finance, Government of India, Chapter 9.

Ministry of Industry (2001) Investment Opportunities: India's Infrastructure, Ministry of Industry, Government of India, http://indmin.nic.in/vsindmin/sectors/default.htm visited on 10.3.2001.

Model Concession Agreement for BOT Roads (2000) Ministry of Surface Transport, Government of India.

PMBOK (1996) *A Guide to the Project Management Body of Knowledge* (ed. W.R. Duncan), pp. 111–122. Project Management Institute, PMBOK, USA.

Policy Impediments to Financing (1999) Key Concern of Lenders. *Indian Infrastructure*, **1**(7), 6–11.

Projects Group Publications (1996) Baker & McKenzie, London. <http://www.bakernet.com/BakerNet/default.htm> visited on 20.1.2001.

Raftery J. (1994) *Risk Analysis in Project Management*, E&FN Spon, London.

Raghuram G., Jain R., Sinha S., Pangotra P. & Morris S. (1999) *Infrastructure Development and Financing – Towards a Public Private Partnership*. Macmillan India Ltd, New Delhi.

Ramesh C.R. (1999) Experiences in execution of Hubli–Dharward bypass by BOT concept. *Civil Engineering and Construction Review*, **11**, 23–29.

Salzmann A. & Mohamed S. (1999) Risk identification frameworks for international BOOT project. In: *Profitable Partnering in Construction Procurement*, pp. 475–486. E&FN Spon, London.

Srinivasan L. (2000) Land acquisition process made easy. *Indian Highways*, **28**(3), 24–30.

Tah J.H.M. & Carr V. (1999) *Towards a Framework for Project Risk Knowledge Management in the Construction Supply Chain*. The Seventh International Conference on Civil and Structural Engineering Computing, Oxford. Civil-Comp Press, Edinburgh.

Tam C.M. & Leung A.W.T. (1999) *Risk Management of BOT Projects in Southeast Asian Countries*. Joint CIB Symposium on Profitable Partnering in Construction Procurement, Thailand, pp. 499–507. E&FN Spon, London.

Thomas A.V., Kalidindi S.N. & Ananthanarayanan K. (2001) Management of privatised infrastructure development in India: key issues. *Journal of the Institution of Engineers (India), Civil Engineering Division*, **82**, 52–56.

Thomas A.V., Kalidindi S.N. & Ananthanarayanan K. (2002a) Identification of risk factors and risk management strategies for BOT road projects in India. *Indian Highways* (in press).

Thomas A.V., Kalidindi S.N. & Ananthanarayanan K. (2002b) Risk perception analysis of BOT road project participants in India. *Construction Management and Economics* (in press).

UNIDO (1996) *Guidelines for Infrastructure Development through Build-Operate-Transfer (BOT) Projects*. United Nations Industrial Development Organisation, Vienna.

Venkataraman K. (2000) *Public Expenditure on Infrastructure Development*. National Seminar on Strategy of Infrastructure Development, Madras. The Institute of Public Enterprises and Public Administration, Chennai, India.

Walker C. & Smith A.J. (1996) *Privatised Infrastructure, the BOT Approach*. Thomas Telford Publications, London.

Wang S.Q., Tiong R.L.K., Ting S.K. & Ashley D. (2000) Evaluation and management of political risks in China's BOT projects. *ASCE Journal of Construction Engineering and Management*, **126**(3), 242–250.

World Bank (2000) PID Report on Project ID INPE71244. *India – Grand Trunk Road Project*. http://www.worldbank.org/pics/pid/in71244.txt visited on 10.9.2001.

Ye S. & Tiong R.L.K. (2000) NPV-AT-Risk method in infrastructure project investment evaluation. *ASCE Journal of Construction Engineering and Management*, **126**(3), 227–233.

PART FOUR

Framework for risks and opportunities management of public-private partnership infrastructure development

16 Multi-party risk management process for a public-private partnership construction project in Asia

Jirapong Pipattanapiwong, Stephen Ogunlana and Tsunemi Watanabe

16.1 Introduction

Many developing countries now see private financing as the solution to their public infrastructure problems. Consequently, there is increasing use of different forms of private financing in many countries. A major reason for the increased use of public-private partnerships (PPPs) is that developing countries are facing heavy debt burdens while development of infrastructures is still a social and economic necessity (Tiong *et al.*, 1992; Setiadi, 1993; Asian Business, 1996). Private sector firms have found it necessary to participate in infrastructure schemes as a commercial necessity to remain competitive in a changing market place (Tiong *et al.*, 1992; Setiadi, 1993). Asian booming infrastructure has been the focus of world attention since the 1980s but the progress in implementing build-operate-transfer (BOT) schemes in Asia is rather slow (Asian Business, 1996).

The major attraction in using private financing is that developing economies can meet their infrastructure needs without having to pay for the projects. As such, it is perceived in many quarters as a great solution to the infrastructure problems of the developing economies (Ogunlana, 1997).

Promoters of privately financed public infrastructure projects tend to focus, almost exclusively, on the commercial and technical aspects. However, experience in transportation projects in Thailand and a power generation project in India show that economic forecasts should be treated as what they are, 'forecasts' (Ogunlana, 1997). The reality may be different if the contractual aspects of the project are neglected. A privately financed public infrastructure project is a contract between two parties: the host government and the concession company. It also involves other parties in a web of contractual relations. If each of the parties focuses exclusively on its own risks without giving consideration to the needs of the others, the project may not run smoothly or it may fail entirely.

In the Second State Expressway System in Thailand, the concession company achieved a major technical breakthrough and the early completion of the project using new technology, but disagreements with the local authority ruined relationships. On the Don Muang Elevated Tollway Project in Thailand, the government was unable to meet its obligations under the contract promptly. As a result, the concession company experienced financial difficulties. The powerplant project being championed by Enron in the State of Maharashtra in India has been on and off for over six years due to problems with the host government. To be successful, privately financed projects in developing countries must focus more on the responsibilities of the host government and on how to deal with government inabilities or unwillingness to meet contractual obligations.

Although many governments may wish to build much of their infrastructure through private financing, not all projects are suitable for privatisation. Projects that are well suited for a BOT agreement should have the following characteristics:

- The country in which the project is situated should have a stable political system
- The legal system in the country should be predictable and proven to be reasonable
- The economy should be promising in the long term with adequate local financial markets. The currency exchange risk associated with the project should be predictable. This is particularly necessary if expected income is to be paid in local currency
- The project itself should be in the public interest, with governmental support being available to it
- There should be long-term demand for the service to be offered by the project
- There should be limited competition from other projects
- Profits from the project must be sufficient to attract investors
- The cash flow from the project must be attractive to lenders
- The risk scenarios should be predictable while still providing acceptable profit and cash flow (Knoblach, 1992; Asian Business, 1996).

Private finance for public infrastructure is a risky venture. The major risks in PPP projects have been identified by several authors (Tiong, 1990; Tiong & McCarthy, 1991). In short, PPP projects are susceptible to all normal risks facing any construction project but are also faced with additional risks of which political, legal and currency fluctuations are very prominent.

Experience shows that many infrastructure projects in Asia are held up at the negotiating table. The reasons have been outlined by the World Bank as the following (Asian Business, 1996):

- A wide gap between the expectations of the government and the private sector
- Lack of clarity about government objectives and commitment, and complex decision making
- Poorly defined sectoral policies (on pricing, competition, public monopolies) and inadequate legal and regulatory policies, including investment codes and dispute resolution mechanisms

- Need to unbundle and manage risks and to raise credibility of government policies
- Inadequate domestic capital markets
- Lack of mechanisms to provide large amounts of long-term finance from private sources at affordable rates
- Poor transparency and lack of competition.

It would seem, from the favourable conditions outlined and the reasons for the slowdown in PPP approvals in Asia, that political climate, regulatory and legal policies and the transparency of government operations are very important in determining the success or failure of PPP schemes. In addition, when managing risks in a multi-party environment, consideration should be given to the needs and constraints of each of the parties. Risk management, with due consideration to the needs of others fulfils two Asian values: (1) the maintenance of harmony in group situations; and (2) the pursuit of profit for all. The procedure for risk management outlined in the rest of this chapter fulfils those values. The purpose of this chapter is to introduce a multi-party risk management system that will reduce the problem of focusing only on self interest which is inherent in the conventional risk management process.

16.2 Risk management on privately procured projects

Risks are inherent in all construction projects. Typically, an infrastructure construction project is large, fraught with uncertainty, and complex in many respects. A project of this nature requires huge investment of human effort and capital with several groups participating. They therefore tend to face more internal and external risks than other types of projects (Tiong, 1990).

Managing risks at the procurement and construction stages of privately financed public infrastructure construction projects is not a straightforward process since many parties such as an executing agency, the promoter, contractors, consultants and the lenders are involved. The involvement of several parties increases the frequency and impact of risk since the various parties have different objectives. The conventional risk management process (RMP) has evolved to assist decision-makers rather than relying solely on intuition. However, the conventional RMP has obvious limitations in that only one party is generally considered and therefore the objectives of the other multiple project participants may be overlooked in the analysis. In the conventional RMP, only one party undertakes the processes of risk identification and risk response. Even when a risk affects several parties, the process of risk analysis and response evaluation from the point of view of the other involved parties is usually absent. Since responses to some risks made by one party may create risks to some other parties, a risk–response–risk chain may be created. However, such a chain among multiple parties is not commonly incorporated in the conventional risk analysis. Therefore, the conventional RMP may not be suitable when several parties are involved in a project. A systematic process of managing risks in a multi-party environment is thus required.

In this chapter a Multi-party Risk Management Process (MRMP) is proposed. The proposed MRMP has been applied to a construction project financed by an international lender to demonstrate its applicability. A bridge and elevated road construction project in Thailand financed by the ADB has been used as a case study. The rest of this chapter is divided into three main parts: the development of the MRMP, the case application of the MRMP, and discussion of its applicability. Details of the methodology are described in the risk identification, structuring, and analysis and response processes in the MRMP development and in the application of the MRMP.

16.3 The development of the MRMP

The proposed multi-party risk management process (MRMP) (Pipattanapiwong, 2000), provides systematic and logical processes for risk management including risk identification, risk structuring, and risk analysis and response. The needs of the different parties involved in a project and their objectives are incorporated in each process. Priorities based on significance of risks and objectives are considered. The MRMP relies on quantitative measurement and analysis as well as utilising decision-makers' experiences and intuition in a systematic and efficient way.

16.3.1 Essence and characteristics of MRMP

In a multi-party environment, several project participants are integrated through contractual agreements or via interactive communication. The proposed MRMP aims to assure decision-makers that risks are managed systematically and efficiently in a multi-party environment. The underlying essence of the MRMP is based on the risk efficiency concept described by Chapman and Ward (1997). Here risk is defined as the deviation of the level of impact from the expected impact of risk associated with the alternative responses. Risk is characterised in terms of impact level and probability of occurrence. Finding an efficient response is the key in the conventional RMP. Efficient responses are portrayed on a risk efficient boundary, which is plotted on a graph demarcated by the degree of risk and the impact level of risk associated with each response. The risk efficient boundary represented by the bold line connecting responses A, B and C in Fig. 16.1, presents efficiency condition for each response. The responses portrayed on the risk efficient boundary provide a minimum level of risk for a given level of impact and vice versa. Thus, as illustrated in Fig. 16.1, responses A, B and C are efficient, but response D is not.

Based on these underlying concepts, a number of promising attributes of the MRMP are developed; including multi-party risk–response–risk, 'objective' evaluation of each party, multi-party response efficiency, and response characteristic evaluations. These deliveries are discussed in the following sections.

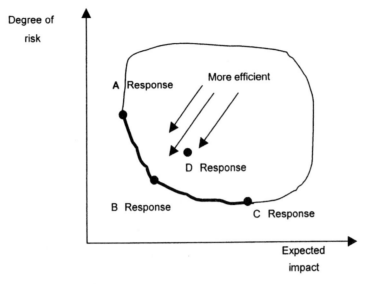

Figure 16.1 *Risk efficient boundary.*

16.3.2 Features of MRMP

When a primary risk is responded to, secondary risks may occur. This risk–response–risk chain, which is discussed by Isaac (1995) and Chapman and Ward (1997), generally involves only one party. In many situations there may be a major risk affecting several parties in a project. When one party makes a response to a major risk this may create risks to other parties, as shown in Fig. 16.2. Explicit incorporation of the multi-party risk–response–risk is the first feature of the MRMP.

When one party undertakes a risk management study with use of the RMP, even if another party observes some deficiency in the party, such an observation may not be included in the study. Since the MRMP can explicitly

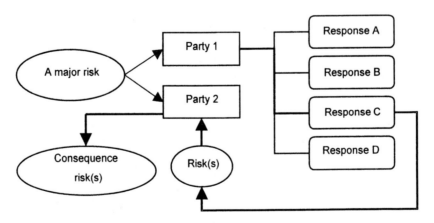

Figure 16.2 *An example of the multi-party risk–response–risk scheme.*

deal with viewpoints of multiple parties, it can incorporate 'objective' observation and evaluation of each party.

When a major risk influences multiple parties it is desirable that the response to the risk be efficient for all the parties involved. In the degree of risk-expected impact map in Fig. 16.3, for example, responses B and C are risk efficient for party 1, and responses A and B are risk efficient for party 2. Responses A and C are not desirable alternatives for party 1 and party 2, respectively. Thus, the most desirable response in this case seems response B. This illustrates 'the multi-party response efficiency evaluation,' which is a direct extension of the risk efficiency concept in the conventional RMP.

Additionally, each party may have his/her preferred disposition towards a particular risk. The possible dispositions to risk include risk aversion, risk neutral and risk seeking (Flanagan and Norman, 1993; Raftery, 1994). Since the conventional RMP is a method developed to systematically obtain risk-efficient responses for a single party, the disposition of other parties towards the response is beyond the scope of the RMP. The disposition issue does not have to be incorporated into the RMP, probably because the disposition of one party is easy to handle in practice. When a risk management study is undertaken from the viewpoint of one party, the most desirable response may be derived without significant difficulty.

When viewpoints of multiple parties have to be incorporated, however, finding and leaving a set of efficient responses to them is probably insufficient. It is not straightforward to incorporate their dispositions to each response and determine a desirable response to the related parties. It is thus necessary to integrate the disposition issue into the MRMP.

The disposition issue can be discussed in the degree of risk-expected impact map. In Fig. 16.4, for example, response B is clearly more desirable than response A for party 2. Response B does not necessarily become a desirable response to party 1. It becomes so only if she/he prefers a risk aversion alternative, response B, to a risk seeking alternative, response A.

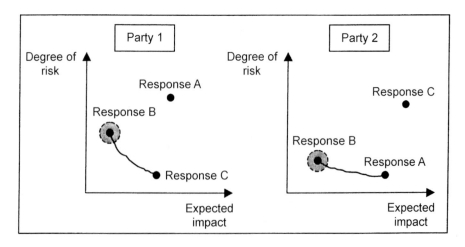

Figure 16.3 *An example of the multi-party response efficiency evaluation.*

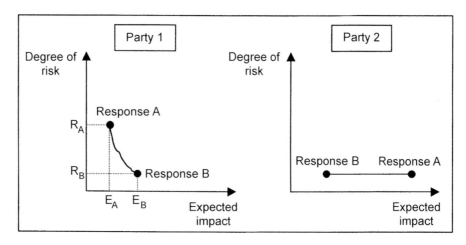

Figure 16.4 *An example of the response characteristics evaluation.*

Such a disposition of party 1 is proved if she/he is willing to pay more than E_B–E_A to reduce the degree of risk from R_A to R_B. Understanding the characteristics of response to a risk perceived by parties is significant in a multi-party environment. This can be easily achieved with the response characteristics evaluation, which is another feature of the MRMP.

The main processes in the MRMP consist of risk identification, risk structuring, and risk analysis and response processes. The outputs of each process are used as inputs in the succeeding processes.

16.3.3 The risk identification process

Risk identification is an important process in risk management because risks identified are the inputs of the structuring, and the analysis and response processes. The main task in the risk identification process is to identify risks affecting the parties involved objectively. Risks are identified based on the objectives of each of the involved parties. The major and minor risks are initially distinguished in the identification process. The risk identification process diagram is shown in Fig. 16.5.

The process assumes that the description and background of the project would have been prepared. Thereafter, the stages of the project are determined and scoped. This is followed by the identification of the parties involved. This work focuses on the executing agency, the contractor and the consultant who were involved in the procurement and construction stages of a publicly managed bridge and elevated road construction project financed by the Asian Development Bank (ADB). Their objectives in each stage of the project are specified based on the transformation system. The transformation system concerns the process of transforming resource inputs into outputs.

In the first stage of the identification process, a preliminary risk checklist was developed from previous risk checklists (Perry and Hayes, 1985; Al-

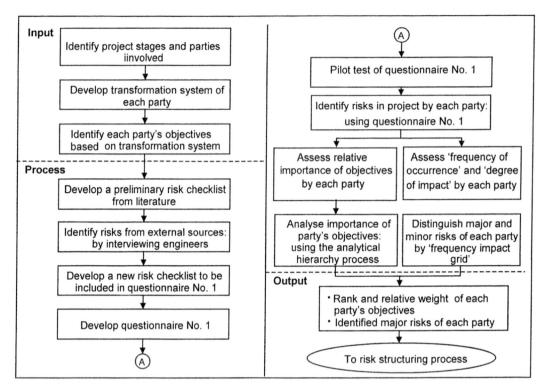

Figure 16.5 *Risk identification process diagram.*

Bahar and Crandall, 1990; Edwards, 1995; Zhi, 1995; Fisk, 1997; Laoh-kongthavorn, 1998) to determine the project risks in general. Next, using the preliminary risk checklist as a guide, unstructured interviews were conducted with 15 experienced engineers, having an average of 15 years experience in order to identify the possible risks related to bridge and elevated road construction projects (whether privately financed or not) in Thailand. The engineers were drawn from the Public Works Department (PWD), the Bangkok Metropolitan Administration (BMA), the ADB, contracting firms, and consulting organisations in Thailand. Afterwards, a new risk checklist was specifically developed for the case study project based on the risks identified in the previous stages. The new risk checklist contained 51 risk items and 124 risk items in the procurement and the construction stages, respectively. This risk checklist is included in the first questionnaire, which is used to identify risks in the case study project.

A structured questionnaire was used to elicit the judgement of the top managers/engineers of each of the parties involved in the project. For this purpose, the chief project engineer from the executing agency and project managers of the contracting and consulting organisations were used. The objective of the first questionnaire is to identify important objectives of the parties involved and risks associated with those objectives. For this purpose, the first questionnaire was distributed to top management engineers for all the parties involved. It can be used to identify risks in general. The Analytical Hierarchy Process (AHP) is employed in analysing each party's important objectives. The relative weights of each objective are obtained from the AHP analysis. The frequency impact grid is used initially to distinguish major and minor risks. The grid consists of two dimensions: frequency of occurrence, and degree of impact. Indeed, the outputs from the risk identification process are the relative weights of objectives of each involved party and the major risks identified. The first three important objectives and risks in order of priority have been selected for further analysis as example in this chapter.

16.3.4 The risk structuring process

The purpose of the risk structuring process is to specify dependencies among risks. The cause and effect relationships among risks associated with specific objectives are also examined. Another task in the structuring process is to identify the most significant risk. This process attempts to improve the understanding of relative importance of different sources of risks. Figure 16.6 is a diagrammatic representation of the risk structuring process.

The MRMP employs an influence diagram, which is a useful technique for developing, structuring and discussing complex risk relationships (McNamee *et al.*, 1990). In order to proceed with the risk structuring process in an efficient and orderly manner, the connectivity matrix technique in graph theory is incorporated with the influence diagram to develop a risk structure diagram. This is defined as the matrix whose (i, j) element is one if the ith risk causes the jth risk and zero if otherwise. The connectivity matrix is used to represent the existence of relationships among risks and objectives.

The second questionnaire is used to investigate the judgement of each

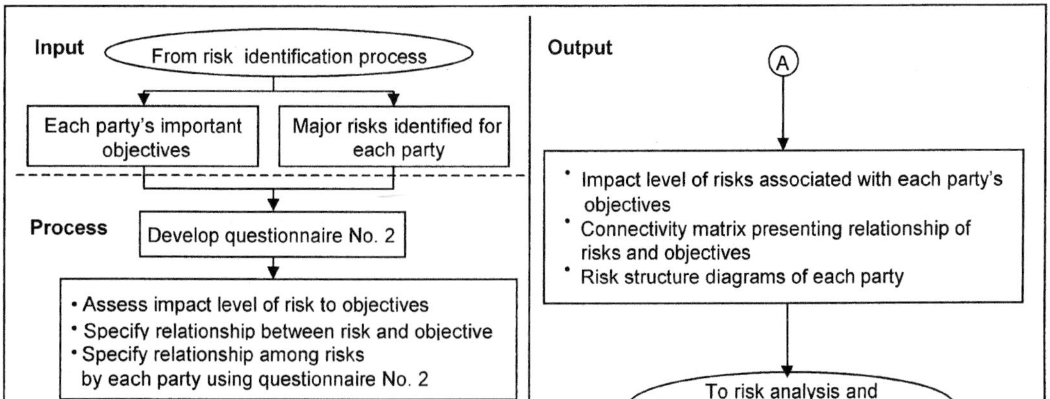

Figure 16.6 *The risk structuring process diagram.*

party's top management engineers. The same group as in the risk identification process was used for the case study. The second questionnaire is developed based on the list of major risks and important objectives drawn from the risk identification process to establish relationships among the risks and the objectives. The questionnaire has been specifically developed for this case study. The significant risks and risk structure diagram based on connectivity matrix are obtained from the analysis of the second questionnaire.

16.3.5 The risk analysis and response process

In the risk analysis and response process, a logical and systematic procedure for evaluating risk response efficiency is provided. The major risks are identified by each of the parties from the risk impact evaluation result in the structuring process. The responses to the major risks are analysed to find the most efficient response for a particular major risk. This is the main task in the risk analysis and response process. The risk analysis and response process diagram is presented in Fig. 16.7. A response is any action or activity that is implemented to deal with a specific risk or a combination of risks. Responses are categorised into three types: accept, pro-active and reactive, based on timing of implementation. A pro-active response is made before the major risk occurs. Its main aim is to prepare for efficient risk management in the project being executed. A reactive response is made after the major risk occurs. Its main aim is to better manage the risk for the rest of the project being undertaken. The 'accept' response can be made either before or after

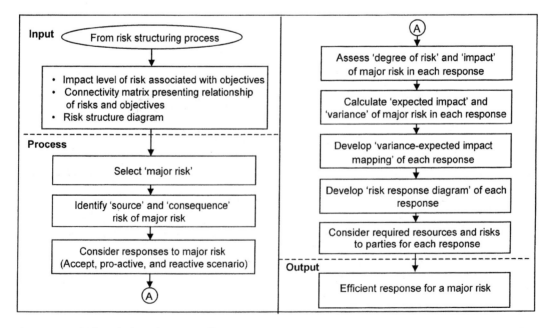

Figure 16.7 *Risk analysis and response diagram.*

the occurrence of a major risk event. It is the baseline for comparison with the pro-active and the reactive responses.

After a major risk has been isolated, the source and consequence risks associated with the major risk are identified. A source risk is defined as the risk that can directly influence and cause the occurrence of the major risk. A consequence risk is defined as the final risk that is directly or indirectly caused by the major risk. The flow of these risks is specified as source risks – major risk – consequence risks as shown in the prototypes of risk-response diagrams in Fig. 16.8. The 'risk analysis and response interview sheet' is used in identifying source and consequence risks, defining probability of risk occurrence and evaluating impact level of a particular major risk considering every alternative response available to each of the parties involved.

The probability of occurrence is one of the components used to characterise risk. Subjective probability has been used in evaluating the probability of occurrence of risks for two reasons. Firstly, most real-world situations are unique, and the possibility of recurrence of the same event under substantially identical conditions is small. Secondly, there is often little possibility of obtaining a large set of relative frequency data. To elicit subjective probability, the direct method was employed. The direct method assumes the existence of a rational decision-maker well aware of the rudiments of probability. Then the method merely consists of asking the subject to assign a number to their opinion regarding the outcome in question.

Another component used to characterise risk is the impact level of risk. The next step is to evaluate the impact level of major, source and consequence risks for each alternative response. Then, the total impact level of a

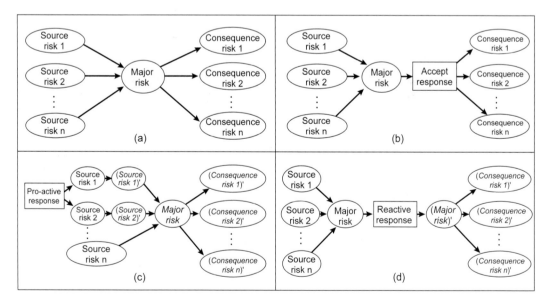

Figure 16.8 *Prototypes of risk-response diagram.*

major risk is calculated and used in deriving the expected impact and variance of the major risk for each alternative response.

The variance of impact is used to measure the degree of risks. Raftery (1994) used variance to represent risk. The expected impact is used to determine the impact level of risk. The calculations of the expected impact and variance is based on the assumption that there are two possibilities for the major risk under each response scenario, i.e. it may 'occur' or it may 'not occur'. If the major risk occurs, the probability of occurrence is assigned. On the other hand, if the risk does not occur, the probability of occurrence is zero. After the expected impact and variance of a major risk under each response scenario have been calculated, they are plotted in a variance–expected impact map. The map is used to determine the risk-response efficiency and characteristics of response in quantitative and graphical format.

16.4 Application of MRMP

The MRMP discussed above was applied to a publicly owned bridge and elevated road construction project financed by the ADB in order to demonstrate its applicability. It should be noted that results of the MRMP have different implications depending on when it is applied. In this case study, although the procurement stage has already been completed, it is assumed that the analysis was conducted towards the end of the procurement stage. The objectives of this analysis are to study whether a major risk could have been managed more efficiently or not and to draw lessons for similar projects in the future. For the construction stage, it was assumed that the analysis was conducted when major risks were occurring.

After going through the steps in the risk identification, the risk structuring, and the risk analysis and response processes, the contractor identified 'executing agency lacks experience in procurement process' as the major risk in the procurement stage. He recommended 'assisting the executing agency by a capable and experienced consultant' as an efficient response. If the executing agency, which considers itself 'almighty', undertakes a risk management study with use of the conventional RMP, such a response of strengthening his/her deficiency would probably not have been proposed. A framework for incorporating 'objective' views is important to derive an 'impartial' response. Incorporation of 'objective' views of each party is a major feature in the MRMP.

In the construction stage, the executing agency, the contractor and the consultant perceived that 'contractor's liquidity and financial problem' was the major risk. Four responses have been proposed to deal with the 'contractor's liquidity and financial problem' risk. The first response is to 'accept' this situation after the major risk occurs. The three remaining responses are reactive responses, being 'new capable contractor joins or takes over the current contractor', 'bank provides financial assistance to the current contractor', and 'the executing agency terminates the contract'.

In this case study, if the 'executing agency terminates contract' response is made by the executing agency, this creates another risk to contractor as shown in Fig. 16.9. The multi-party risk–response–risk evaluation is another feature of the MRMP.

Desirable responses in this case study seem to be the 'acceptance' and the 'new capable contractor joins or takes over current contractor' response as shown in the variance-expected impact map in Fig. 16.10. This illustrates 'the multi-party response efficiency evaluation'. The risk-response diagram for the efficient response, 'new capable contractor joins or takes over current contractor' response, is presented in Fig. 16.11.

In Fig. 16.10, the second response of 'new capable contractor joins or takes over current contractor' was evaluated to be a little more risk seeking than the first response of 'acceptance' by all parties. If all the parties are willing to

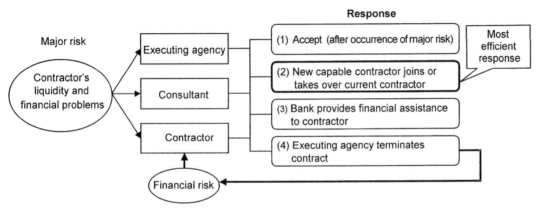

Figure 16.9 *The multi-party risk–response–risk scheme.*

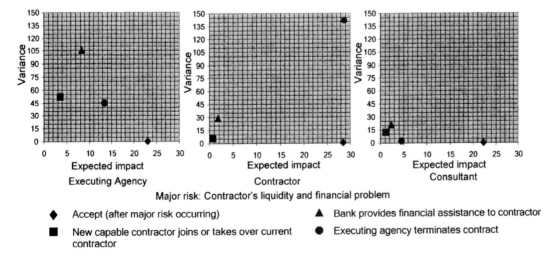

Figure 16.10 *Variance-expected impact map for the major risk in construction stage.*

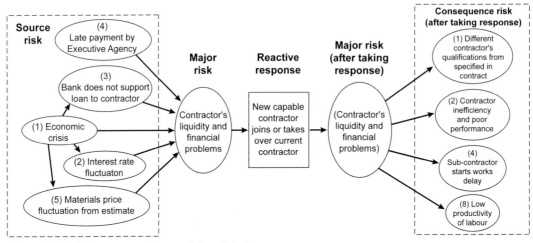

Figure 16.11 *Risk-response diagram of the efficient response.*

take this risk, the second response will be the best option to implement. Understanding the characteristics of response to a risk perceived by parties is significant in a multi-party environment. This can be achieved with the response characteristics evaluation, which is another feature of the MRMP. The findings of the MRMP application in the case study are summarised in Table 16.1.

Table 16.1 *Findings of the MRMP application in the case study.*

Party	Important objective	Identified major risk	Efficient response	MRMP contributions
Procurement stage				
Executing agency	Selecting capable contractor	Delay in awarding contract	Preparing clear bid document	Response efficiency evaluation (same as in the conventional risk management process)
Contractor	Contract price	Executing agency lacks experience in the procurement process	Capable and experienced consultant assists the executing agency in the procurement process	'Objective' evaluation of other party
Construction stage				
Executing agency	Schedule, budget, quality	Contractor's liquidity and financial problem	New capable contractor joins or takes over the current contractor	Multi-party response efficiency evaluation
Contractor	Schedule			Multi-party risk–response–risk evaluation
Consultant	Schedule			Response characteristics evaluation

16.5 Summary

The core concept of the MRMP is to understand how to handle a risky situation, to reduce the impact of risk, to decide on the required resources and to minimise the effects on all related parties when responding to a risk. A number of components of the MRMP have been developed from the conventional RMP to assist decision-makers to better make decisions on risk management efficiently in a multi-party environment. These are multi-party risk–response–risk, 'objective' evaluation of each party, multi-party response efficiency, and response characteristics evaluations.

First, risks to one party occurring from a response made by another party can be notified. This is the multi-party risk–response–risk chain. Second, the opportunity is given to evaluate another party 'objectively'. Any party can notify the deficiency regarding the experience, technical or managerial skills, of another party involved in the project during the identification of risks. Third, the multi-party response efficiency evaluation is provided. From this premise, and in order to manage risk more efficiently, it is desirable to find a response that is risk-efficient to all related parties. Finally, the response characteristics (i.e. risk avoiding, risk neutral and risk seeking) associated with a major risk can be specified from the presentation of a variance-expected impact map. This feature can assist decision-makers to find and select the most preferable response for all the parties. These illustrate the advantages of involving multiple parties in the risk management process.

References

Al-Bahar J.F. & Crandall K.C. (1990) Systematic risk management approach for construction projects. *Journal of Construction Engineering and Management*, **106**(3), 533–546.

Asian Business (1996) Special report on Asia's infrastructure boom. *Asian Business*, March, 60–69.

Chapman C. & Ward S. (1997) *Project Risk Management: Process, Techniques and Insights*. John Wiley & Sons, Canada.

Edwards L. (1995) *Practical Risk Management in the Construction Industry*. Thomas Telford Publications, London.

Fisk E.R. (1997) *Construction Project Administration*, 5th edn. Prentice-Hall, Upper Saddle River, New Jersey.

Flanagan R. & Norman G. (1993) *Risk Management and Construction*. Blackwell Scientific Publications, Oxford.

Isaac I. (1995) Training in risk management. *International Journal of Project Management*, **13**(4) 225–229.

Knoblach P. (1992) *Project Management on Major Infrastructural Projects*. Seminar presented at the Asian Institute of Technology, Bangkok, 18 November (Unpublished).

Laokhongthavorn L. (1998) *Recurrent Risk Assessment for Bangkok High-rise Building Projects*. Master Thesis No. ST-98-31, Asian Institute of Technology, Bangkok.

McNamee P., Celona J. & the Strategic Decisions Group (1990) *Decision Analysis with Supertree*, 2nd edn. The Scientific Press, San Francisco.

Ogunlana S.O. (1997) *Build-Operate-Transfer Procurement Traps: Lessons from Transportation Projects in Thailand*. Proceedings of the CIB W92 Symposium on Procurement, Montreal, Canada, May, pp. 585–594.

Perry J.G. & Hayes R.W. (1985) Construction projects – know the risks. *CME* (Chartered Mechanical Engineer), **32**(2), February, 42–45.

Pipattanapiwong J. (2000) *Multi-party Risk Management Process (MRMP) for a Construction Project Financed by an International Lender*. Master Thesis, Asian Institute of Technology, Thailand.

Raftery J. (1994) *Risk Analysis in Project Management*. E & FN Spon, London.

Setiadi M. (1993) *A Study of Build-Operate-Transfer Approach to Selected Infrastructure Projects*. AIT Thesis No. ST-93-21.

Tiong L.K.R. (1990) BOT projects: risks and securities. *Construction Management and Economics*, **8**, 315–328.

Tiong L.K.R. & McCarthy S.C. (1991) Financial and contractual aspects of BOT contracts. *International Journal of Project Management*, **9**(4), 222–227.

Tiong L.K.R., Yeo K.M. & McCarthy S.C. (1992) Critical success factors in winning BOT contracts. *Journal of Construction Engineering and Management*, **118**, 2.

Zhi H. (1995) Risk management for overseas construction projects. *International Journal of Project Management*, **13**(4), 231–237.

17 Private finance initiative uptake in UK local authorities

Matthias Beck and Caroline Hunter

17.1 Introduction

When the private finance initiative (PFI) was launched in 1992, it largely focused on the involvement of private sector companies or consortia in infrastructure projects which were sponsored by central government departments. As a result, up until the mid 1990s, local authority (LA) involvement in PFI was at best peripheral. Accordingly, Harding *et al.* (2000) report that, during the budget years 1992/93 and 1993/94, only about £500 million was raised for PFI projects originating from organisations such as the London Docklands Development Corporation, and, perhaps to a lesser degree, from hospital trusts. As a rule, most of the organisations who participated in the first wave of PFIs had already been involved in some form of public/private collaboration and, therefore, could build on prior experience when entering the PFI domain.

This relatively slow uptake of PFI during the first half of the 1990s was seen as a disappointment by government officials, who had expected that, by 1996/97, a total of £10 billion, almost one quarter of the Government's capital investment, would be raised through PFI (Harding *et al.*, 2000). With few or no LAs being involved in PFI, there was an indication that, apart from legal obstacles, LAs lacked the expertise to enter into these arrangements. As a response, in 1996, the Public-private Partnership Programme Limited (or the 4Ps) was established by LA associations, with a view towards supporting partnership arrangements. Despite these initiatives, by the time of the election only one single LA had signed a PFI contract (Harding *et al.*, 2000).

When the new government took power in 1997, it was clear that either a situation had to be accepted where LAs did not play a major role in PFI, or that measures to facilitate LA involvement in PFI had to be expedited. Choosing the latter strategy, the New Labour administration quickly adopted a number of initiatives aimed at eliminating obstacles to LA involvement in PFI, as well as at creating explicit and implicit incentives for LA participation. As concerns the removal of obstacles, the Local Government (Contracts) Act, which came into force in December 1997, clarified the powers of LAs to enter into long-term service contracts with the private sector. In addition to expanding the authority of LAs, the Act made provisions for the compensation of private sector partners in the event that a contract was set aside by judicial or audit review. In practical terms, this measure had become necessary on account of a ruling by the City of London's Financial Review Panel which suggested that, where an action of a local authority was subsequently deemed illegal, creditors, such as merchant banks, might not be able to recover debts incurred in that arrangement. Retroactively, however, the government's swift and comprehensive response to this obstacle must also be seen as reflecting an overarching concern with eliminating sources of apprehension among potential private sector partners of LAs.

The Local Government (Contracts) Act clarified the powers of LAs to enter into contracts, introduced certification procedures which allowed LAs to enter into major long-term contracts and, lastly, guaranteed compensation to private sector parties in case a contract was set aside. Despite this extension of the authority of LAs with regard to PFIs, the government reserved a significant degree of control over LA-PFI activities.

Today, the Treasury and other government departments are still required to approve the business plan of all PFI schemes that require central government revenue support. Under current arrangements, this means that LAs must put forward their PFI scheme to the relevant central government department, such as the Department of Education and Skills in the case of a school project. If the department approves of the scheme, it forwards the business case to the Project Review Group, which is chaired by the Treasury Task Force. The Project Review Group evaluates the scheme against agreed criteria such as affordability, value for money and, if appropriate, includes the scheme in a list of approved projects which is published in the official journal. Underlying this procedure is the goal to grant Treasury support to those projects only which are likely to achieve success, both in terms of marketability at the contracting stage and in terms of future operability. More broadly, it could also be argued that the level of government control imposed on LAs has been aimed, especially at the early stages of PFI development, at fostering commercial confidence in these arrangements.

Along similar lines, where a PFI scheme does not require central government revenue, the general framework relating to LA-PFI projects in England and Wales has been amended to account for a fuller range of projects (Akintoye *et al.*, 1999). In this context, government authorities still maintain some degree of implicit control through value for money requirements and by prescribing, *inter alia*, the discount rate for calculating the cost of credit for arrangements entered into by LAs.

In addition to clarifying and facilitating the legal and contractual arrangements for LA-PFIs, these have been actively promoted by a number of government departments. These initiatives commenced with the announcement of 30 pathfinder projects in 1997, which included all types of urban services ranging from schools, social services and libraries to waste management (Harding *et al.*, 2000).

Although there are – due to definitional and jurisdictional problems – no completely reliable estimates of the number of PFI projects, most studies would indicate that, partially as a consequence of these initiatives, the number of LA-PFIs had increased to well over 100 by 1998/99. More specifically, according to estimates by the 4Ps, by mid October 1998, there were 184 projects which involved 79 local authorities (Akintoye *et al.*, 1999). An alternative estimate, based on PFI schemes approved by ministers, indicates that by 1999, 103 PFI schemes had been approved. This suggests an approximate ratio of central government-supported to other LA PFI projects of around less than 1 : 1. As concerns the mix of LA-PFI projects by sector, Akintoye *et al.* (1999) have suggested that education-related projects are by far the most prevalent (29%), followed by public/administrative buildings (15%), transport-related projects (15%), housing (11%) and IT (9%).

While some researchers have suggested that this rise in PFIs can be taken as evidence for the success of the Government's concerted effort to introduce PFI to LAs (Ball *et al.*, 2000), it can be argued that, given the scale of infrastructure investment requirements facing LAs, PFI uptake has in fact been slow. This argument of a slow uptake of PFI is underscored by recent estimates which indicate a possible decline in PFI capital spending between fiscal years 2000/01 and 2001/02 (Robinson, 2000).

Based on a quantitative analysis of the incidence of PFI in English LAs, the following sections will examine potential causes for the arguably slow uptake of PFI by LAs. In this context, several alternative scenarios are examined, including the possibility of a political bias against PFI among LAs, a preference for other forms of procurement, especially among richer authorities, and, lastly, technical obstacles. Our analysis commences with a brief discussion of the political background of PFI and its position within the broader context of different models of urban governance. Section two then examines some of the potential technical obstacles encountered by LAs wishing to engage in PFI projects. Section three presents the results of a quantitative analysis, which indicates that PFI uptake in LAs is largely determined by their technical capabilities, which correlate to size factors.

17.2 The politics of PFI

Contrary to public belief, the origins of public-private partnership can be traced to the Labour administration of the late 1970s. At the time, the US was experimenting with public-private partnership as a means of regenerating depressed communities. When Peter Shore, then Environment Secretary, visited the US in 1978, a decision was made to emulate US approaches and plans were made to set in motion the creation of similar groupings in the UK (O'Brien, 1997).

Following Labour's defeat in the 1979 election, the Conservatives took the concept forward. In this context, PFI was no longer seen solely as a tool for regenerating depressed areas through partnership, but rather as a means for adjusting the balance between the in-house provision of public services and private sector provision (Glaister, 1999). When in 1981, the National Economic Development Council noted that external financing limits prevented greater levels of investment, the first steps were undertaken to explore the possibility of real-world public-private partnerships.

Perhaps due to the complexity of PFI transactions, the Government's focus on partnership initially centred on other models of co-operation such as the use of private sector operators in public sector contracts (e.g. via compulsory competitive tendering), or the outright privatisation of previously state-owned service providers (such as utilities). PFI, as a model of public-private partnership, was eventually introduced by the Conservative Chancellor of the Exchequer in 1992, amidst announcements that the use of this facility would allow public investment outside public sector borrowing constraints.

As previously discussed, despite government exhortation it took until the mid 1990s for the PFI option to have any significance for LAs. Around that time, advocacy for PFI was accompanied by vociferous rhetoric which portrayed PFI as a means of introducing market discipline into the way local authorities delivered services and financed infrastructural developments (Harding *et al.*, 2000). In addition, on account of its focus on the service contract, as compared to the creation of infrastructure *per se*, PFI was heralded as a means of tapping into, and releasing the innovative potential of the private sector. With hindsight, it can be argued that much of this rhetoric was politically unhelpful, if only because it placed PFI firmly within a free market ideology.

Early advocacy of PFI in many ways associated the scheme with Conservative ideology, which was so convinced of the advantages of the private sector, that it sought either to pass on responsibilities to the private sector, or to coerce government authorities to act like private sector actors. The fact that PFI today embraces a strong focus on clients and the provision of Best Value to the public has not yet fully broken this link. Opposition to PFI often continues to be based on the association of PFI with a radical pro-market ideology. In this context, PFI opponents view PFI with apprehension, be it on account of its alleged links to new public management (Mayston, 1999) or its perceived adverse effect on employers (Ruane, 1999).

Estimates on how many potential PFI projects have not materialised because of the political aversion of policy actors, community groups or other interest groups, are, of course, impossible to gather. However, there is a very real possibility that one reason for the slow uptake of PFI in LAs is its political pedigree, whether it is correctly or falsely understood.

17.3 The technicalities of PFI

Today there is ample empirical evidence that LA officials often view involvement in PFI as daunting. PFI schemes involve contractual, financial and/or organisational arrangements which typically fall outside the experience of LA managers, and which, as a consequence, require a significant investment in time and resources. While some of these problems may be alleviated through the involvement of organisations such as the 4Ps, and/or the availability of information on previous PFI schemes of the same or other LAs, the sheer complexity of PFI schemes is likely to continue to place high demands on the client.

Research into the preconditions of success of public-private partnership confirms many of these concerns. Thus Grant (1996) found public-private partnerships to be most successful where:

- The partners are financially strong and organisationally stable
- The partners are willing to commit their best human resources to the project
- The project provides opportunities for all partners
- There is shared authority and responsibility.

Quite clearly some of these factors cannot be altered over a short time by individual public sector managers. Stable and efficient organisational structures are often the outcome of organisational learning and evolution. Likewise, at least in the public sector, the availability of human and technical resources usually closely corresponds to the size of the authority; putting smaller LAs with less specialised expertise at an inherent disadvantage. The first and second Bates Review and related studies have recognised some of the technical problems that LAs experience as clients of PFI projects and suggested a number of possible remedies. These include the bundling of smaller projects, closer collaboration of LAs, and the improved dissemination of PFI-relevant information. However, it is unlikely that collaboration, improved information or even standardisation of PFI projects alone will eliminate the technical difficulties faced by some LAs.

On account of these difficulties, it is conceivable that the technical challenges involved in the planning and implementation of a PFI scheme are continuing to present barriers to a greater PFI uptake in LAs. In other words, some LAs may be reluctant to pursue a PFI arrangement, not because they prefer other forms of procurement, but rather because they feel that they lack the technical capability and the resources to enter into such a scheme.

17.4 Quantitative analysis

To date, the 'PFI Map of Great Britain' (2001), published on behalf of the Bank of Scotland and Allen & Overy, represents the most up-to-date and comprehensive listing of PFIs in the UK by geographical region. Based on data from the Office of Government Commerce and other government departments, the PFI map lists PFI projects completed by 2000 or signed in 2000, by location, type and value. This allows both for the identification of projects led by LAs, as well as their designation to the respective LA.

The core of this study is based on an analysis of the incidence of PFIs amongst the 116 LAs of England which have been identified on the basis of the PFI map. These LAs fall within three categories, notably, county councils, unitary authorities and metropolitan borough councils. While each of these authorities has a slightly different legal status with different responsibilities for services, this does not, on the whole, affect this analysis. For the purpose of this analysis, PFIs in Scottish LAs were excluded on account of difficulties arising from local government reorganisation, while Welsh LAs were excluded because some data used in the analysis was not available at the LA level. In addition, London Boroughs could not be included on account of the differential local government structure.

17.4.1 Incidence of PFI projects

By the end of the year 2000, a total of 40 authorities out of 116, or 34.5%, had either signed or completed a PFI project. Among these 40 authorities, three had shared in a PFI project with other authorities, without having completed a PFI project on their own. The majority of authorities who had acted as PFI client, 21 in total, or 18.1% of all surveyed authorities, had participated in only one PFI project. A total of nine authorities had been involved in two projects, while six authorities had been party to more than two projects (Figs 17.1 and 17.2).

Although this data indicates that the penetration of LAs by PFIs is still relatively moderate, there is an indication that some authorities may have already acquired substantial expertise in PFI procurement and are now routinely procuring infrastructure and services via this route. This analysis would be supported by the fact that three councils have participated in more than four PFI projects, as well as by the fact that almost as many LAs have particiated in two or more PFIs as have in one single PFI project.

Whether this, by itself, is indicative of a breakthrough for LA-PFIs, however, is questionable on at least two accounts. Firstly, rather than being an indication of the ease in which LAs now engage in PFIs, the concentration of several PFIs in a few councils may in fact be a reflection of these LAs having gathered special expertise, while it has remained difficult for other councils

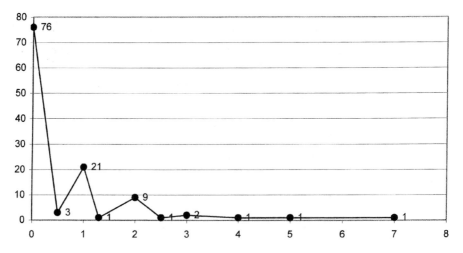

Figure 17.1 *Number of LAs having signed or completed one or more PFIs by end of 2000.*

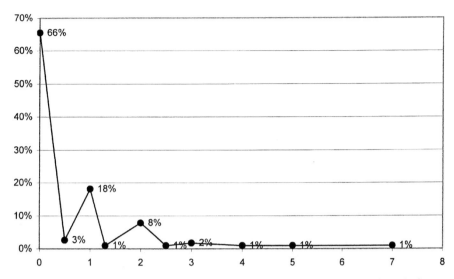

Figure 17.2 *Percentage of LAs having signed or completed one or more PFIs by end of 2000.*

to enter the field. Secondly, there is a possibility that experienced PFI participants have become sought after partners for the private sector which, in turn, offers them attractive packages based on their past record.

In any case, even at this early stage of LA involvement in PFI, there appears to be a notable concentration of PFI activity amongst a limited number of councils, while a large number of LAs have yet to become involved (Table 17.1).

These findings are mirrored, to a lesser degree, by the most significant functional types of PFIs. Following earlier findings by Akintoye *et al.* (1999), the most widespread PFI projects of the 116 LAs surveyed in this analysis included, firstly, educational services and infrastructure, followed by public/administrative buildings, transport-related projects and housing. Specifically, of 116 authorities, 16 LAs, or 13.8%, had participated in school-related PFI projects (Figs 17.3 and 17.4).

Again, there is some, albeit much less pronounced, indication of concentration, whereby four out of a total of 16 school PFIs occur in LAs with more than one school PFI. Both LAs with multiple school PFIs, incidentally, are also amongst the LAs with more than three PFIs.

Table 17.1 *Number and percentage of PFIs in LAs with one or more PFI (shared PFIs counted as fraction of 1).*

In LAs with	Up to 1 PFI	More than 1	More than 2	More than 3	More than 4	More than 5
Number LAs	24	16	6	3	2	1
Number of PFIs	22.5	43.8	21.5	16	12	7
Percent of total PFIs	33.9%	66.1%	32.4%	24.1%	18.1%	10.6%

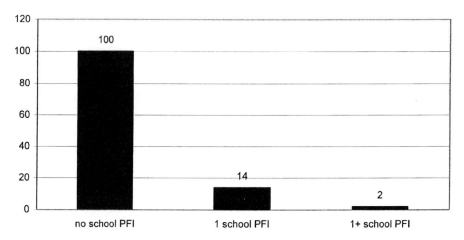

Figure 17.3 *Number of LAs having signed or completed one or more PFI school projects by end of 2000.*

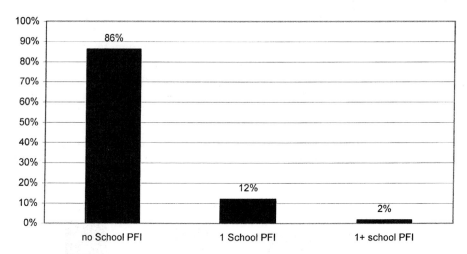

Figure 17.4 *Percentage of LAs having signed or completed one or more PFIs by end of 2000.*

17.4.2 The politics of PFI examined

Some of the literature on PFI would suggest that certain political groups inherently favour PFI projects, while others equally strongly oppose them. Accordingly, it should be expected that Conservative-dominated councils, on the whole, support PFIs primarily on account of a preference for market over government-based solutions. Conversely, it should be expected that traditional Left, or Old Labour-dominated councils, as well as those with heavy links to unions, should oppose PFI schemes on a number of grounds. Specifically, Left opposition to PFI would be expected on account of the association of PFI with an intrusion of the private sector into public service provision, private sector profit making, the potential weakening or marginalisation of organised labour and similar reasons.

Data on the political composition of LAs in England is readily available through various sources, including the *Local Authority Handbook*. This data allows for the identification of relevant political majorities in terms of the three leading parties – Labour, Liberal and Conservative. Needless to say, none of this information permits for a more detailed analysis of the respective political majority, such as its classification as Old versus New Labour dominated; although it must be said that such detailed classifications would, in any case, run the danger of imposing undue rigour on potentially unstable political majorities.

The influence of political leadership on PFI participation can be easily depicted in terms of an odds rate, which describes the percentage likelihood of a council, under different leadership regimes, having participated in a PFI project. (This approximated odds rate should not be confused with the statistical notion of a genuine odds ratio, which typically is calculated on the basis of four rates.) According to this rate, LAs under Liberal leadership were the most likely to be engaged in one or more PFI projects (37.5%), followed by Labour-led authorities (35.7%). Conservative LAs, meanwhile, where least likely to participate in PFI projects, with only 27.8% having engaged in any PFI activity in the past (Fig. 17.5).

Obviously, this analysis hardly supports the myth of PFI being favoured by Conservatives and opposed by the Left. The reasons for this can be manifold. Firstly, it is quite possible that political factors have indeed very little influence on PFI adoption, with PFI participation being driven by other considerations. Thus, with Conservative-led councils being either rural and/or, on average, more wealthy than Labour-led councils, there is every possibility that these authorities perceive less of a need to pursue PFI partici-

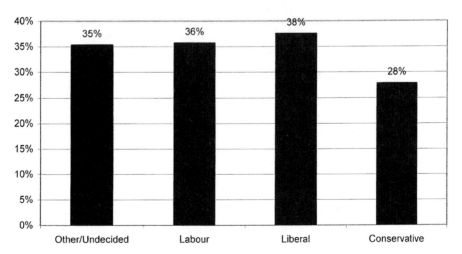

Figure 17.5 *Percentage of LAs under different political leadership which have been involved in PFI.*

pation in the first place. In addition, the analysis could be affected by a size factor, whereby Conservative councils, which are often smaller in terms of population, are less likely to engage in PFI; not on account of their political preferences, but on account of their need and/or technical abilities. Alternatively, the above analysis could be taken to reflect a preference for PFIs among New Labour councils which may have a specific interest in implementing policies favoured by central government in their authority. None of this, however, does explain the significant portion of Liberal-led councils which have engaged in PFI, as that party has repeatedly expressed its preference for direct public spending. In any case, statistically speaking there is no evidence for a significant link between the political orientation of a council to its proneness in adopting PFI.

17.4.3 The technicalities of PFI examined

The previous analysis has suggested that there is a concentration of several PFI projects in some LAs, while a significant number of other LAs have not yet participated in any PFI projects. In this context, it has been argued that there may be substantial obstacles to some LAs entering the PFI arena.

An obvious source of obstacles to PFI participation relates to the technical resources and competencies of councils. While it can be surmised that large councils, which are accustomed to handling very substantial budgets, as well as having significant experience in project management, will be well equipped to enter PFI partnerships, this is not necessarily the case in the context of smaller and/or rural authorities.

As part of this analysis, the size-competency factor has been operationalised alongside two variables; notably LA budgets and the size of population within an LA. Both variables correlate closely, which allows for a confirmation of analytical results which are derived from their use. For the purpose of this analysis budget size has been approximated by the Net Revenue Expenditure (NRE) which is published annually for all English LAs on the website of the Department of Environment, Transport and the Regions (DETR). The NRE can be roughly defined as the total amount spent by an LA on services, plus capital charges, but minus specific grants. Population data likewise has been obtained from the DETR website, which bases its own estimates on the Office of National Statistics.

For the purpose of this analysis, LAs were divided into three roughly equal size groups according to their NRE. These categories included LAs with an NRE of less that £200 million, LAs with an NRE of £200–300 million and, lastly, LAs with an NRE of over £300 million. Within these categories, the odds rate of PFI adoption amongst the three size groups differed dramatically. Thus, amongst the 50 authorities belonging to the smallest budget category (less than £200 million), only 12.0% (six LAs) had participated in a PFI project. Amongst the 29 councils from the second largest category (£200–300 million), 41.3% (or 12 LAs) had been involved in PFIs. Lastly, within the largest budget category (+£300 million), of 37 LAs, a majority of 59.4% (27 LAs) had been involved in PFI development (Fig. 17.6).

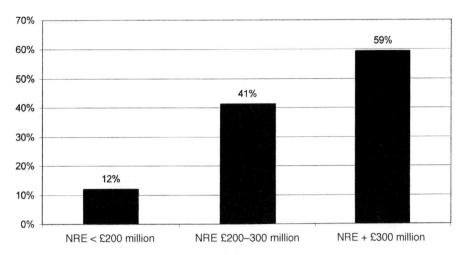

Figure 17.6 *Percentage of LAs with different budgets which have been involved in PFIs.*

In statistical terms this relationship could be identified as highly significant both in terms of the proposed 2 (adopter; non-adopter) × 3 (budget size category 1 to 3) tabulation (chi square 22.014), as well as in terms of more refined statistical tests. This pattern of LAs with larger budgets being more likely to become involved in PFIs could be confirmed when looking at the largest functional category of PFI schemes – that of schools. Although the total number of signed or completed PFI school projects in this sample was comparably small, there was again a clearly identifiable differentiation of PFI involvement by size. Thus amongst the 50 authorities belonging to the smallest budget category (less than £200 million), only 2.0% (one LA) had participated in a school-related PFI project. Amongst the 29 councils from the second largest category (£200–300 million), 20.6% (six LAs) had been involved in school PFIs. Lastly, within the largest budget category (+£300 million) of 37 LAs, 24.3% (nine LAs) had been involved in school-related PFIs (Fig. 17.7).

Following the previous approach, in the second stage of this analysis the sample was again divided into three roughly equal groups, this time based on population size. The first groups included LAs with a population of less than 200 thousand; the second LAs with 200–400 thousand inhabitants; and the third LAs with over 400 thousand inhabitants. Again, the percentage share of LAs which had adopted PFIs in each size category differed dramatically. Amongst the 41 authorities belonging to the smallest category (less than 200 thousand), only 9.8% (four LAs) had participated in a PFI project. Amongst the 39 councils from the second largest category (200–400 thousand), 33.3% (or 13 LAs) had been involved in PFIs. Lastly, within the largest budget category (+400 thousand) of 36 LAs, a majority of 63.8% (23 LAs), had been involved in PFI development (Fig. 17.8)

In statistical terms this relationship could again be identified as highly significant both in terms of the proposed 2 (adopter; non-adopter) × 3

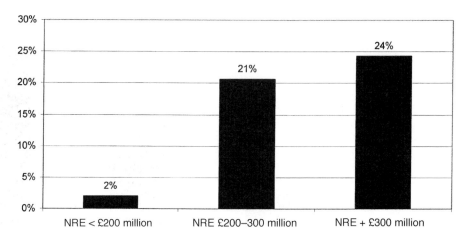

Figure 17.7 *Percentage of LAs with different budgets which have been involved in school PFIs.*

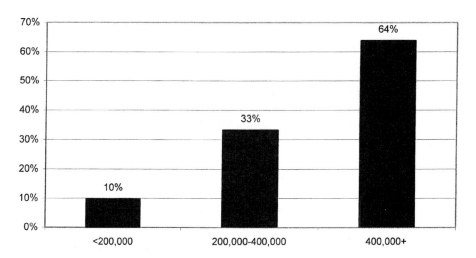

Figure 17.8 *Percentage of LAs with different populations which have been involved in PFIs.*

(population size category 1 to 3) tabulation (chi square 24.897), as well as in terms of more refined statistical tests.

This pattern was again closely mirrored by the largest functional category of PFI schemes – that of schools. Thus, amongst the 41 authorities belonging to the smallest population category (less than 300 thousand), only 2.4% (one LA) had participated in a school-type PFI project. Amongst the 39 councils from the second largest category (200–400 thousand), 10.3% (or four LAs) had been involved in school PFIs. Lastly, within the largest budget category (+400 thousand) of 36 LAs, 30.6% (11 LAs), had been involved in PFI-based development of school facilities (Fig. 17.9).

Overall, both the budget and population-based analyses indicate a strong link between size or resource factors, and LA proneness for PFI adoption.

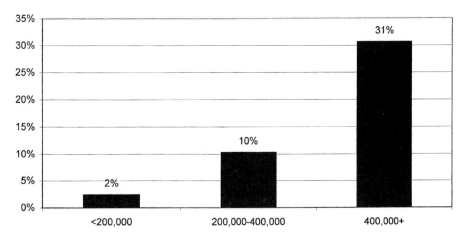

Figure 17.9 *Percentage of LAs with different populations which have been involved in school PFIs.*

Larger councils with larger budgets appear to be amongst the most ardent participants in PFI ventures, whereas small LAs appear to be very reluctant to become involved in PFIs. As long as there are ample opportunities for PFI development in larger LAs, this situation may be largely unproblematic from the perspective of commercial partners or even government officials. If, however, PFI saturation amongst larger LAs is reached, there could be a possibility of a bottleneck, where smaller authorities are unable to ensure a continued growth in PFI schemes.

17.5 Conclusions

Statistical analyses of real-world phenomena often raise more questions than they answer, and in many ways this chapter is only a partial exception. At the outset of this paper, the question was posed about which factors are responsible for the relatively limited uptake of PFIs in UK LAs. At this stage, the speculation was put forth that PFI adoption might be hampered by political reservations especially amongst Left-dominated authorities. Alternatively, it was proposed that the reason for the reluctance of many, especially smaller councils, to participate in PFIs may well arise from the fact that they lack the resources and expertise to engage in such complex projects.

The quantitative analysis allows us partially to refute the first, or the 'political', hypothesis. It also allows us partially to accept the second hypothesis, namely that PFIs are concentrated among large, well-resourced LAs which have the technical and knowledge infrastructure to cope with the challenge of PFIs. Statistically, this, albeit brief, analysis has found great difficulty in identifying a credible link between the political make-up of English LAs and the adoption of PFIs. This finding, as it is, does not imply that politics does not matter, but rather that the politics of UK LAs may have changed to a stage where the old simplistic dividing lines of Left/anti-market and Right/pro-market no longer apply.

By contrast, there appears to be convincing evidence that PFI adoption is strongest where budget and populations sizes are large. While these findings appear to lend support for the 'technical' hypothesis, it has perhaps a more complex subtext. Clearly, it has to be expected that larger LAs, which are accustomed to handling large budgets, will find it easier to cope with the challenge of a PFI. However, this does not mean that PFI is necessarily driven only by the 'ability' of an LA to participate. Large councils also are amongst those most in 'need' for investment and, in many cases, amongst those least able to afford it via means other than PFI. Whether 'need' or 'ability' drives PFI participation may be commercially irrelevant, but politically this is by no means a moot question. In an ideal world, public-private partnership has to grow out of a voluntary coming together of the public and private sectors, whether this involves a large or a small public sector client. At this stage, as far as UK local authorities are concerned, the verdict appears to be, on both counts, undecided.

Acknowledgements

The authors would like to thank Dr Peter Falconer and Professor Bobby Pyper for their helpful contributions to this paper. Any errors are, of course, solely the responsibility of the authors.

References

Akintoye A., Fitzgerald E., Hardcastle C. & Kraria H. (1999) *Local Authority Risk Management and the Private Finance Initiative*. Report on behalf of the Royal Institution of Chartered Surveyors, London.

Ball R., Healey M. & Kulg D. (2000) Private Finance Initiative – a good deal for the public purse or a drain on future generations? *Policy and Politics*, **29**(1), 95–108.

Glaister S. (1999) Past abuses and future uses of private finance and public private partnership in transport. *Public Money and Management*, July/Sept, 29–36.

Grant T. (1996) Keys to successful public-private partnerships. *Canadian Business Review*, **23**, 3, 27–28.

Harding A., Wilks-Heeg S. & Hutchins M. (2000) Business, Government and the business of governance. *Urban Studies*, **37**(5/6), 975–994.

Mayston D. (1999) The Private Finance Initiative in the National Health Service: an unhealthy development in new public management. *Financial Accountability Management*, **15**(3/4), 249–274.

O'Brien S. (1997) London's team spirit. *The New Statesman*, 13 June.
Robinson P. (2000) The Private Finance Initiative. *New Economy*, **20**, 148–149.
Ruane S. (2000) Acquiescence and opposition: the private finance initiative in the National Health Service. *Policy and Politics*, **28**(3), 411–424.

18 A framework for the risk management of private finance initiative projects

Akintola Akintoye, Matthias Beck, Cliff Hardcastle, Ezekiel Chinyio and Darinka Asenova

18.1 Introduction

The planning and implementation of a private finance initiative (PFI) project require a broad range of risk identification, assessment and management activities by individual participants, as well as by different groups of participants interacting with each other. A conceptual framework, which identifies these needs, can provide a basis for pro-active rather than reactive risk management (Al-Bahar & Crandall, 1990).

This chapter focuses on a broad-based framework, developed for identifying and addressing PFI risk on the basis of best practice in the PFI domain. This chapter also reviews the gateway produced by the UK's Office of Government Commerce (OGC), as a prelude to the presentation of the PFI risk assessment framework. This enables the framework to be put into perspective, in the sense that risk management process and activities are required at all the six stages of the gateways (gateway 0–5).

Although the PFI risk management framework proposed herein is not developed on the OGC gateway basis, the framework is applicable to the realisation of the gateway risk management objectives.

The goal of the framework is to provide a basis for:

- Improved communication between the different PFI participants
- Greater consistency in PFI risk assessment and management practices
- A harmonisation of the risk management terminology used by different participants
- Improved guidance on tools available for risk management, and when they may be used.

Previous research has argued that a 'risk management process' (RMP) must be overt, pro-active and farsighted (Sells, 1994). In addition, in most contexts a form of routine risk audit is advisable, to check that the RMP is proceeding according to plan (Carter *et al.*, 1994).

Each organisation has stakeholders, which may include owners, customers, employees, funders, regulators and the greater community. The attitudes, beliefs, values and expectations of these stakeholders usually have an impact on risk considerations (Baldry, 1998). Public organisations are expected to be accountable, which can render them relatively risk averse. Project consultants, meanwhile, may be more willing to suggest innovative solutions that foster their standing in the professional community, without necessarily focusing on the full range of risks this may pose to all participants.

Specific sources of uncertainty must be identified for each project individual (Reutlinger, 1970). Therefore risk analysis can be quite complex. It may be wise to keep it simple, employing sophistication only when necessary (Chapman, 1997). To employ an appropriate level of resources requires each participant not only to be aware of its own risk portfolio but also the way it is affected by the risk portfolios of others.

18.2 OGC gateways for PFI risk management

Succeeding the Treasury Task Force and Partnerships UK is the OGC, which has become the custodian of the Treasury's standard contract terms (Wilson, 2000). The OGC Gateway Process was launched on 20 February 2001. Its task is to review PFI projects based on six gateways processes (gateway 0–5), as shown in Fig. 18.1. It is expected that a team of experts, independent of the project team, will conduct these six reviews per project. These reviews take place at critical points of the project life, as shown in the figure.

OGC reviews are aimed at reducing project delay and ensuring that risks have been adequately assessed. The key goals of the reviews include that the best skills and expertise have been utilised for projects, and that more predictable time and cost targets will be achieved.

Based on the gateway framework review, reports are produced for the project sponsors, to enable them to progress with their schemes. According to OGC, pilot reviews are said to have yielded valuable additions to the respective projects at minimal expense. Each gateway is associated with risk management as shown in Table 18.1.

18.3 Framework for 'Project Risk Analysis and Management' (PRAM)

A framework for the risk management of PFI projects has been developed, which has taken into account the specific characteristics of the PFI process. This framework draws on the findings of non-project specific interviews with 68 organisations involved in PFI projects (contractors, facilities management organisations, financial institutions, clients and consultants) and eight case studies (two projects each for hospital, waste management, school and housing).

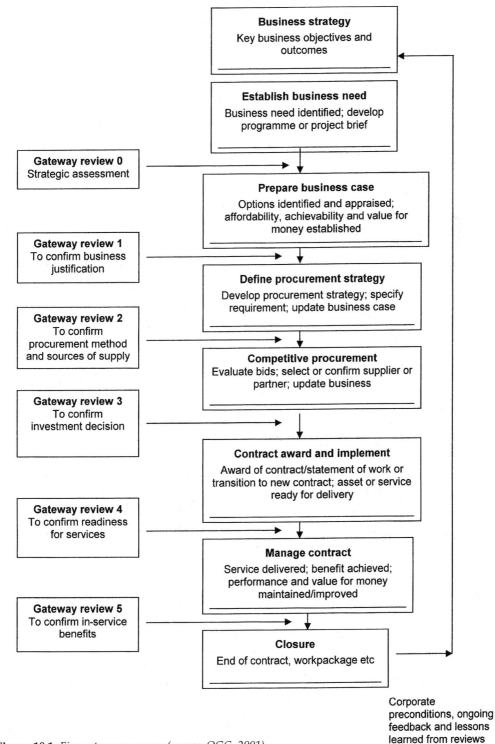

Figure 18.1 *Five gateways process (source: OGC, 2001).*

Table 18.1 *OGC gateway risk management review.*

Gateway	Risk management	Risk management activities
(1) Business justification	Establish processes to identify, assess and monitor current, anticipated and emerging risks	List of risks, including business, technical, operation and innovation risks; establish risk management strategy; and distribute risk managing responsibilities to individual
	Evaluate the options for each risk	Risk classification based on probability, severity, ownership, effect and countermeasure
	Assessment risk for the preferred option	Analysis of preferred option, plus management strategies
	Assessment of costs of risks in worst case	Assessment of risk financing or contingency funding available
	Identification of the category of costs in managing risks	Separately identify the costs of handling risks. Combine the costs of managing risks into base estimate or as a contingency funding
	Identification of project – whether it is a pioneer in its area	Examination of leading edge projects to assess their impact on the business, users and customers. Collection of evidence from similar projects or activities from which lessons may be drawn. Assessment of the innovative solution by experts, such as IT projects
	Economic assessment of the project	Market sounding taken
(2) Procurement strategy	Determination of the procurement strategy in terms of risk identification and understanding, financial evaluation and consideration of the major risks, and establishing a risk management plan	(1) Logging major risks (2) Assessment of risk financially and including it in a business plan (3) Establishing a risk management plan for each risk (4) Regularly review risk log (5) Assessment of all technical risks, including those associated with innovation and new technology documents (6) Agreement on risk transfer strategy

(3) Investment decision	Update risk and issue management plans	(1) Checking risks and issuing management strategies and systems in place
		(2) Updating risk management plans and risk log
		    • Risks associated with project resourcing and funding
		    • Team competencies
		    • Users and stakeholders
		    • Owners of risks/issues assigned

continued

Table 18.1 *continued.*

Gateway	Risk management	Risk management activities
	Arrangement to minimise the risks to the business in the event of major problems during implementation and rollout	(1) Agreeing a business continuity and contingency approach with users and stakeholders (2) Developing business or client continuity and contingency plans (3) Assessment of supplier's continuity and contingency plans
	Assessment of contract, whether reflecting standard terms and conditions and the required level of risk transfers	(1) Tenders comply with standard terms and conditions (2) Changes assessed for their impact, legality and acceptability (3) Analysis of risk transfer proposed by supplier or partner versus expectations or the original rationale for project
(4) Readiness for services	Proper management of the risks and issues that arose in the contract award and implementation phase	(1) Settlement of risks without outstanding claims (2) Quantification and planning of remaining risks associated with commissioning and service delivery
	Identification of unresolved issues and the risks of implementing rather than delaying	(1) Assessment of all remaining issues and risks (2) Evaluating report on the risk and impact of delaying or proceeding with implementation that considers: (a) option and management plans for both scenarios and a recommendation; and (b) ratification of the recommendation to delay or proceed with implementation by project or programme board
(5) Benefit evaluation	None	

The process of the development of the framework involves six stages as follows:

Stage 1 – investigation of the approaches used by the private sector (contractors, operators and PFI project consultants) in the assessment of PFI risks and establishment of how and why these specific methods of risk assessment have been chosen

Stage 2 – investigation of the prevalent practices of risk assessment and risk management of public sector client groups

Stage 3 – investigation of the framework used by financial institutions to standardise risk assessment and management across the financial sector

Stage 4 – analysis of the compatibility between these approaches adopted by contractors, facilities management organisations, financial institutions, clients and consultants

Stage 5 – identification of best practice in the risk assessment and risk management approaches used by these parties

Stage 6 – development on the basis of the above, of a standardised state-of-the-art assessment and management of risk model for PFI schemes.

This chapter only presents Stage 6. The findings on Stages 1–5 have been documented in the DETR/EPSRC final report. In essence, the framework presented in this chapter represents measures which have been identified as the best practice in risk management in the PFI domain.

The basis on which the framework has been developed can be summarised as follows:

(1) The public sector initiates a PFI project in which it identifies and proposes how risks should be allocated in the scheme.
(2) After *Official Journal of the European Community* (OJEC) notice, the private sector is involved where the risk allocation of the client is appraised. Initially, subjective assessments, which rely on experience, suffice. However, as each consortium is (further) shortlisted in the procurement process, more detailed risk analysis has to be performed. Due to the high costs involved, a detailed analysis is postponed until a consortium has become the preferred bidder.
(3) When the number of bidders has declined to three, more elaborate computations are performed to evaluate the project risks. At this stage, the probabilities of risk occurring, and the potential impact of the cost, are estimated. In the absence of historical data, previous experience is called upon. External experts are consulted for an informed view on the risk assessments. Sensitivity analyses are performed to improve the competitiveness of bids, while still targeting profits.
(4) During the negotiations, due diligence is practised, while a financial model is used by both the public and private sector to study and resolve the best distribution and pricing of the project risks.
(5) The level and intensity with which risks are assessed increase until the financial close.

Figure 18.2 shows a miniature version of the framework. At a macro level, the framework can be conceived as four building blocks (or grids). Figures 18.3–18.6 illustrate these main grids in relation to the framework and the relevant PFI stages. For convenience, the grids are used as a basis for explaining the framework. The following sections focus on the main activities performed within each grid. The bulk of activities within Grid 1 are undertaken by the public sector, while the private sector is more functional within Grids 2 and 3. The final Grid, No. 4, concerns a much more communal effort between the two sectors.

18.3.1 Grid 1: risk introduction

This grid pertains to the preliminary PFI project phase, and involves a range of activities performed by the public sector client. This phase involves activities performed by the public sector, from project conception up to the

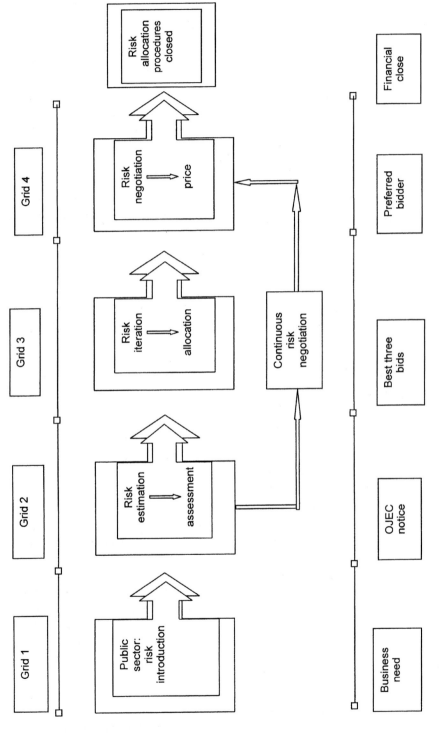

Figure 18.2 *Summarised framework for risk analysis and management of PFI projects.*

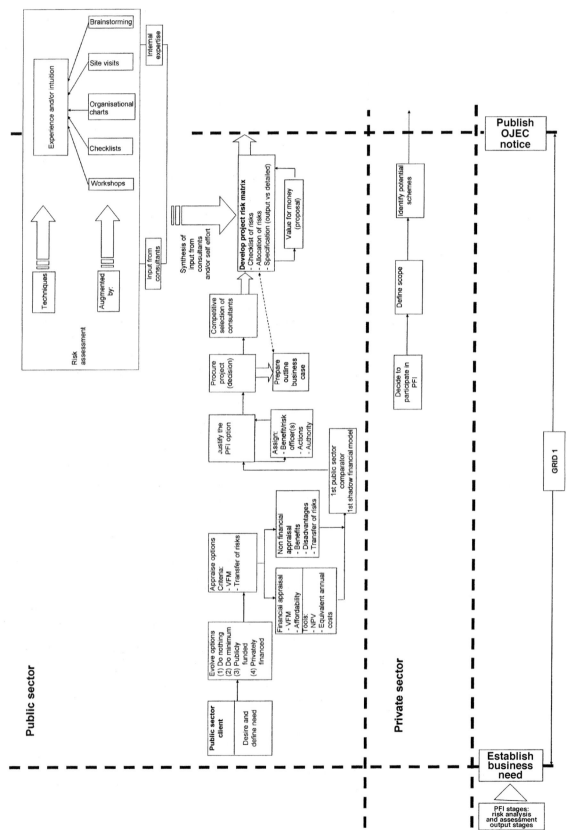

Figure 18.3 *Grid 1 – public sector risk introduction and initial allocation.*

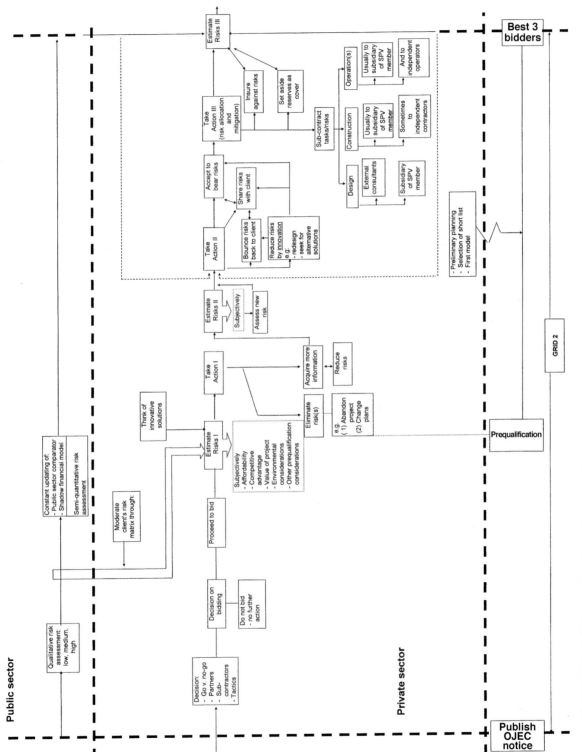

Figure 18.4 *Grid 2 – initial risk estimation and assessment.*

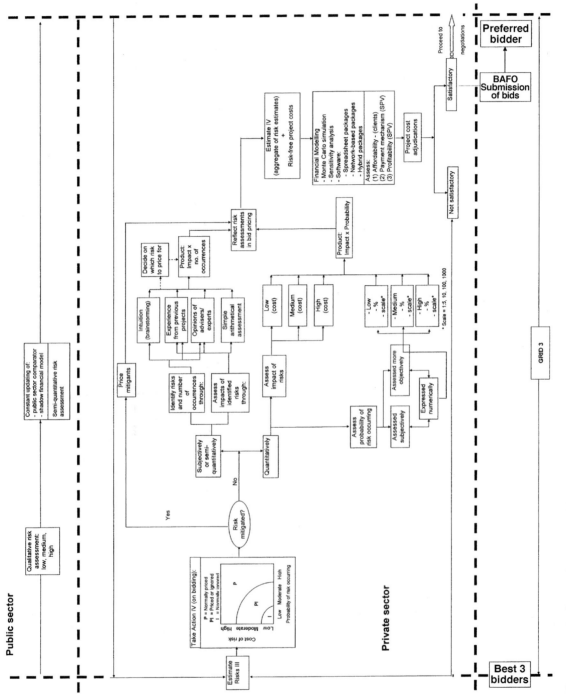

Figure 18.5 *Grid 3 – risk iteration, allocation and pricing.*

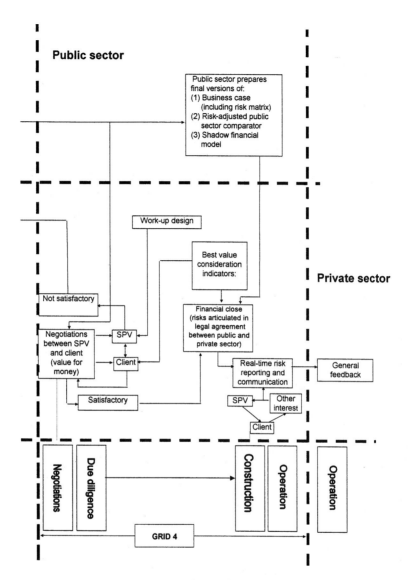

Figure 18.6 *Grid 4 – risk negotiation and financial close.*

issue of an OJEC notice. While the public sector has to decide on crucial issues concerning the future project, the private sector plays a rather passive role during this phase, other than identifying potentials for involvement in a PFI project in terms of resources available to the organisation and the sector in which it would like to be involved.

The tasks performed by the public sector side within Grid 1 include the following:

- Defining the precise scope and nature of the public services required
- Appraising the options in terms of affordability, the 'value for money'

(VFM) criteria, and the impact of risk allocation under different options

- Conducting a feasibility study to select (or justify) the PFI option
- Investigating the competencies available within the organisation
- Selecting external consultants (legal, financial, engineering etc.) on a competitive basis
- Assembling project team members, who have the critical mass expertise
- Developing an initial view on the desirable risk allocation in the project
- Using the project team to develop the first version of the 'public sector comparator' (PSC) and a shadow financial model
- Continuously refining the PSC and the shadow financial model, in an iterative manner
- Conducting a final assessment of the PFI option
- Developing the risk matrix, using the VFM criteria, and preparing the outline business case
- Preparing an Information Memorandum, which is a summary of the main project features, to be given to the private sector.

The main outputs arising in this phase on the public sector side, include:

- The preparation of the PSC as a benchmark for the evaluation of bids
- The development of the initial project risk matrix
- The development of the outline business case
- The generation of 'Information Memorandum' to explain the project details to potential bidders.

The techniques that facilitate risk assessment within Grid 1 typically include the use of:

- Experience acquired from similar or previous projects
- Brainstorming
- Workshops
- Checklists
- Site visits.

Towards the end of this phase, the public sector should be in a position to have an in-depth understanding of the type of project risks and their desirable contractual distribution. Strategies for achieving Best Value should also be identified at this stage.

Meanwhile, pro-active private sector bidders might use different forms of enquiries to identify projects that are about to come on stream, even if an OJEC notice has not been placed. For example, a public sector client might talk to their bankers concerning the funding of a potential PFI project. Such a discussion would indicate to the bank that a project might be forthcoming. Given this experience with PFI schemes, some banks can assist in preparing to bid for incoming projects.

By associating themselves with (previous) project partners, contractors and operators can gain information about projects that are about to come on stream. In doing so, private sector participants may be able to identify a PFI project before it is highlighted through the OJEC notice. However, the formal participation of any party only commences once the OJEC notice is published where 'expressions of interest' are called for.

18.3.2 Grid 2: risk estimation

The response to an OJEC notice marks the official beginning of a private sector party's involvement in the PFI process. At this stage, activities centre round creating a bidding consortium that will win the project. Each consortium will initially go through the pre-qualification stage during which six to eight consortia will be shortlisted to proceed to the next stage of bidding. The shortlisted consortia will then submit detailed proposals that will inform the selection of the three best bidders. Grid 2 ends at this stage.

The actions taken by the private sector organisations at this stage typically include the following:

- Form a definite opinion on the suitability of the project
- Conduct an initial assessment of opportunities and risks
- Choose partners to team with
- Select a bidding strategy
- Establish lines of responsibility and consolidate the team
- Appoint external consultants
- Conduct an initial estimation of the risks, usually in a subjective manner
- Refine the risk estimation through an initial iterative process supported by the external experts. Usually this involves about three iterations being suggested
- Decide on possible risk mitigating measures, e.g. insure risks, transfer risks or set up a reserve account.

The tools/methods employed by the private sector usually include:

- The use of a preliminary financial model
- Consultations with experts.

Towards the end of this stage, the private sector consortia should be fully acquainted with the project and its anticipated risks. The private sector companies should also have formed a view on the desirable risk distribution from their point of view. All anticipated risks should be assessed in terms of their probability and impact. Most of the risks that constitute deal breakers should be investigated. Apart from this, there will be little by way of formal output on risk issues.

On the public sector side, the main task of the project team is the comparative analysis of bids, both during pre-qualification and subsequent shortlist of bidders. At this stage, the public sector side should use its PSC as a benchmark for the comparison of bids. This process should be supported by projections made from the shadow financial model. Towards the end of Grid 2, the public sector should be in a position to summarise its findings and select the best three bidders.

18.3.3 Grid 3: risk iteration

This stage is a continuation of the previous one, where the public and the private sector partners devote additional effort towards achieving a more favourable risk distribution. The public sector may oblige all three (best) consortia to produce quotations, which reflect different risk allocation scenarios.

A full range of qualitative and quantitative risk assessment techniques often informs the estimates returned by the private sector consortia. These bidders usually re-analyse their own financial models and conduct sensitivity analyses, to evaluate risks in greater detail.

The techniques employed at this stage by the private sector, depending on the type of project and the level of expertise available in-house and outside the organisation, should include the use of:

- Experience and intuition
- Probabilistic analysis
- Simple arithmetic analysis
- Eliciting the opinions of advisers
- Sensitivity analysis through a Monte Carlo simulation.

The outputs derived from this stage should include:

- Establishment of firmer project costs/price
- Assessment of clearer profit margins (especially by the private sector)
- Assurance that the project is still within the client's affordability
- Initial identification of the payment mechanism.

At this stage, the public sector would normally rely on their shadow financial model and updated version of the PSC to assess the private sector's evolving risk profiles. In this process, the client would be mainly interested in meeting the affordability criteria and optimising VFM, while the private sector consortia would be concerned with the balance between profitability and risks borne by them. The end of this stage is marked by the selection of a preferred bidder by the client, and the commencement of negotiations with that consortium.

18.3.4 Grid 4: final risk negotiation

This stage leads towards signing the contract and financial close. As such, it marks the final stage as far as risk allocation is concerned. Most of the project risks would have been assessed or allocated during the earlier stages. Any remaining deal-breaking issues will be sorted out, and missing details will be clarified. The parties scrutinise their earlier estimates using the same methodology and tools, and arrange the final distribution of risks through the negotiations.

At this stage, the senior debt provider also engages external consultants to perform due diligence. The purpose of the due diligence is threefold:

- Double-check the reliability of all estimates
- Investigate any possible legal shortfalls
- Provide assurance that there are no discrepancies.

The appointed consultants investigate the legal, technical and financial aspects of the whole project, and also audit the financial model. The main tool used at this stage is the financial model, which is used to re-evaluate different scenarios as negotiated decisions are made. When consultants give their assurance that the project is acceptable, the private sector participants become confident enough to enter the project agreement.

The outputs arising from the final negotiations should include:

- The project agreement between the main parties
- The agreed risk matrix
- The risk-adjusted version of the public sector comparator.

Financial close does not mark the end of risk management. The agreement between the two parties is monitored both during the subsequent construction and operation phases. Any risks that materialise during these later stages are addressed in the most appropriate manner. Reports on such incidences are fed back into the project documentation, and noted for future use.

18.3.5 Grids 2–4: continuous risk negotiations

It is expected that continuous risk negotiation should take place between the time that the initial risk estimation is undertaken and the time that the final risk negotiation is reached. The essence of these activities is to ensure that there is risk feedback and feedforward between the private sector and the public sector. This should guide against deal-breaking risks, which could arise if the risk negotiation was only undertaken at the end of the process.

The structuring of the framework into grids is not an indication that negotiations are held in discrete stages. Risk management evolves progressively and is done in an iterative manner.

18.4 Application of the PFI projects risk assessment framework

The process of the framework is discussed to illustrate its applicability. A hypothetical project is used for this purpose, and thus assumptions are made where relevant. Although this hypothetical project is not directly related to the case studies and projects investigated in the research, it takes its best practices from them. Many PFI projects have been procured, and many of these projects are now fully operational. It may thus not be surprising if this hypothetical project shares similar features with some schemes in the market. Any semblance of the illustration to any PFI project is merely coincidental (see case study).

Case study

The details of the hypothetical project are as follows:

Feature	Description
Client	A hospital (NHS) Trust
Type of scheme	Hospital
Main type of work	New build: an existing old hospital will be demolished and a new one built as a replacement
Projected price	• The capital price of construction work is estimated to be £115 million • The charge for the premises, is about £850k a year, to cover water, electricity, utilities, ground maintenance, etc.
Construction period	20 months
Concession period	25 years for the management aspects
Type of finance	Expected to consist of about 90% debt and 10% equity

The procurement of this hospital project is now discussed in the light of the framework. The four Grids are used as the basis for this illustration.

18.4.1 Grid 1

Project conception

The existing facilities of the NHS Trust (client) consist of wards, operating theatres, out-patient departments, pharmacy, etc. In addition to using these facilities to service its ever-expanding community, it has been identified that waiting lists are not coming down and that the hospital, which was built in the Victorian era, is becoming inadequate for this community. The client needs to prepare to face any new challenges that will be posed to its functions in the twenty-first century.

The client is highly concerned about the physical state of its buildings, which are over 70 years old. In general, its facilities are either getting old or less efficient. This project was then conceived because this particular client feels that it needs to carry out a major overhaul of its facilities.

The initial intention of the client was to carry out a major refurbishment programme. However, the money this client receives from the Government is not sufficient to undertake such major work. The client therefore decided to explore other ways of achieving its aim.

Optioneering

A decision has been made that something must be done, and the client must then explore how it is to be done. The cost of refurbishing existing facilities, and maintaining those facilities afterwards, has been considered. The figures that evolved led the client to perceive that it was cheaper in the long run to construct and maintain new facilities, rather than to continue with the old stock.

As PFI is available for procuring public projects, this client should explore its usage. It is expected that PFI should be used when it is viable to do so, and that it should not be applied blindly. Therefore, this client should assess the possibility of using different options before a final choice is made.

At least four options should be carefully considered, with the prime objective of achieving best value for money. The other criteria to consider in the comparison of these options should include: flexibility of facilities with respect to future changes, disruption of the client's normal services, meeting health and safety standards, ability to deliver services with modern facilities and standards and the demand on the client's staff and resources.

Table 18.2 shows that both quantitative and qualitative criteria should be considered. Initial capital and subsequent maintenance costs should be considered quantitatively, while risks and benefits may initially be considered qualitatively. Many and detailed criteria should be considered in the comparative analysis, many more than reflected in Table 18.2. The client may need to appoint a project manager at this point, if not before, to head the process.

Justifying the PFI option

The evaluation of Table 18.2 should be complemented with a consideration of soft issues, as shown in Table 18.3. Information in Tables 18.2 and 18.3

Table 18.2 *Comparison of different developmental options.*

Criteria	Weighting (1–10)	Do nothing	Do minimum	Refurbish	New build
Demand on staff resources	4	0	3	6	10
Disruption to normal service	4	0	2	7	10
Other considerations					
Meeting health & safety standards	10	8	8	9	10
(Sub-total)		80	100	142	180
Costs (£000)					
Cost of land	3	0	0	0	0
Cost of decanting	5	0	100	500	1200
Building costs	6	0	50	650	5000
Maintenance costs	9	4500	2500	1750	850
Other costs		X1	X2	X3	X4
Total (NPV p.a.)*		4000	3300	2150	1750

NB: *Totals are approximated and should depend on X1, X2, X3 and X4.

Table 18.3 *Prospects and risks of the options.*

Option	Risks	Prospects
Do nothing	The existing facilities are either old or outdated, and thus not suitable for long-term use	Should be discounted for strategic reasons
Do minimum	Hospital will not operate efficiently, due to over strain on resources. There is the attendant frequent and high cost of maintenance and health and safety risks to staff and patients	Given the age of the facilities, this option is insufficient to address the modern needs of the client, let alone prepare it for the future expansion envisaged in that locality
Major refurbishment (publicly funded)	The dynamic and modern requirements of the hospital will not be addressed. Also, the capacity of the hospital will not improve much	This alternative is feasible, but expensive. Meanwhile, refurbishment would have extended the life of the facilities for just a few years
New build via PFI option	Disruption to services during decanting, and health and safety risks arising thereby	The PFI options offer the client the chance to acquire new facilities that will last much longer than refurbished facilities

suggests that a new-build hospital scheme by means of PFI is cheaper for the client in the long run. Although its initial capital costs are higher, it offers a better choice strategically.

Satisfied that the PFI option offers best value for money at this time, the hospital Trust may then approach the NHS for project approval and commitment to funding in respect of this project. The estimates prepared in the analysis should be used as a basis for determining the level of funding required. In other words, affordability should be established.

Procuring the project

Having obtained approval for this project, the client should then proceed to let the project on a PFI basis. At this juncture, design and engineering consultants should be engaged to assist with expert advice. As the hospital may not have in-house expertise in construction and PFI procurement, consultants should be engaged at an earlier stage. A discussion of a design solution with the consultants will, in part, inform the refinement of the estimated project costs.

The onward procurement of the project should involve the preparation of the outline business case, project brief and risk matrix. If the client's project team is sufficiently capable, it should prepare the business case and get the consultants to vet it for a second opinion. Otherwise, the consultants should assist with the preparation of the business case.

Developing the risk matrix

The client should fully perceive the risks that attend its proposed scheme before it attempts to distribute such risks. In this respect, the client could see risk through the limited window of 'something that can go wrong'.

The significant risks to be carefully considered by the client in a scheme like this should include:

- Design solution that meets the requirements of the client
- Construction delays
- Cost overruns
- Political risk (i.e. if the private sector made a mistake, the scheme would get negative publicity)
- Efficient performance in terms of facilities
- Ability of equipment and appliances to meet the future demands of health provision.

These risks should be thought through carefully, especially the major issues. The client's project team or consultants should meet regularly to discuss the project and its risks. As the procurement process evolves, the client should also engage and use the advice of lawyers to draft the contract documents.

Public sector comparator

Forming part of the business case is the 'public sector comparator' (PSC), which gives an overview of how much it would cost the public sector to undertake the works through traditional procedures. In the PSC, risks like cost and time overruns should be priced. There are other aspects like wage increases, inflation, disruption, etc. which the client may not be able to price precisely, but should be identified in the PSC.

What the client organisation should do, in conjunction with its consultants, is to put a figure on the possibility of a risk occurring. They should do this by looking back through other historic schemes and through the experience and knowledge of their team members that have been acquired elsewhere. Suppose, for example, that information had been generated from 40 hospital schemes, and that claims amounting to £50 000 had been made for bad weather and £120 000 for bad ground conditions. Taking an average of these 40 schemes, the client may divide the amounts paid in the past by 40, to obtain the expected value of each risk. By this analogy, the expected risk estimates will be as in Table 18.4

The total in Table 18.4 is an average mechanism, and does not suggest that all the risks will occur in this particular job. The assessment of risks in this way spreads their impact over many projects, and assumes that bidders will see and assess risks in a likewise manner.

In preparing the PSC and assessing risks, at least three scenarios should be considered:

Table 18.4 *Risk estimation.*

Number	Risk	Estimation	Allowance (£)
1	Bad weather [*Other risks*]	£50 000 ÷ 40	1250
n	Bad ground conditions	£120 000 ÷ 40	3000
Total			4250

- Best/dream case – where nothing would go wrong
- Medium case – where some things will go wrong
- Worst case – where everything will go wrong.

The precise detail and number of scenarios to be considered should be decided upon between the client and its consultants.

The assessment of risks in accordance with the different scenarios should guide the client in reaching a decision as to the allowance to be made for the occurrence of risks. This should also help in moderating the affordability estimate produced by the client organisation.

Advertising the project

Having developed the project thus far, and barring any impediments, the client should proceed to advertise the scheme in the bidding market. The project brief and documentation should be worked upon, and the client's consulting team should be relied upon for major assistance. The lawyers should, in particular, help in drafting the project documentation. The existing standard NHS contracts for PFI schemes should make the client's task in this regard easier.

When the documentation is complete and the client is satisfied with it, the project should be advertised through an OJEC notice. Expressions of interest should be elicited in the OJEC notice. The OJEC notice marks the transition from Grid 1 to 2, and the official beginning of the involvement of the private sector in the project. From this stage, the activities of the public and private sectors should be interrelated, and are subsequently discussed that way.

18.4.2 Grid 2

Following the publication of the OJEC notice, it is assumed that 12 consortia have expressed their interest in this scheme. They have each used the client's proforma to supply information of a general nature, such as:

- Calibre and experience of personnel for the job
- Experience of participating organisations

- Experience of health or allied schemes
- Perception of risks involved in the project.

Initial risk assessment

The project documentation prepared by the client should provide the main basis on which the consortia will bid for the project. The private sector bidders, on their part, should give consideration to:

- The client's affordability
- The value (size) of the project
- Environmental considerations
- Ability to raise finance for the scheme
- Risks inherent in the project, as documented in the risk matrix, and initial overview of how to deal with most of these risks
- The identities of their likely competitors and what they will be pricing
- The competitive advantage(s) they might have over their competitors.

The foregoing considerations will inform the way the bidders answer the client's questionnaire. Each consortium will endeavour to convince the client that it is able to deliver the scheme, and thus worthy of further consideration.

First shortlisting of bidders

The client organisation should critically assess the information supplied in its questionnaire by the private sector bidders. In the absence of structured data, confidence ratings should be used to assess the 12 expressions of interest. The assistance of the consultants would be valuable for this exercise. Occasionally, a client might be familiar with some consortia through previous interactions. Such familiarity should not preclude the client from assessing the bidders thoroughly, as a new team that can offer better value for money can only be identified that way.

The bidders would at this stage be rated in broad terms; the confidence towards them being expressed subjectively in terms of high, medium and low. These yardsticks should be used to rate the ability with which each consortium can achieve the primary and secondary objectives of the client. Firstly, the consortia should be assessed on general ability to deliver, followed by the price offered and the expected quality to be achieved. Thereafter, other criteria should be considered, such as market share issues, stand-by suppliers, design linked to construction, commissioning and hand-over, operation, etc.

The client's project team should evaluate the consortia and involve staff from various departments such as design, construction, site and planning, operations, etc. In this case, medical personnel from different disciplines should be involved, or consulted, to make sure that their requirements are not overlooked. The client's project team should also seek input from its advisers.

The project team should collate the respective assessments and opinions sought. The exercise should culminate in the overall rating of each bidder. The client may choose to rank the bidders at this stage, or impose a yardstick to be used for the initial shortlisting of bidders. In this example, it is assumed that six bidders have been pre-qualified and asked to proceed to submit more detail bids.

Second stage bidding

The six selected consortia should proceed to prepare more detailed bids that will include design solutions and a more detailed assessment of risks. At this stage, the risk assessment iterations should begin.

The consortia should consider the allocation of risks between the public and private sector. They should also consider the allocation of risks between each consortium and its potential sub-contractors. The evolving design solution will be priced, as well as the identified risks with each scheme and the implications of their allocation. For instance, if a consortium decided that the client was going to be responsible for obtaining planning permission, if a delay was envisaged in the process, then the cost consequence of such a delay would be estimated and added to that consortium's bid price.

As the project evolves, the consortia should identify the different risks or combinations of risks that impact on the project, and work out their cost implications. Sometimes, the cost implications of the risks would be high, leading a consortium to have a re-think on how to circumvent risks with such high consequences. It is during such times that members of a consortium should brainstorm on how to estimate, or what to do with such risks. It is also during such times that serious thought should be given to innovative solutions that will offer the client value for money and keep the project price down.

The quotations of sub-contractors should also be analysed by the consortia. A typical format used by sub-contractors while assessing risks is shown in Table 18.5. The design, construction and operation sub-contracts and their risks should all be considered.

Any aspect that has a price implication should be analysed to inform the establishment of a bid price that is ultimately submitted to the client. Along this line, the shortlisted consortia should develop an initial model to assist them in assessing the different project and/or risk scenarios. The impact of different combinations of risks on the project (price) should be assessed with the model, but detailed use of the model may not be made at this stage.

Second stage shortlisting

On the basis of their project proposals and risk assessments, the six bidders may be able to submit quotations for this hospital project. The client will then scrutinise these offers, by going through the risk library that the bidders have submitted, and decide whether all these risks apply to the current scheme or not. Some items may be deleted from the bidder's checklist or brought to their attention for clarification if they are considered unnecessary.

Table 18.5 *Evaluating FM risks in a PFI project.*

Risk feature	Description	Frequency	Rate	Cost
Move-outs	Clearing, cleaning and tidying-up occasioned by moving out of a building. (50%* × 150 offices) = 75; 75 offices ÷ 25 years = 3 offices p.a.	3	£150*	£450
[*Other features*]				
Ground maintenance	Cutting and re-laying grass	5000 sq.m*	£25	£125 000

* Assessed from benchmarks and moderated by means of aggregated experience.

At the same time, risks that had not been identified before, but have been flagged up by the bidders, should be added to the risk matrix.

The client may use the services of its consultants to help with the evaluation of bids. The client should, at this stage, use its shadow financial model to perform a sensitivity analysis or Monte Carlo simulation (if possible, although advisable), to assess the risk profiles of the bidders. The bidding consortia should again be ranked in order of their suitability for the job. The ability to achieve the client's objectives and offer value for money should be used to rank the bidders.

If the assessment of bids is elaborate, three consortia should eventually be shortlisted at this stage, to work-up and submit their proposals. The project should move onto Grid 3, after the three best bidders have been selected.

18.4.3 Grid 3

At this stage, the competition between the consortia should heighten, and the consortia should concentrate their efforts on allocating risks more optimally. The consortia should also be making final concessions and firming-up their price offers. In determining the optimal allocation of risks, the consortia should pursue the agenda of 'what is comfortable for us to bear'. A consortium should consider treatment of all other risks, e.g. should they either be passed down to sub-contractors or re-allocated to the public sector client.

For a bid to be competitive, consortia should choose to make concessions, by not pricing for all risks in their bids. They should rather look at other ways in which some of these could be mitigated, apart from pricing them into their bids. The general principle to be followed is illustrated in Fig. 18.7, where risks with a low chance of occurrence and a low impact should generally not be priced. Some risks with a moderate chance of occurrence and a moderate financial or time impact should be ignored too, if the competition is very stiff.

The actual number of risks in this category varies with each consortium.

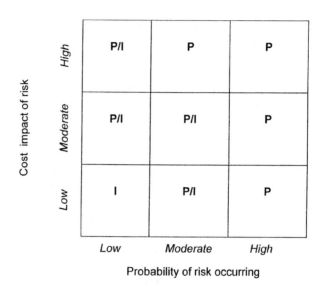

Figure 18.7 *Action taken in view of risk characteristics. NB: P means priced, P/I means either priced or ignored, while I means ignored.*

The extent to which each consortium is willing to ignore the pricing of certain risk features should, in part, depend on the extent to which it wants to be competitive.

Profit margins too would have an influence on the number of risks priced, and each consortium should overlay its risk concessions with the effect on the eventual profit to be made from the project. In essence, each consortium should determine what risks to price for and the price to put on them at board level. Notably, the board ultimately adjudicates the bid price the consortium is comfortable with.

Generally though, all risks with a high impact and high chance of occurrence are priced. Some organisations may in addition choose not to price risks that have a high impact but low chance of occurring. Thus, concessions should concentrate on the moderate and minor risks.

The actual choice of which risks to price and which to ignore is circumspect. All things being equal, contracts should be won and lost on the basis of adequate risk evaluation. As consortia differ, what each of them would feel comfortable with accepting differs. Therefore, individual consortia should evolve their policies to guide their actions in terms of risk inclusion or exclusion in the pricing of bids. Having identified and agreed on the risks to be priced in the bid, albeit circumspectly, there are two main ways in which each consortium should go about achieving this objective.

The first approach is to adopt a full quantitative approach, where both the probability of occurrence of each risk and its associated impact are assessed numerically. In this context, the probability of occurrence should be assessed on a point scale of 1 to 5, 10, 100 or 1000, depending on each organisation's choice. Previous records and experiences should be used to score the probability of each risk. The cost impact of each risk is assessed in monetary terms. The two values of probability and impact of each risk are then multiplied to obtain its average estimate. That is:

Average estimate of risk = (probability × impact).

The second approach is semi-quantitative. While the monetary impact of each risk can be assessed in terms of this category, information is usually insufficient to warrant an assessment of their likelihood of occurrence. Therefore, subjective decisions are made with respect to probabilities, underpinning the outcome of such decisions on the opinion of advisers, brainstorming activities, intuition and projections from non-similar projects. The resulting risk estimate is, however, obtained as in the quantitative approach.

The monetary impact of a risk can always be ascertained. In this context, any implication of a risk, such as time delay or employee turnover, can be translated into monetary equivalent. It therefore means that, once probabilities can be assessed, a quantitative evaluation of risks can be performed. In essence, where data is available to support the assessment of probabilities, a quantitative approach should be employed.

In retrospect, a bid should consist of three constituent parts:

- The cost of facilities (or construction)
- The cost of operating the facilities (FM)
- The life cycle (maintenance) costs.

In addition to these three, ancillary items such as cost of finance, development fees, overheads, etc. should be accounted for. Each of these constituent costs should have an in-built risk factor.

Modelling of risks

Whichever way the risks have been priced, a financial model should be utilised to smooth out the effect of risk pricing on the project cash flow. The constituent costs should be fed into the financial model to identify sharp rises in expenditures at several points in time. Such spikes, that render a bid less attractive, should be swept away.

In financial modelling, sensitivity analyses should be performed on the assessed risks, to determine those that are most volatile. The output should then be adjusted, to minimise the unitary charge to the client. At this stage, the client's affordability should be weighed *vis-à-vis* the profitability of the SPV. Where it is felt that the sub-contractors' prices are excessive, a negotiation with them should be sought to reconsider the prices. The SPV should be negotiated with each participant until the most competitive price possible is obtained.

The financial model should be able to reflect how the project finance would perform over its life. Using the model, lenders should check if there is enough leeway in the overall financing and that the whole financial structure of the project is viable. This is especially important as a substantial sum (up to £100 million) is being borrowed for this hospital project on the basis of assumptions concerning interest rates and running (future) costs of the hospital for 25 years. The lenders should thus check to see that sufficient money is available to cover certain changes in circumstances that may occur over the concessionary period. Without such a sufficient leeway, any hiccup may affect the SPV profitability of the project.

Some key financial ratios that are useful for the lenders' assessments include:

- The annual debt service cover ratio (ADSCR), which is calculated as the cash revenue available for debt servicing, divided by the amount of debt in the corresponding year. This ratio gives an indication of how close the consortium is to not being able to pay its debt
- The loan life cover ratio (LLCR), which is the NPV of the sum of all future income for the loan life, divided by the consortium's outstanding debt at a particular point in time.

Other appropriate formulae should be used to produce a number of outputs, like the profit and loss account, the balance sheet and the cash flow. The outputs from the financial model should demonstrate the level of retained cash, corrected total cost curve, ratio maps against base cover ratios, shareholders' return graphs, etc.

Other different scenarios should be considered, for example:

- What would happen if inflation fell to 1% or went up to 15%?
- What would happen if maintenance costs were advanced, say by six months or one year?

The key objective of the sensitivity analyses should be the determination of a strong cash-flow position, as it allows for the repayment of loans. The lenders should also examine the strength of the management team and ownership of the SPV, to be satisfied that if things went wrong, appropriate personnel would be there to address them.

The model should continually be refined during the negotiation phase. The sculpting of the model should be done several times over, and the adjustments should continue until the SPV is satisfied with a final bid price, which is submitted as the final offer.

Third shortlisting

When the consortia have finished pricing their schemes and are reasonably satisfied with their bids, submissions are made to the client, which provide the basis for selection of a preferred bidder. A preferred bidder should be selected by the client on the basis that the consortium concerned will offer better value for money to the client than the other two bidders. The client organisation should use its shadow financial model and updated version of the business case to compare the submissions of the bidders.

Consultants should also be utilised when necessary by the client. The financial consultants should, in particular, check the effect of each proposal on the payment mechanisms, to ensure that the affordability for the client is not undermined. The technical consultant should make sure that the respective solutions are sound and workable. A scoring of the proposals should be made (see Table 18.6 for an example), and the consultants should make a recommendation to the client on the suitability of the bidders.

In contrast, any downsides with the consortia should be identified and listed against them. A decision on the suitability of a consortium would hinge on its strengths and weaknesses. Therefore, a SWOT analysis should be conducted to compare the Strengths, Weaknesses, Opportunities and Threats of the bidders before a preferred bidder is selected. Having qualified twice before, the three bidders should have little or no significant downsides. This should enable the client to concentrate on identifying the potential gains to be made through each bidder.

On the basis of the relative strengths of the consortia, Table 18.6 suggests that bidder No. 2 stands out as having the potential to offer the best value to the client, and should be considered as the preferred bidder. Bidder No. 1 should be named as reserved bidder in view of this consortium's qualities, which virtually match those of No. 2.

The selection of a preferred bidder ends the activities of Grid 3. The transition to Grid 4 is characterised by the client sending an invitation to this preferred bidder for the two to enter into negotiations.

18.4.4 Grid 4

At this stage, negotiations are held between the public sector client and the preferred bidder (consortium). Detailed issues that have not been sorted should be settled during the negotiations. The negotiations should consider:

- The final distribution of risks, especially those that are shared between the two sectors
- The penalty system and how it should be applied
- The workability of the final programme of works

Table 18.6 *Third stage screening of bidders.*

Criteria	Weighting	Consortium		
	(1–10)	No. 1	No. 2	No. 3
Meeting affordability	10	10	10	10
Offer of value for money	10	8	9	6
Experience of similar schemes	7	6	7	6
Experience of PFI schemes	7	8	8	8
Innovative solution	8	6	8	6
[*Other desirable criteria*]				
Firmness of funding arrangement	8	8	8	8
Total	100	87	90	70

- The soundness of the project agreement between the main parties
- The agreed risk matrix
- The risk-adjusted version of the public sector comparator
- Other relevant issues.

The foregoing issues should be negotiated until a settlement is reached on each of them. Consider an example of rates on properties, where the parties decide the party who should have responsibility for this risk. The client should get an assessment of the likely evaluation of the new buildings from the rates office, which may, for example, amount to £60 000 a year. On that basis, the client may propose that the consortium bears this risk. The consortium may in return say 'Okay, we will charge you £75 000 a year and we will take whatever happens with rates in the future'. The client may say '£75 000 is unacceptable to us as a public body'. The two sides should then negotiate in order to reach a settlement. One side might convince the other to change its position, or they might reach a middle ground. Whichever way it goes, the decision should be reflected in the updated risk matrix.

The negotiations should seek to ensure that all identifiable risks have either been mitigated or covered in some way. Each sector should check its insurance to ensure that no aspect of the scheme has been left uncovered.

When a consortium is selected as preferred bidder, it should begin to work on the contract agreement. Issues that are resolved during the negotiations should be reflected in the draft agreement. When the two sides are ultimately satisfied with their resolutions, the contract documentation should be finalised, to reflect their consensus decisions. Lawyers on the two sides should scrutinise the agreement, to make sure that no unwarranted risk is allocated to their clients. If any contractual clause poses a new risk to one of the sides, the lawyers should spot and point it out to the relevant party.

Due diligence

If it has not been done before, the senior debt provider should, at this stage, initiate proceedings for conducting due diligence, in order to make sure that risks have been priced fairly. The aim of the due diligence is to support the risk identification and assessment processes by involving top experts in each particular field. Additionally, due diligence can be considered as a tool for off laying risks to the companies performing it. For example, if there are unidentified problems with the financial model, the auditors have to take responsibility for such problems. Thus, the purpose of the due diligence is threefold:

- Double-check the reliability of all estimates
- Investigate any possible legal shortfalls
- Provide some insurance in case there are any big discrepancies.

In this case, the financial model should be run to determine the effect of price increases on equity returns. In general, any mismatch between the revenue stream and the way the costs behave under different inflation scenarios should be investigated as they may have a negative impact on the equity returns. Different risk scenarios should be run through the model to examine relationships between any two or more risks.

During due diligence, the bank should be investigating the design and construction solutions to confirm that they are 95% confident that the project will be completed on time. In addition, the bank should put in place other protections in terms of recourse to the contractor's resources if they do not deliver within the time limit specified in the contract. External financial consultants can be used to conduct the due diligence, wherein they should investigate the legal, technical, and financial aspects of the whole project, and also audit the financial model.

Due diligence should continue up to financial close, but the risk assessment should not stop there. Every six months, the financial model should be re-run, where 'look-back' and 'look-forward' regimes should be considered. An ongoing monitoring of the project's risk profile should also be made.

Financial close

Financial close should be reached when the two sides are happy with the agreement and ready to commit themselves to the deal. After financial close, the construction works should officially commence, to be followed by the operation of facilities for service delivery. Real-time risk reporting and mitigation also continue throughout the concession period.

18.5 Conclusion

The framework discussed in this chapter offers PFI participants a state-of-the-art basis for identifying, assessing, mitigating and reporting the risks of PFI projects. Leading PFI practitioners and academics assisted in shaping and finalising the model. Therefore its practical usage should ensure that risks are adequately assessed, and optimum value is achieved in the delivery of PFI schemes.

The implementation of the framework was illustrated; however, a report of this nature cannot be exhaustive. Different consortia and projects have their peculiarities. The application of the framework under different circumstances has been discussed on many occasions and modified by the research project's industrial partners and has continued to be discussed with the PFI participants at various workshops.

Acknowledgements

We wish to thank the Engineering and Physical Science Research Council (EPSRC) and the Department of Environment, Transport and the Regions (DETR, now Department of Trade and Industry) for funding the research on

'Standardised framework for risk assessment and management of Private Finance Initiative projects' which informs this paper. We are also profoundly grateful to the project industrial partners who have supported the research. Many organisations have either participated in the research or supplied data. Their co-operation is fully appreciated.

References

Al-Bahar J.F. & Crandall K.C. (1990) Systematic risk management approach for construction projects. *Journal of Construction Engineering and Management*, **116**(3), 533–546.

Baldry D. (1998) The evaluation of risk management in public sector capital projects. *International Journal of Project Management*, **16**(1), 35–41.

Carter B., Hancock T., Morin J. & Robins N. (1994) *Introducing Riskman Methodology: The European Project Risk Management Methodology*. Blackwell, Oxford.

Chapman C.B. (1997) Project risk analysis and management – PRAM the generic process. *International Journal of Project Management*, **15**(5), 273–281.

OGC (2001) *OGC Best Practice and Operational Guidance: OGC Gateway Process*. Office of Government Commerce, http://www.ogc.uk/index, HMSO, London.

Reutlinger S. (1970) *Techniques for Project Appraisal Under Uncertainty*. International Bank for Reconstruction and Development, World Bank Staff Occasional Paper Number Ten, Washington, DC.

Sells B. (1994) What asbestos taught me about managing risk. *Harvard Business Review*, **72**(2), 76–90.

Wilson, T. (2000) Farewell to the taskforce. *The Private Finance Initiative Journal*, **5**(2), 6–7.

Index